Fundamentals of Electronic Devices

Second Edition

Merrill's International Series in Electrical and Electronics Technology

Samuel L. Oppenheimer,
Consulting Editor

Fundamentals of Electronic Devices

Second Edition

Ronald J. Tocci
Monroe Community College

Charles E. Merrill Publishing Company
A Bell & Howell Company
Columbus, Ohio

To Cathy, Mike, Mark and Carrie—

*for their undeserved tolerance of me
during the writing of this book*

Published by
Charles E. Merrill Publishing Co.
A Bell & Howell Company
Columbus, Ohio 43216

International Standard Book Number: 0–675–08771–6

Library of Congress Catalog Card Number: 74–83047

Printed in the United States of America
6 7 8—82 81 80 79

PREFACE

The process of educating highly qualified electronics personnel has become complicated by the fact that both traditional and modern aspects of electronics have to be mastered in the amount of time previously devoted to traditional aspects alone. For this reason, it is essential that students in a two-year technology program start training in mathematics *and* in their technical specialty during the first semester.

Having realized this situation, most technical schools have constructed their electronics technology programs so that the student is introduced to the basics of electricity in the first semester; more often than not, this is the only electrical course studied in the first semester. Although this coverage has been sufficient in the past, the electronics faculty at Monroe Community College believes that the combined effects of an expanding technology and the increasing requirements of modern industry dictate a change from this traditional approach if the limited time available is to be used effectively.

At Monroe Community College the electronics program is organized so that students study both basic electricity (DC circuits) and electronic devices during the first semester. The electronic devices course is concerned with the operation, characteristics, parameters, limitations, and applications of electronic devices. This approach requires that courses be closely coordinated to give a maximum amount of subject coverage in a minimum amount of time. As soon as a mathematical tool is developed, it is applied in the electrical courses; as soon as a sufficient foundation is laid in the basic electricity course, this background is used to promote understanding of electronic devices. This process is continued through the second semester, when a second term of basic electricity (AC circuits) is studied along with a second term of electronic devices. The second course in devices is a continuation of the first in that it uses some of the same devices as the first course in amplifier circuits as well as introducing new ones. There are many important advantages to this approach:

(1) The student enters the program to study electronics and immediately starts training in this specialty, thus providing motivation.
(2) Introduction of electronics in the first semester makes it possible to achieve the depth of understanding of basics needed for advanced work later in the program.
(3) The student sees immediate application for the principles studied in the mathematics and basic electricity courses.

The material presented in this textbook makes up the major portion of the two freshman electronic devices courses at Monroe Community College. Although originally intended for use in two-year technician programs, the

first edition of this textbook was also used by several vocational school programs and even by a number of four-year baccalaureate programs. This wide
spectrum of users can probably be attributed to the fact that a minimum
amount of math (simple algebra) is all that is required to follow the material.

In preparing this second edition, many passages were rewritten to improve
clarity or to provide more extensive explanations of difficult concepts. A
number of illustrative examples were added where I felt they were needed, and
many new and practical homework problems were added to most of the
chapters. In addition to these general improvements, many specific changes in
the coverage of topics in the second edition were made. Many of these modifications were incorporated at the suggestion of first-edition users who were
kind enough to give constructive criticism. Still other changes were dictated by
the desire to include devices which have become important in the last five
years. In brief, the major modifications include:

(1) The coverage of photoelectric devices (Chapter 8) has been expanded
    to include more on the nature of light energy and the addition of
    LEDs and photodiodes.
(2) The transistor switch (previously Chapter 12) has been added to the
    chapter on basic transistor operation (Chapter 9).
(3) The monumental coverage of transistor amplification (previously
    Chapter 11) has been divided into three chapters to make it more
    palatable. In addition, this material has been extensively rewritten,
    using a generalized approach to amplifier analysis and a simple
    equivalent circuit for the transistor.
(4) The material on transistor technology (previously Chapter 13) has
    been eliminated.
(5) The coverage of PNPN devices (Chapter 14) has been expanded to in-
    clude material on PNPN devices in addition to the four-layer diode
    and the SCR.
(6) Chapter 15 on unijunction transistors has been expanded to include
    coverage of complimentary UJTs and programmable UJTs.
(7) Chapter 16 on field-effect transistors has been expanded to include
    more on biasing and amplification.
(8) The chapter on special devices (previously Chapter 17) has been
    eliminated.
(9) Chapter 17 on integrated circuits has been updated.
(10) Appendix VI has been added to cover the use of the ohmmeter in
    performing simple checks on semiconductor devices.

Once again I express my gratitude to Ms. Kathy Langworthy for her typing of
the new portions of this edition. In addition, I would like to thank all the
people who have provided valuable suggestions for improving this edition,
especially Professor Albert Camps and my worthy colleague Professor Joseph
G. Baker. I only hope that my revisions have come close to fulfilling their
recommendations.

R.J.T

CONTENTS

4 SEMICONDUCTOR PRINCIPLES 37

5 THE P-N JUNCTION 55

6 ZENER DIODES 97

7 TUNNEL DIODES   117

8 OPTOELECTRONIC DEVICES   131

13 COMMON-COLLECTOR AMPLIFIERS AND OTHER TOPICS 287

14 PNPN DEVICES 311

15   UNIJUNCTION TRANSISTORS   357

16   FIELD-EFFECT TRANSISTORS   385

1

Basic Atomic Theory

1.1. Introduction

Any comprehensive study of semiconductor devices must begin with a study of atomic theory. The extent to which it helps explain semiconductor phenomena warrants at least a brief look at atomic structure, based on the simple Bohr model of the atom. Although modern science points out the limitations of the Bohr model, for a nonrigorous study like ours, we can stick with it as long as we are prepared to accept some rather strange additional rules governing atomic structure. The concept of energy bands shall be introduced to help classify matter electrically, and to explain certain electrical properties of matter.

1.2. The Bohr Atom

The most fundamental unit of all matter is the atom. We shall begin consideration of materials leading to semiconductors by studying the single, isolated atom. Investigation of the properties of solid materials containing many combined atoms shall then follow.

For purposes of this study we can consider the atom to consist of three distinct types of particles: neutrons, protons and electrons. Neutrons and protons, the heaviest particles, make up the nucleus, or core, of the atom. Neutrons have no electrical charge, while protons are charged positively. Electrons, which are negatively charged, have a weight of about 1/1800 that of a neutron or proton. An atom contains an equal number of electrons and protons.

The Bohr model of an isolated atom depicts the nucleus, containing the heavier protons and neutrons, as the center of a miniature solar system about which the atom's electrons revolve. Figure 1.1 shows a symbolic representation of the Bohr model for hydrogen, which consists of one proton making up the nucleus, and one electron orbiting about the nucleus.

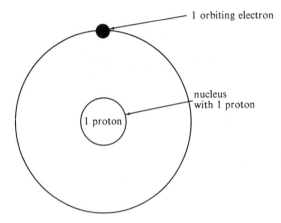

Figure 1.1.   BOHR MODEL FOR HYDROGEN ATOM

The analogy between the atom and the solar system is, in many instances, a good one. The forces and relationships between the nucleus and orbiting electrons are similar to those existing between the sun and its planets. There are, however, two major differences between atoms and our solar system. The more obvious is the extreme difference in physical size: whereas the solar system is billions of miles in diameter, a typical atomic diameter is 10^{-11} inches (10 trillionths of an inch). The other difference is in the nature of the attraction forces: our solar system is acted upon by gravitational forces; the forces existing in the atomic system are electrostatic, pertaining to Coulomb's law which specifies that like charges repel and unlike charges attract. The positive charges in the nucleus exert an electrostatic force on the orbiting electrons. (It should be noted that gravitational forces act only as attracting forces and never as repelling forces.) Noting these differences, the basic model of the two systems can be considered otherwise the same.

1.3.   Atomic Number and Atomic Weight

The different types of atoms make up the many different elements. These elements differ in the number of each particle (electron, proton, neutron) that make up their atoms.

The *atomic number* of an atom or element is given by either the number of protons in its nucleus or by the number of orbiting electrons. Copper, for example, has 29 protons in its nucleus (along with 34 neutrons) and 29 orbiting electrons. Thus its atomic number is 29.

The *atomic weight* of an atom or element is given *approximately* by the number of protons and neutrons in its nucleus, expressed in atomic mass units. Copper has an atomic weight of about 63. It should be noted that protons and neutrons have approximately the same weight, while electrons are much lighter (1/1800 of a proton or neutron).

1.4.   Electron Orbits and Energy

In accordance with the analogy to our solar system, the atom is made up of a heavy nucleus around which one or more electrons revolve in orbits. For each isolated atom, however, there are only a certain number of orbits available. These available orbits represent energy levels for the electrons; that is, each orbit corresponds to a certain value of total electron energy and, furthermore, no more than two electrons may exist in any one level, or orbit. This last rule is known in modern physics as *Pauli's exclusion principle*. It is very important to note that these energy levels, or orbits, exist at discrete levels for the atom.

At this point, it is advisable to introduce the energy unit which we will employ in all our work on atomic theory and semiconductors. The *electronvolt* is defined as that amount of energy gained or lost when an electron moves with or against a potential difference of one volt. In terms of *joules*, a common unit of energy, an electronvolt (abbreviated eV from now on) is equivalent to 1.6×10^{-19} joules. This is a very small amount of energy, indeed, when one realizes that a 200-lb man climbing one stair exerts almost 300 joules of energy. In terms of an electron, however, it is a great deal since an electron has so small a mass.

Returning to the subject of atomic energy levels, it is necessary to understand the difference between having a continuum of energies and having discrete values of energy available for electrons. To illustrate: If a continuum of energies were available, an electron could have *any* value of energy between, say, 1 eV and 2 eV. However, modern physics tells us that only discrete values of electron energies are possible. Thus an electron could not have *any* value of energy between 1 eV and 2 eV, but only certain permissible values, such as 1.00000000000000000001 eV or 1.5 eV, for example; no electron may exist at any energy level, or orbit, other than a permissible one. In addition, no more than two electrons of the same atom can have the same energy level, or orbit, at the same time.

For purposes of considering the chemical behavior of the atoms, a group of permissible levels, or orbits, are combined into electron "shells." The differences in energy levels within a shell are smaller than the difference in energy between shells. Thus the shells can be regarded as appropriate groupings of the possible orbits. The number of electrons existing in the various shells determines the chemical properties of the atom.

Electron shells in an atom are usually denoted by the letters *K*, *L*, *M*, *N*, ..., *Z*, the *K*-shell being the closest to the nucleus. Figure 1.2 shows the symbolic atomic structures of hydrogen, lithium and silicon.

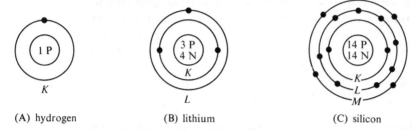

(A) hydrogen   (B) lithium   (C) silicon

Figure 1.2.   EXAMPLES OF VARIOUS ATOMIC STRUCTURES

We say that a particular atom is more complex than another if it contains more orbiting electrons. In other words, the silicon atom, since it has three shells and 14 electrons, is more complex than the lithium atom, which fourteen has only two shells and three electrons.

Electrons revolving in orbits far from the nucleus are influenced less by the electrostatic force of the protons in the nucleus than are electrons in closer orbits.* Accordingly, these more distant electrons possess a greater total energy.† In general, we can state that, as the orbital radius (the distance of the orbit from the nucleus) increases, the energy of electrons travelling these orbits increases. In other words, as an electron moves further from the nucleus, its energy increases. An electron orbiting very close to the nucleus in the *K*-shell is tightly bound to the nucleus by the electrostatic force and possesses only a small amount of energy. It would be difficult to knock this electron out of its orbit. On the other hand, an electron orbiting further from the nucleus would have a greater energy and could more easily be bumped out of its orbit. This is analogous to trying to knock someone off the top of a concrete wall. If the person is running along the wall, using lots of energy, it is much easier to knock him down than if he were standing rigidly still, using no energy.

1.5.   Valence Electrons

Many of the chemical properties of the different atoms are determined by the behavior of the electrons existing in the shell furthest from the nucleus. These electrons, being the most energetic, are the ones involved in chemical bonding

---

*Electrons in the inner orbits act as a shield from the nucleus; also, recall that coulomb force varies indirectly as the square of distance.

†Total energy includes kinetic and potential energies.

and chemical reactions. The outer shell electrons are called *valence electrons*. An atom with one electron in its outer shell, such as hydrogen, is said to have a *valence* of 1. In general, an atom with *n* electrons in its outer shell has a valence of *n*. Referring again to Figure 1.2, we can see that hydrogen and lithium have valences of 1, while silicon has a valence of 4.

When the outermost shell of an atom is completely filled, that atom is said to be inert and incapable of combining with other atoms of another element to form a compound. Conversely, if the outer shell is not completely filled, the atom is capable of forming compounds with other atoms. In forming these compounds, two atoms may share their electrons so that the combined electrons will fill the outer shells of both. Or one atom may donate its valence electrons to a second atom in order to fill the second atom's outer shell. These two types of compounds will be discussed in more detail when we talk about atomic bonding.

1.6.   Excitation of the Atom

The energy of an orbiting electron has two parts. The electron has *kinetic energy* due to its motion, and *potential energy* due to the electrostatic attraction of the nucleus. When energy is added to an electron, both its kinetic and potential energies are increased. It can easily be understood that, since an electron can only exist at discrete energy levels, the amount of energy it can receive must be discrete in value.

To re-emphasize the significant points concerning orbits, or energy levels, of the electrons in an isolated atom: the atom contains many energy levels corresponding to orbits in which it is permissible for an electron to exist, and there are many more of these orbits, or levels, than there are electrons. Every atomic electron must exist in one of the permissible energy levels. Since there are many available orbits, it is possible for an electron to move from orbit to orbit as it gains or loses energy. If an electron gains energy, it must move to a higher energy level in an orbit further from the nucleus (Section 1.4). An electron that loses some of its energy must drop to a lower energy level in an orbit closer to the nucleus (Section 1.4).

Since the electrons must exist in certain discrete energy levels, it is clear that when energy is absorbed or lost by an electron it is done so in discrete quantities. These discrete quantities, called *quantums of energy,* correspond to the differences in energy between the permissible levels for the electrons.

One of the most convenient mechanisms by which the energy of an atom may be increased is the application of heat. If we can visualize a single atom isolated in space, the application of heat would result in some electrons in the atom being raised to higher energy levels (orbits further from the nucleus). Those electrons with the highest energy to start would be the first to have their orbits changed. Thus it is the valence electrons in the outermost shell that would have their orbits raised first upon the application of external energy.

Another way to increase the electron energy of an atom is to direct light onto the atom. Light represents the visible portion of the entire electromagnetic spectrum and consists of the same radiation as radio waves, but the frequency is higher for the visible portion of the spectrum. If an atom absorbs electromagnetic radiation, the radiation will be converted into energy. Thus light represents a source of energy for raising the electron levels of an atom.

Light may be considered as possessing little packets, or quantums, of energy. As the frequency of the light increases, the energy increases. In order for the radiation (light) to raise the level of an electron, its quantum of energy must be greater than that needed to raise the electron to the next highest energy level. It is for this reason that only a portion of the electromagnetic spectrum is capable of giving energy to the electrons of solids.

When an atom absorbs any form of energy, then, the energy levels of some of the electrons (usually the valence electrons) are raised and the atom is said to be *excited*. It must be understood that in an excited atom the permissible energy levels do not change; the electrons simply jump from one level to another.

To illustrate what has been said thus far about excitation of an atom, Figure 1.3 shows the process of excitation for the hydrogen atom. In Figure 1.3A, the one valence electron is depicted as existing in the first permissible orbit ($n = 1$) before external heat is applied. The application of heat is sufficient to raise this valence electron to its second permissible orbit ($n = 2$), in which it is further from the nucleus and at a higher energy. Further application of heat could raise the electron to the third orbit ($n = 3$). A sufficient amount of heat may be applied so that the valence electron can be completely removed from the influence of the atom ($n = \infty$). This process is called *ionization*, and it leaves the atom with an excess positive charge since it has lost an electron. The positively charged atom is now called a *positive ion*. If the valence electron in Figure 1.3 were to escape from the atom's influence, a positive hydrogen ion, denoted H^+, would result. Negative ions are produced in the reverse manner. An external electron attaching itself to a normal atom causes the atom to acquire a negative charge and become a negative ion.

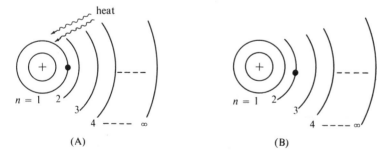

Figure 1.3.   (A) HYDROGEN'S VALENCE ELECTRON IN ITS FIRST PERMISSIBLE ORBIT (ENERGY LEVEL) ABOUT THE NUCLEUS BEFORE APPLICATION OF HEAT; (B) VALENCE ELECTRON IN SECOND PERMISSIBLE ORBIT AFTER APPLICATION OF HEAT

When a positive ion such as H^+, is formed, the liberated electron is virtually free of any electrostatic force from the nucleus, and thereby becomes susceptible to any externally applied force. As we shall see, these liberated electrons, along with the valence electrons, play an important role in semiconductor operation.

1.7.   Combinations of Atoms

The discussion of the properties of isolated atoms leads us naturally to consideration of combinations of atoms as they occur in nature. Here we shall be interested in combinations of atoms of the same element and of different elements. To this end, our attention will be focused on solid materials, ignoring for a time liquids and gases.

When atoms link together or interconnect in some fashion, they form *molecules* of matter. This linkage or interconnection, called *bonding*, takes place through the interaction of the valence electrons of each atom.

It is now pertinent to discuss the forces that bind atoms together. These forces are electrostatic in nature and can be divided into three categories: (1) *ionic* binding forces, (2) *covalent* binding forces, and (3) *metallic* binding forces. Although our chief concern will be with the covalent type, each will be discussed briefly.

When atoms bond together to form molecules of matter, each atom attempts to have eight electrons in its outer shell. The outer shell of each atom is considered filled when it contains eight electrons. In *ionic* bonding, the binding forces occur when the valence electrons from one atom join together with those of another atom to fill the latter's outer shell. If an atom has four or more valence electrons, it has a tendency to acquire additional electrons when joining with other atoms in order to fill its outer shell. When this occurs, the outer shell contains eight electrons and the binding forces are termed ionic since the atom acquiring the electrons becomes a negative ion and the atom losing electrons becomes a positive ion. A common example of an ionically bound molecule is salt, or sodium chloride, shown symbolically in Figure 1.4.

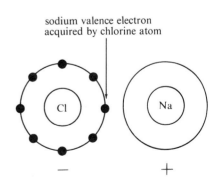

sodium valence electron
acquired by chlorine atom

Figure 1.4.   SODIUM CHLORIDE NaCl IONIC BOND

Only the valence electrons are shown. Sodium (Na), having one valence electron, donates it to the chlorine atom with seven valence electrons. The chlorine atom thus acquires a full outer shell of eight electrons. As a result of this electron transfer, the sodium atom becomes positively charged and the chlorine atom becomes negatively charged. Each atom has become an ion and the forces which bind them together are ionic forces.

Covalent bonding occurs when the valence electrons of neighboring atoms are shared among the atoms. An atom which contains four valence electrons may share one electron with each of four neighboring atoms. This concept of sharing electrons should be distinguished from the ionic situation of gaining or losing valence electrons. The important difference is that no ions are formed in a covalent bond. The covalent forces are established when two electrons coordinate their motions so as to produce an electrostatic force between the electrons. An illustration of covalent bonding between silicon atoms is given in Figure 1.5. Again, only the valence electrons are shown. The center silicon atom shares one of its valence electrons with each of the four surrounding silicon atoms, each of which in turn shares one of its valence electrons with the center atom. This sharing of valence electrons produces pairs of electrons that

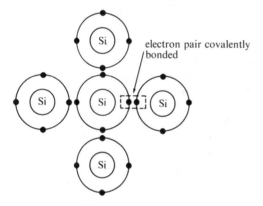

Figure 1.5.   COVALENT BONDING OF SILICON ATOMS

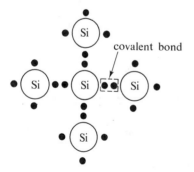

Figure 1.6.   COVALENT BONDING OF SILICON ATOMS

are covalently bonded, giving each silicon atom an apparent eight valence electrons in its outermost shell. Since in the following chapters we shall be looking
frequently at covalently bonded structures, the abbreviated type of symbol
shown in Figure 1.6 will henceforth be used.

The *metallic* binding forces are neither ionic nor covalent in nature, but are
rather thought of as occurring when positive ions float in a "cloud" of electrons. There is an electrostatic force between the positive ions and the negative
electrons which form a cloud about the ions. Since in a metal the electrons have
such a great mobility, they cannot be associated with any particular atom. The
atoms are held together by the attraction of these electrons and the resulting
positive ions. Figure 1.7 illustrates this for the metal copper, Cu.

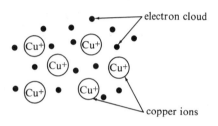

Figure 1.7.   METALLIC BONDING IN COPPER

1.8.   Electron Energies in Solids: Energy Bands

Referring to Section 1.4, it was stated that a single atom possesses many permissible energy levels in which its electrons can exist. Recall, also, that no electron can exist at an energy level other than a permissible one. For a single atom,
a diagram can be constructed showing the different energy levels available for
its electrons. Figure 1.8 is the energy-level diagram for the hydrogen atom. The
energy levels available to the single electron of the hydrogen atom are numbered $n = 1, 2, \ldots$, in increasing order of energy. There are an infinite number of energies between the various levels shown. However, the hydrogen
electron can exist only at one of the permissible levels. The higher the energy

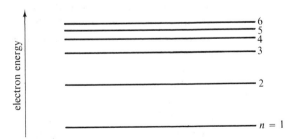

Figure 1.8.   ENERGY LEVELS OF A HYDROGEN ATOM

level, the further away from the nucleus is the electron's orbit. Note that the permissible energy levels are closer together as *n* increases.

When atoms bond together to form molecules of matter, this simple diagram of electron energies is no longer applicable. In a solid, the atoms are so close to each other that certain important changes occur in the state of the energy levels. When atoms are brought into close proximity, as in a solid, the energy levels which existed for single isolated atoms split up to form *bands* of energy levels. Within each band there are still discrete permissible energy levels rather than a continuum. Figure 1.9 shows the *energy-band diagram* for an atom in a silicon crystal.

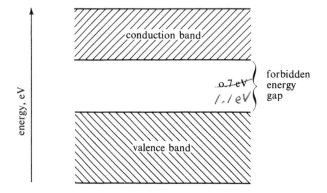

Figure 1.9.   ENERGY-BAND DIAGRAM FOR SILICON

Figure 1.9 shows only the two upper (highest-energy) bands of energy levels. There are many bands below the level of the *valence band* shown in this figure; however, only the two upper bands are of interest in considering electrical properties. The uppermost energy band is the *conduction band*. Separating these two bands is a *forbidden energy band*, which may not be occupied by any electron in the silicon crystal.

It must be stressed that the different energy bands contain many discrete energy levels which are much closer in energy than the levels in a single atom. Thus, in a solid structure, there are many more energy levels available for an electron, and these levels are grouped in bands.

The *valence band* of electron energies contains all the energy levels available to the valence electrons in the structure. These valence electrons (Section 1.5) are more or less attached to the individual atoms and are not free to move about as are the electrons in the conduction band. Every valence electron has an energy in the valence band.

The *conduction band* is a band of energies in which the level of energy of the electrons is high enough so that electrons in these levels are not attached or bound to any atom but rather are mobile and capable of being influenced by an external force. Electrons do not normally exist in the conduction band. The electrons that move to the conduction band are those electrons from the valence band that gain sufficient energy, through some form of excitation (Section

1.6), to be elevated to the conduction band. In order to do this, they must jump the forbidden energy band between the valence band and the conduction band. Referring to Figure 1.9 for silicon, an electron existing at an energy level near the top of the valence band needs to gain 0.7 eV of energy in order to jump the gap and reach the bottom of the conduction band. It is this energy difference across the forbidden energy band that determines whether a solid behaves as a *conductor*, *insulator* or *semiconductor*.

Electric current can be defined as the movement of charges. Since electrons are negatively charged particles, then it is logical to conclude that the ability of a material to conduct electricity depends upon the availability of free, or conduction-band, electrons within the material.

A *conductor* is a solid containing many electrons in the conduction band at room temperature. In fact, there is no forbidden region between the valence and conduction bands on a good conductor's energy-band diagram. The two bands actually overlap as shown in Figure 1.10A. Since the valence-band energies are the same as the conduction-band energies for a conductor, it is very easy for a valence electron to become a conduction (free) electron. Thus conductors have many conduction electrons available to conduct electric current without the need for applied energy, such as heat or light.

An insulator material has an energy-band diagram with a very wide forbidden energy band, such as that in Figure 1.10B. The forbidden energy band is so wide that practically no electrons can be given sufficient energy to jump the gap from the valence band to the conduction band. In the ideal insulator, all the levels of the valence band are occupied by electrons and the conduction band is empty. Thus a perfect insulator has no conduction electrons and will not conduct an electric current; practical insulators, of course, have a few conduction electrons and conduct a very small current.

A *semiconductor* is a solid which has a forbidden energy band as shown in Figure 1.10C. Its forbidden energy band is much smaller than that for an insulator but larger than that for a conductor. Normally, since it has this forbidden energy band, it would be considered that a semiconductor has no electrons in its conduction band. However, the energy provided by the heat of

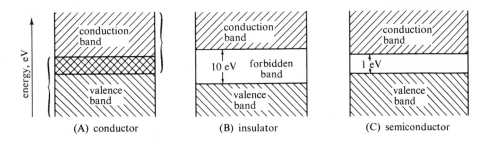

(A) conductor   (B) insulator   (C) semiconductor

Figure 1.10.   ENERGY-BAND DIAGRAMS FOR THREE ELECTRICAL CLASSIFICATIONS OF MATTER: (A) CONDUCTOR; (B) INSULATOR; (C) SEMICONDUCTOR

room temperature is sufficient to overcome the atomic bonding forces on a few valence electrons so that some can jump the gap into the conduction band. Therefore, at room temperature, semiconductors are capable of conducting some electric current. It is this class of materials which shall be of interest to us.

GLOSSARY

Atomic number: equal to the number of protons or electrons of an atom.

Atomic weight: equal approximately to the total number of protons and neutrons in an atom's nucleus.

Electronvolt: unit of energy equivalent to 1.6×10^{-19} joules.

Permissible energy levels: those values of energy available for the electrons of an atom.

Electron shells: groupings of permissible energy levels.

Valence electrons: electrons existing in an atom's outermost shell.

Atomic excitation: application of external energy of some form to an atom.

Ionization: process of an atom either losing or gaining an electron so as to leave it with a nonneutral charge.

Ion: charged atom (either $+$ or $-$).

Molecules: atoms bonded together.

Bonding: linking or interconnecting of atoms.

Ionic bonding: type of atomic bonding whereby one atom donates valence electrons to another.

Covalent bonding: type of atomic bonding whereby the atoms share one another's valence electrons.

Metallic bonding: type of electrostatic bonding between positive nuclei and negative electron clouds occurring in metals.

Energy bands: groupings of energy levels which occur when atoms bond together in solid structures.

Valence energy band: band of energies available for valence electrons.

Conduction energy band: band of energies available for conduction or free electrons.

Forbidden energy band: band of energies between valence and conduction bands containing nonpermissible energy levels.

Electric current: movement of charges.

Conductor, semiconductor, insulator: electrical classification of matter in decreasing order of ability to conduct electric current.

PROBLEMS

1.1   Describe the Bohr model of an isolated atom of silicon. (See Figure 1.2.)

1.2   What is the atomic number and weight of lithium? (See Figure 1.2.)

1.3   Explain the difference between discrete energy levels and a continuum of energy levels.

1.4   Will an electron in the *M*-shell of an atom have a higher or lower energy than one in the *L*-shell?

1.5   What is the valence of an atom with three electrons in its outer shell?

1.6   Why can't an orbiting electron have its energy increased to any and all values?

1.7   What happens to an electron's orbit if it loses energy?

1.8   Name two forms of energy which can excite an atom.

1.9   What happens to an atom when an electron receives enough energy to free itself from its atom? What is this called?

1.10 Name the three types of atomic bonding. Briefly describe each.

1.11 When atoms share electrons, what is their bonding called?

1.12 Explain the differences in conductors, semiconductors and insulators using the energy-band diagram concept.

1.13 What will happen to the number of electrons in the conduction band of a semiconductor as temperature of the material is increased? Decreased?

REFERENCES

Anderson, D. L., *The Discovery of the Electron*. Princeton, N. J.: D. Van Nostrand Co., Inc., 1964.

Branson, L. K., *Introduction to Electronics*. Englewood Cliffs, N. J.: Prentice-Hall, Inc., 1967.

Foster, J. F., *Semiconductors, Diodes and Transistors*, Vol. 1. Beaverton, Oregon: Programmed Instruction Group, Tetronix, Inc., 1964.

Riddle, R. L. and M. P. Ristenbatt, *Transistor Physics and Circuits*. Englewood Cliffs, N. J.: Prentice-Hall, Inc., 1958.

Romanowitz, H. A. and R. E. Puckett, *Introduction to Electronics*. New York: John Wiley & Sons, Inc., 1968.

$$K = 9 \times 10^9 \ \frac{N m^2}{c^2}$$

2

Basics of Current Flow*

2.1. Introduction

The study of electronics is basically a study of the movement of charges. This movement of charges constitutes *current* flow. In this chapter we shall briefly discuss the concepts of current, resistance of different materials to the flow of current and the conditions under which a current will flow. The terms *drift current* and *diffusion current* will be introduced as necessary background for the work in semiconductors to follow.

2.2. Electric Current

If a potential difference or voltage is applied across a conductor — for example, by connecting it across a battery — the free electrons in the conductor will move or drift toward the positive terminal of the battery. This movement of electrons is an electrical current. In a lightning stroke, mother nature's brilliant generation of great electrical currents, both negative and positive charges move

---

*The popular term *current flow* is actually redundant since charge is what flows and current is charge flow. However, we shall use this term throughout the text.

— in opposite directions. In high-energy accelerators (cyclotrons), charges are made to move about, and again this movement or flow of charges constitutes an electrical current. The flow of electrons in conductors and semiconductors, and between the electrodes of vacuum tubes, are examples of currents encountered in electronics.

Since current is a flow of charge, the unit current is given in terms of a given amount of charge flowing past a certain point per second. The unit of charge used is a *coulomb*, which is equivalent to the charge on 6.25×10^{18} electrons or protons. The unit of current is called the *ampere*. Thus

$$1 \text{ ampere} = 1 \text{ coulomb/second} \tag{2.1}$$

In terms of electron flow,

$$1 \text{ ampere} = 6.25 \times 10^{18} \text{ electrons/second} \tag{2.2}$$

Figure 2.1 illustrates the ampere of electric current flowing in a conductor.

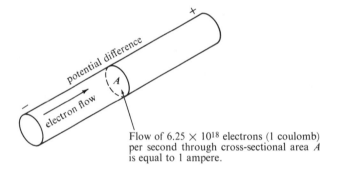

Flow of 6.25×10^{18} electrons (1 coulomb) per second through cross-sectional area A is equal to 1 ampere.

Figure 2.1.   ELECTRIC CURRENT

In conductors or semiconductors, the only charge that moves or flows is the charge carried by electrons. Even the movement of holes* (which are thought of as positive charges) in semiconductors is really caused by the jumping of valence electrons from hole to hole. In light of this, the direction of current flow will henceforth be defined as the direction of electron flow. This conflicts with the traditional definition of current flow as being in the direction of positive charge flow. However, since in material bodies electrons are the particles which are free to move about, it will aid the understanding of semiconductor operation if current flow is synonymous with electron flow. Thus, in Figure 2.1 we shall consider the current as flowing from left to right.

Table 2.1 gives the fractional parts of an ampere which shall be used in this text and with which the student should become familiar.

---

*"Holes" will be discussed in detail with semiconductor theory.

TABLE 2.1

AMPERE UNITS

1 ampere	= 1 ampere (A)
10^{-3} ampere	= 0.001 ampere = 1 milliampere (mA)
10^{-6} ampere	= 0.000001 ampere
	= 1 microampere (μA) (μ is the Greek letter mu)
10^{-9} ampere	= 0.000000001 ampere
	= 1 nanoampere (nA)
10^{-12} ampere	= 0.001 nanoampere
	= 1 picoampere (pA)

2.3. Resistivity and Conductivity

In the previous chapter we defined conductors, insulators and semiconductors in terms of their energy-band structures. In this section we shall consider an equivalent definition of electrical properties of materials which will augment, but not replace, the previous definition.

Resistivity is a measure of the degree to which a material opposes or resists the flow of electrical current; *conductivity* is the degree to which a material allows current to flow. A material which exhibits high opposition to current flow is said to have a high resistivity. The same material is said to have a low conductivity. The symbols ρ (the Greek letter rho) and σ (the Greek letter sigma) will be used for resistivity and conductivity, respectively.

Different materials have different values of resistivity. The resistivity of rubber, which is an insulator, is far greater than that of copper, which is a conductor. The resistivity of silicon, which is a semiconductor, is between that of rubber and copper.

The resistivity of a material is determined by a number of factors, including the atomic structure of the material, its temperature and the density of free charge carriers available to move under an external force such as an electric potential. Resistivity is a measurable quantity and is independent of the geometrical shape of the material. It is an intrinsic property in the same sense that color is, for example. A long, thin wire of copper has the same color and resistivity as a large, thick copper brick.

Resistivity has the units of *ohm-centimeters* in the CGS (centimeter-gram-second) system of units. *Ohms* are the familiar units of electrical *resistance*. Resistance is also a measure of opposition to current, but it depends on both the resistivity of a material and on its geometry (length and cross-sectional area) and is therefore useless as a means of classifying materials electrically. Conductivity is the exact opposite of resistivity and, mathematically, is equal to the reciprocal of ρ. That is,

$$\sigma = \frac{1}{\rho} \tag{2.3}$$

As such, the units for σ are 1/ohm-cm, or mho/cm, since a mho is the reciprocal of an ohm.

Table 2.2 lists several materials and their resistivities. Conductors are considered to have resistivities below 10^{-3} ohm-cm; semiconductors are in the range of 10^{-3} to 10^6 ohm-cm; and insulators are materials with resistivities greater than 10^6 ohm-cm. Thus, in Table 2.2, all the materials listed above and including porcelain are classified as insulators. Silicon, germanium and carbon are semiconductors and the metals platinum, aluminum, copper and silver are conductors. This classification of materials is in exact agreement with the energy-band classification, since materials with a very narrow forbidden band have the lowest resistivities and materials with a large forbidden band have the highest resistivities.

TABLE 2.2

RESISTIVITIES IN OHM-CM

Insulators	Fused quartz:	10^{19}
	Hard rubber:	10^{18}
	Nylon:	4×10^{14}
	Glass:	17×10^{12}
	Porcelain:	3×10^{11}
Semiconductors	Silicon (pure):	2×10^5
	Germanium (pure):	65
	Carbon:	4×10^{-3}
Conductors	Platinum:	10^{-5}
	Aluminum:	2.8×10^{-6}
	Copper:	1.7×10^{-6}
	Silver:	1.6×10^{-6}

2.4.   Conditions Producing Current

In the study of semiconductors, two current-producing conditions, or mechanisms, are encountered. One is the already familiar electric potential, or voltage. The other, less familiar, is called a *concentration gradient* and is very important in semiconductor work. Both of these mechanisms occur in many of the semiconductor devices which we shall study in this text.

Drift current

Figure 2.2 illustrates the mechanism of current flow produced in a conductive material when a potential difference is applied across it. The free electrons in the material are accelerated by the potential difference toward the positive end of the material. These conduction electrons, however, move erratically through the material as they encounter collisions with atoms in the structure. We can

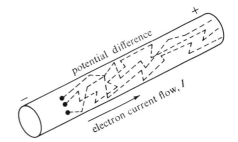

Figure 2.2.   ELECTRON DRIFT CURRENT

ignore this erratic motion and consider that the electrons drift toward the positive end at a much slower speed than if they were moving in a vacuum with no atoms impeding their movement. This drifting of electrons caused by the potential difference constitutes an electric current in the direction shown. This current is often referred to as *drift current* and is the most common type. Whenever a voltage is applied to a wire or resistor, the current that flows is actually a drift current.

Diffusion current

It is possible for a current to flow in a material even in the absence of an applied voltage if a *concentration gradient* exists in the material. A concentration gradient occurs when the concentration of charges is greater in one part of the material than in some other part. Figure 2.3 depicts this situation in a bar of silicon. When a concentration gradient of charge carriers exists in a material, the charge carriers tend to move or *diffuse* from the region of high concentration to the region of lower concentration, much as the molecules of a gas diffuse to fill a container. The current produced by this movement of charge is called *diffusion current* and is of major significance in semiconductor electronics. In the silicon bar, the electrons will tend to move from left to right from the region of high electron concentration to the region of lower electron concentration in an effort to reach an equilibrium situation of uniform concentration

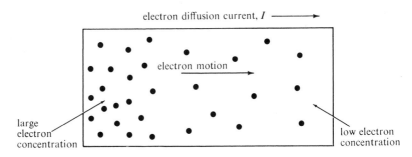

Figure 2.3.   DIFFUSION CURRENT RESULTING FROM A CONCENTRATION
GRADIENT OF ELECTRONS

throughout the bar. The direction of the diffusion current is thus left to right as shown. The same action takes place for positive-charge concentration gradients as is shown in Figure 2.4. Here the positive charges are symbolized by little zeros, or holes, for reasons which will become clear later. Note that the diffusion current direction in this case is *opposite* to the flow of the positive charges, since we have defined current flow to be in the direction of electron flow, that is, negative charge flow. Positive charges moving in one direction produce the same current effect as negative charges moving in the opposite direction.

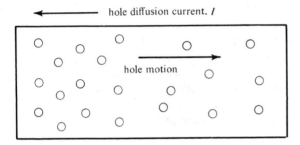

Figure 2.4.   DIFFUSION CURRENT RESULTING FROM A CONCENTRATION GRADIENT OF HOLES (POSITIVE CHARGES)

To summarize, *drift current* occurs whenever a potential difference (voltage) is applied to a body, causing any free charges in the material to move toward the opposite polarity; *diffusion current* occurs whenever an imbalance of charges (concentration gradient) exists in a material so that the charges repel each other from an area of high charge concentration to one of low concentration. For either type of current, it should be remembered that the ease with which the charges move depends on the resistivity of the material.

GLOSSARY

Electric current (I):   movement of charges.

Coulomb: amount of charge on 6.25×10^{18} electrons or protons.

Ampere: unit of electric current equivalent to one coulomb per second.

Resistivity (ϱ):   measure of the ability of a material to resist the flow of electric current.

Conductivity ($\acute{o}$):   opposite of resistivity.

Drift current: current produced by movement of charges under influence of a potential difference.

Concentration gradient: a difference in charge concentration within a material.

Diffusion current: current produced by movement of charge due to a concentration gradient.

PROBLEMS

2.1 Two copper wires, with cross-sectional areas of 1 cm² and 2 cm², respectively, both have charge flowing through them at the rate of 0.5 coulomb per second. Which one has the larger current flowing through it?

2.2 Will any platinum wire always have a greater resistivity than any copper wire? Will it necessarily have a greater resistance to current?

2.3 Explain the difference between resistivity and resistance.

2.4 What condition is necessary to produce a drift current?

2.5 Describe the condition necessary to produce a diffusion current.

2.6 Answer *true* or *false:*
(a) No current will flow when a voltage is applied to a piece of fused quartz.
(b) As the conductivity of a material increases, its resistance decreases.
(c) The movement of positive ions such as H^+ does not constitute *electrical* current.
(d) A block of germanium will *always* have a smaller resistivity than an identical block of silicon.
(e) Diffusion current does not require an applied voltage.
(f) Drift current only takes place in wires and resistors.

2.7 For each of the diagrams in Figure 2.5, indicate the direction of electrical current as defined in this text. The dots represent electrons and the circles represent holes (+ charges).

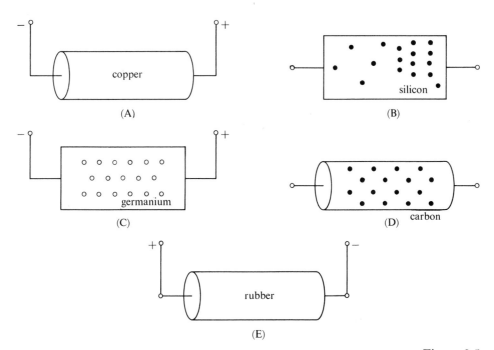

Figure 2.5.

REFERENCES

Branson, L. K., *Introduction to Electronics*. Englewood Cliffs, N. J.: Prentice-Hall, Inc., 1967.

General Device Analysis

3.1. Introduction

This chapter constitutes a general introduction to device analysis. In the discussion, no distinction will be made between the various electronic devices; the treatment will be generally applicable to any type of device. A device will be treated as a "black box" with consideration given only to its terminal characteristics and how these terminal characteristics are used to analyze the device's operation. Two- and three-terminal devices will be discussed in detail and the results extended to devices with more than three terminals.

3.2. The General n-Terminal Device

Figure 3.1 depicts a general n-terminal device as a box with n terminals or leads emanating from it (n can be any number greater than one). The terminals are physically and electrically connected by conductive material to points within the box. The box, or black box,* as it is commonly called, represents the actual

---

*The term "black box" usually refers to a physical device, circuit or system whose internal makeup is not known but must be deduced from external measurements.

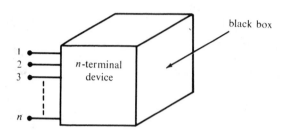

Figure 3.1.   SYMBOLIC REPRESENTATION OF A GENERAL *n*-TERMINAL
DEVICE

physical makeup of the device, which in general is not box shaped. Each terminal of the device is available for electrical connection to external points, such as the terminals of other devices, battery terminals, ground terminals or another of its own terminals.

The black box contains the internal physical and chemical properties of the device that determine its particular electrical characteristics. In this chapter we will not be concerned with what is happening inside the black boxes, but rather with what sort of information we can obtain and use simply by taking external measurements at the device terminals. After all, this is the information available to the circuit designer or analyst and is the data needed to determine how a device will operate in a circuit in conjunction with other devices and electrical components.

Figure 3.2 is an example of a familiar two-terminal device, the resistor. Its physical appearance is shown in part A and its standard electronic symbol is shown in part B. In the resistor symbol, the terminals are denoted *a* and *b*. The squiggly line with the letter *R* represents the physical resistance. Without examining or knowing the composition of the resistor, we can find out all we need to know simply by measuring the resistance between the terminals or by determining the relationship between the current flow in the resistor and the voltage applied across its terminals. This is illustrated in Figure 3.3. For the resistor, this is a straight-line relationship as indicated in the graph of resistor current, I, versus resistor voltage, V. This graph is called the static *I-V characteristic* of the resistor since it is obtained by applying a constant (static) voltage and measuring the resultant constant current, or vice versa. Throughout the text, consideration will be given to the static *I-V* characteristics of many devices. In doing so, the word *static* will be dropped and *I-V* characteristics

Figure 3.2.   THE RESISTOR, A TWO-TERMINAL DEVICE: (A) PHYSICAL
APPEARANCE; (B) SYMBOL

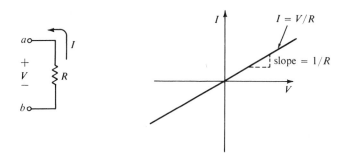

Figure 3.3.   CURRENT-VOLTAGE   RELATIONSHIP   OF   RESISTANCE R BE-
TWEEN TERMINALS a AND b

will be referred to with the understanding that they are static characteristics. In practice, a device's static I-V characteristic is sometimes obtained by applying a varying voltage to the device and noting the variation in the device current. This practice is valid as long as the applied voltage varies slow enough so that the device responds the same as it would to a static applied voltage.

The procedure of obtaining the I-V characteristics between any two terminals of a device will be followed for every device we study in this text, regardless of the total number of terminals each has. However, the procedure is slightly more complicated for multiterminal (more than two terminals) devices than for two-terminal devices like the resistor above. The difference is that for a multiterminal device the I-V characteristic between any two terminals depends on what is occurring at the other terminals. The next two sections describe the general procedure for determining the complete static electrical characteristics of two- and three-terminal devices. This procedure can then be easily extended to devices with four or more terminals.

3.3.   Two-Terminal Device Characteristics

A general two-terminal device is shown in Figure 3.4. Its terminals are denoted a and b. The voltage between terminals a and b is given by V_{ab}, which is defined as the voltage at a relative to the voltage at b. For instance, if we placed a 6-volt battery such that its positive terminal was at a, then V_{ab} would be $+6$ volts since terminal a is 6 volts positive with respect to terminal b. This is illustrated in part B of Figure 3.4. If we were to reverse this 6-volt battery, then V_{ab} would be -6 volts since terminal a would be 6 voltages negative with respect to terminal b. Thus V_{ab} is positive if terminal a is more positive than terminal b, and negative if terminal a is more negative than terminal b.

The current flow I_a through terminal a is shown as coming out of terminal a. This direction is chosen since a positive value of V_{ab} will cause current to flow in this direction, electrons flowing toward the positive voltage. Similarly, I_b

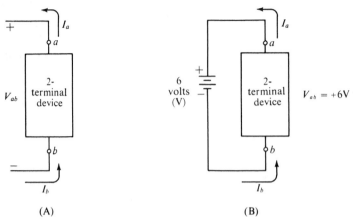

Figure 3.4.   (A) GENERAL TWO-TERMINAL DEVICE; (B) DEVICE WITH 6-
VOLT BATTERY APPLIED BETWEEN ITS TERMINALS

flows into terminal b, away from the negative voltage. Note that in a two-
terminal device I_a and I_b are always equal since the current into one end of the
device must equal the current coming out of the other end.

For a two-terminal device, the only I-V characteristic to determine is the
relationship between I_a and V_{ab}. To re-emphasize a point, the information obtained from the I-V characteristic is needed in order to be able to determine the
device's voltage and current as it operates in a circuit with other circuit elements. The I-V characteristic of a two-terminal device is obtained simply by

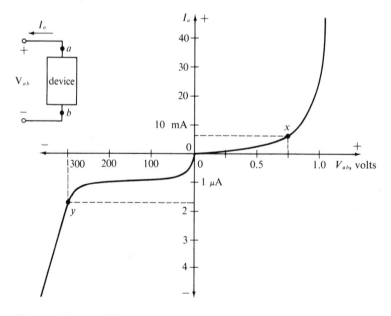

Figure 3.5.   *I-V* CURVE FOR A TYPICAL TWO-TERMINAL DEVICE

applying various values of V_{ab} and measuring the corresponding values of I_a. Figure 3.5 shows a typical *I-V* characteristic for one of the two-terminal devices that we will study in subsequent chapters. Note that the curve is *nonlinear*. This means that the curve is not a straight line like the *I-V* characteristic of the resistor in Figure 3.3. A device with a nonlinear *I-V* curve is called a nonlinear device. Also note in Figure 3.5 that the scale values for positive current and voltage are different from the scale values for negative I and V. This is a common characteristic of many electronic devices.

▶ EXAMPLE 3.1

Use the *I-V* curve in Figure 3.5 to determine the device current, I_a, for $V_{ab} = +0.75$ V and for $V_{ab} = -300$ V.

The value of I_a for $V_{ab} = +0.75$ V is obtained from the *I-V* curve by drawing a vertical line (shown dashed in the figure) through the $V_{ab} = +0.75$ V coordinate. This line intersects the *I-V* curve at point x. A horizontal line is drawn through x and intersects the I_a axis at $+6$ mA. Thus $I_a = +6$ mA for $V_{ab} = +0.75$ V. These are the *I-V* coordinates of point x.

A similar procedure is followed for $V_{ab} = -300$ V, resulting in point y on the *I-V* curve. At point y, the device current is $I_a = -1.7$ μA. The negative value of current indicates that, in this case, the current through terminal a of the device is actually flowing in a direction opposite to the direction shown on the diagram. This is because the applied voltage $V_{ab} = -300$ V makes terminal a negative, thereby repelling electrons into the device at terminal a.   ◀

3.4.   Three-Terminal Device Characteristics

We saw in the previous section that for two-terminal devices only *one I-V* characteristic is needed to describe their operation. For three-terminal devices, *two I-V* characteristics are required, but the procedure is complicated somewhat by the effect of the third terminal on the *I-V* characteristics between the two terminals of interest. A general three-terminal device is shown in Figure 3.6 with its

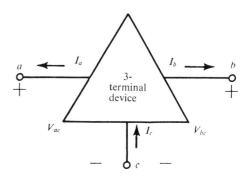

Figure 3.6.   GENERAL THREE-TERMINAL DEVICE

terminals denoted a, b and c. Between terminals a and c, the voltage is denoted as V_{ac}, the voltage at a relative to c. Similarly, V_{bc} is the voltage at b relative to c. V_{ab}, which is the voltage between a and b, is not shown since only two I-V characteristics are needed for a three-terminal device. A third I-V characteristic would give no new information. The assumed current directions indicated for each terminal are such that electrons flow away from the negative and toward the positive voltage.

In Figure 3.6, terminal c is common to both voltages V_{ac} and V_{bc}. For this reason, the two I-V characteristics obtained using these voltages, namely I_a *versus* V_{ac} and I_b *versus* V_{bc}, are called "common-c characteristic curves." We could choose a or b as the common terminal and obtain the common-a or common-b characteristic curves. However, since this is a general discussion, we can concern ourselves with the common-c characteristics without loss of generality. In order to distinguish between the two common-c I-V characteristics, we can arbitrarily refer to a as the *input* terminal and b as the *output* terminal (see Figure 3.7). The relationship between I_a and voltage V_{ac} can be called the *common-c input characteristic*, and the relationship between I_b and V_{bc}, the *common-c output characteristic*. In practice, the choice of input and output terminals is determined by the particular device and its mode of operation.

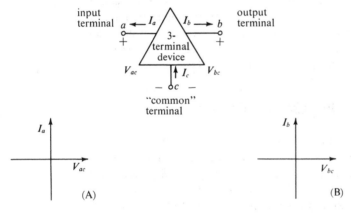

Figure 3.7.   THREE-TERMINAL DEVICE: (A) LABELED AXES FOR INPUT CHARACTERISTICS; (B) LABELED AXES FOR OUTPUT CHARACTERISTICS

Refer to the device diagram in Figure 3.7. It may appear inconsistent that the *input* current I_a is shown flowing *out* of terminal a rather than into a. On the contrary. The direction of I_a is consistent with the polarity of the input voltage V_{ac} as shown. It should not be assumed that input current always flows into the device and output current flows out of the device. The actual current directions depend on the device voltage polarities. For example, if the input voltage were such that the input terminal a were negative relative to common terminal c, then the input current direction would be into the device. Similarly, the output current I_b would flow into terminal b if the output voltage were such that b was negative relative to c. It is best to think of the input current as the current flowing in the input terminal (either direction) and the output current as the current flowing in the output terminal.

Families of *I-V* characteristic curves

We have seen that for a two-terminal device the complete *I-V* characteristic consisted of one curve relating the device current to its voltage. A three-terminal device has two *I-V* characteristics, as is the case in Figure 3.7. Each of these *I-V* characteristics consists of a set, or *family*, of curves. This is because for most three-terminal devices the relationship between current and voltage at a given pair of terminals, say a and c, depends on the value of current and/or voltage at terminal b due to the physical properties of the device. This means that for each value of current (or voltage) at terminal b there will be a different *I-V* curve between a and c. This is illustrated in Figure 3.8, where a family of input characteristic curves for a typical three-terminal device is shown.

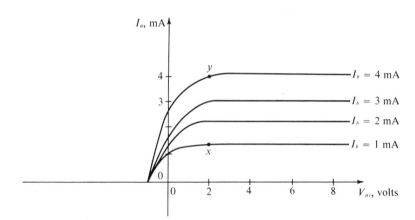

Figure 3.8.   ILLUSTRATION OF INPUT CHARACTERISTICS FOR A HYPOTHETICAL THREE-TERMINAL DEVICE

Each curve shows how the input current I_a varies with the input voltage V_{ac} for a given *fixed* value of current I_b, a different curve resulting for each value of I_b. The laboratory procedure for obtaining the family of curves requires that I_b be held fixed at a given value while V_{ac} is varied and the resulting values of I_a are recorded.

▶ EXAMPLE 3.2

For the device characterized in Figure 3.8, determine the input current when the input voltage is 2 V and $I_b = 1$ mA. Repeat for $I_b = 4$ mA.

For $I_b = 1$ mA we must use the *I-V* curve labeled as such. Using this curve, we can determine that for $V_{ac} = 2$ V the current is approximately 1.2 mA (point x on curve).

For $I_b = 4\,\text{mA}$, the upper curve is used, resulting in $I_a = 4\,\text{mA}$ for $V_{ac} = 2$ V (point y). ◀

The family of input *I-V* curves in Figure 3.8 are examples of *current-controlled* characteristics, because the *current* at terminal *b controls* the relationship between I_a and V_{ac}. A possible family of *voltage-controlled* input *I-V* curves is shown in Figure 3.9, where the voltage at terminal *b* relative to common terminal *c* (V_{bc}) controls the I_a-V_{ac} relationship. Here different input *I-V* curves result for different fixed values of V_{bc}. The nature of the device usually dictates whether the current-controlled or voltage-controlled curves are used.

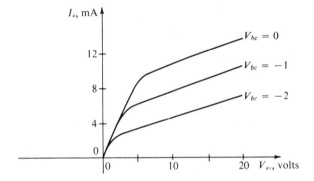

Figure 3.9.   ANOTHER POSSIBLE SET OF INPUT CHARACTERISTICS

The output *I-V* characteristics of a three-terminal device also consist of a family of curves relating the output current to the output voltage for various *fixed* values of input current or voltage. Figure 3.10 illustrates. In part A of the figure, a family of *current-controlled* output *I-V* curves are shown where the I_b-V_{bc} relationship depends on the value of input current I_a. In B, a family of *voltage-controlled* output *I-V* curves are shown where the input *voltage* V_{ac} determines the I_b-V_{bc} relationship.

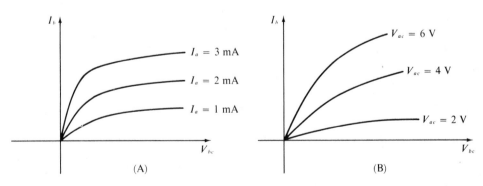

Figure 3.10.   OUTPUT   CHARACTERISTICS: (A)   CURRENT-CONTROLLED;
                  (B) VOLTAGE-CONTROLLED

To completely describe the characteristics of any three-terminal device requires *one* set of input *I-V* curves (voltage- or current-controlled) and *one* set of
output *I-V* curves (voltage- or current-controlled). Figure 3.11 shows the complete common-*c* *I-V* characteristics for a particular device.

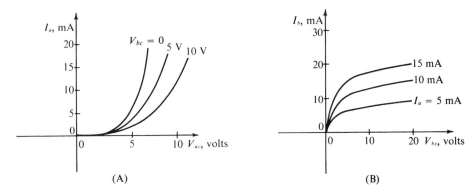

Figure 3.11.   COMMON-*c* CHARACTERISTICS: (A) INPUT CHAR-
ACTERISTICS; (B) OUTPUT CHARACTERISTICS

It must again be pointed out that the *I-V* relationship between any two
terminals depends on the third. For example, in the device shown in Figure
3.11, the relationship between I_a and V_{ac} depends on the value of V_{bc}. In analyzing a circuit, it is necessary to know the value of V_{bc} in order to determine which
curve to use. Similarly, the value of I_a must be known in order to determine
which output curve to use. This interdependency may present difficulty at this
time. However, with most devices, approximations can be made which greatly
simplify the circuit analysis.

In summary, the added complexity of three-terminal device characteristics
over that of two-terminal devices is a result of the physical interaction among
the three terminals. This causes the *I-V* characteristics between any two terminals to consist of a family of curves rather than one curve as for two-
terminal devices.

3.5.   Devices with Four or More Terminals

The procedure for determining the complete characteristics of a device with
more than three terminals is the same as in the three-terminal case. A common
terminal is chosen and the *I-V* characteristics between each terminal and the
common terminal are found. Figure 3.12 illustrates one possible terminal desig-
nation for a four-terminal device. Here d is the common terminal, and the
complete characteristics consist of the relationships I_a versus V_{ad}, I_b versus V_{bd}
and I_c versus V_{cd}. Each of these relationships consists of a family of curves, a
different curve for different conditions on the other two terminals. For example, Figure 3.13 shows an I_b versus V_{bd} family of curves for a particular
device.

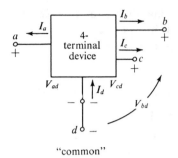

"common"

Figure 3.12.   CONFIGURATION FOR DETERMINING COMMON-d CHAR-
ACTERISTICS FOR A FOUR-TERMINAL DEVICE

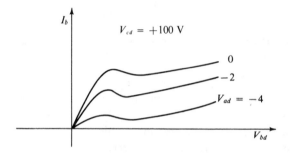

Figure 3.13.   EXAMPLE OF CURVES RELATING I_b TO V_{bd}

In this particular example, a different curve relating I_b to V_{bd} is obtained for
different sets of conditions on terminals a and c, namely V_{ad} and V_{cd} (I_a, or I_c or
both could also be used).

It is important to remember that, no matter how many terminals there are,
the conditions on each terminal must be specified (current or voltage relative
to common), and the *I-V* characteristics between each terminal and the common terminal constitute a complete set of characteristics. Using these characteristics can become extraordinarily complicated for more than two terminals unless some approximations can be made. We shall see that this is
usually the case.

GLOSSARY

n-terminal device: electronic device with n terminals or connections emanating from
it.

Black box: representation of the physical properties of a device used when one is
interested only in its terminal characteristics.

(Static) *I-V* characteristic: graph of current versus voltage between two terminals of a device (Section 3.2).

V_{ab}:   voltage at a with respect to b (Section 3.3).

Common-*c* characteristics: complete *I-V* characteristics of a device using the c terminal as the common terminal.

Input (output) characteristic: *I-V* characteristics between the designated input (output) terminal and the common terminal (Section 3.4).

Current-(voltage-)controlled *I-V* curves: *I-V* curves controlled by value of current (voltage) at other terminal.

PROBLEMS

3.1   In the circuit of Figure 3.14, what are the values of V_{ac}, V_{bc} and V_{ab}?

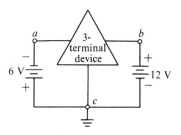

Figure 3.14.

3.2   How many sets of *I-V* characteristics are needed for a two-terminal device? A four-terminal device?

3.3   In the circuit of Figure 3.15, is the three-terminal device in the common-*a*, common-*b* or common-*c* configuration? What is the direction of I_b?

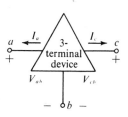

Figure 3.15.

3.4 For the circuit of Figure 3.15, choose one of the terminals as input and the other as output; then draw and label the axes needed to plot the input and output characteristics.

3.5 Figure 3.16 shows the output *I-V* characteristics between terminals *a* and *c* for a particular three-terminal device. What is the value of I_a when $V_{ac} = 20$ volts and $V_{bc} = 0$ volts? Repeat for $V_{bc} = -1$ volt; -2 volts. The value of V_{bc} determines which curve relating I_a to V_{ac} is to be used. A device whose output *I-V* characteristics are controlled by a voltage at another terminal is said to be a *voltage-controlled* device.

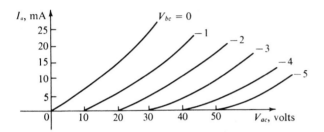

Figure 3.16.

3.6 Figure 3.17 shows the output *I-V* characteristics between terminals *a* and *c* for a particular three-terminal device. What is the value of I_a when $V_{ac} = 10$ volts and $I_b = 1$ mA? Repeat for $I_b = 2$ mA; $I_b = 3$ mA.

The value of I_b determines which curve relating I_a to V_{ac} is to be used. A device whose output *I-V* characteristics are controlled by a current at another terminal is said to be a *current-controlled* device.

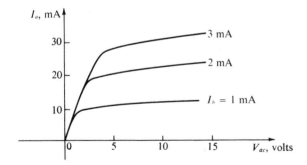

Figure 3.17.

3.7 The device in Figure 3.18 has the *I-V* characteristics shown in Figure 3.11. Using these characteristics, determine the values of I_a and I_b for the voltages shown in the circuit.

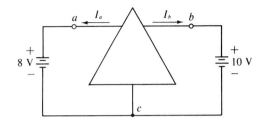

Figure 3.18.

4

Semiconductor Principles

4.1. Introduction

Although the use of semiconductor materials in electronics is not new, semi-conductors played only a minor role in electronics for many years, and only a token effort was directed toward developing an understanding of their unique characteristics. In recent years, a remarkable concentration of effort has been given to the field of semiconductors. This emphasis on semiconductors was brought about principally by the development of the transistor by Schockley, Bardeen and Brattain in 1948. From its initial applications in hearing aids and pocket radios, the transistor has spearheaded a complete revolution in most areas of electronics, particularly in the design of computers and space vehicles. A whole new family of related semiconductor devices — diodes, zener diodes, controlled rectifiers, unijunction transistors, photo devices, field-effect devices, etc. — has been developed since the appearance of transistors, culminating in the recent development of integrated circuits that consist of complete function-al circuits produced on a single minute chip of semiconductor material.

The brief description of electronic processes in solids presented in the previ-ous chapters is sufficient for comprehending the basic electronic properties of solid materials. However, considerable elaboration is necessary to bring about a full appreciation of material behavior. For the purposes of understanding semiconductor devices, the following discussion will emphasize semiconductor principles and properties almost to the exclusion of conductors and insulators. In particular, we will study the types of atomic bonding peculiar to semiconductors, the two types of current carriers in semiconductors, and how they are produced. The effects of heat on the electronic behavior of semiconductor material will also be studied. Throughout this discussion, energy-band diagrams will be utilized in an effort to relate new material to what has already been presented.

4.2.   Conductors, Semiconductors and Insulators

In Chapter 1, solid materials were classified according to their energy-band structures as conductors, semiconductors or insulators. Exactly what determines whether a material has the electrical characteristics of a good, fair or poor conductor? Obviously the number of free charge carriers available within a material is a major factor. This number is determined by the complexity of the atom, the number of valance electrons and the type of bonding between atoms.

Recall that an atom is considered more complex than another if it contains more orbiting electrons. More complex atoms have their valence electrons farther from the nucleus than simpler atoms. Electrons in the valence band of more complex atoms possess more energy than those in the valence band of simple atoms. The farther the electron is in its orbit from the nucleus, the more loosely bound it is to the nucleus, since it possesses more energy (Section 1.4). Atoms with few valence electrons tend to give them up more readily than atoms with many valence electrons. Atoms possessing few valence electrons have them more loosely bound to the nucleus than the atoms with many valence electrons. An atom with two valence electrons will more easily give up a valence electron than an atom with a valence of five. A complex atom with a valence of one will lose its valence electron more readily than a simple atom with a valence of one. This follows from the fact that a complex atom's valence electrons are less tightly bound to the nucleus. This all means that complex, low-valence atoms have narrow forbidden energy bands between the valence and conduction bands and thus are good conductors. On the other hand, simple, high-valence atoms have wide forbidden energy bands, making them poor conductors.

If we look at the periodic table of elements in Appendix I, we can confirm some of these ideas. Silver and copper are both excellent conductors. Each is a fairly complex atom: silver has 47 electrons, copper has 29 electrons, and each has a valence of one. Since silver is more complex than copper, it is a better conductor. Indicated in the valence IV column are the two most important semiconductor materials, silicon and germanium. Germanium has an atomic number (the number of electrons) of 32 and silicon has an atomic number of 14. Thus, both have the same number of valence electrons. However, since germanium is more complex, we can expect it to be a better conductor than silicon.

It is, although neither is a good conductor. On the other hand, both are found to be good semiconductors, with germanium possessing a forbidden band gap of 0.7 eV and silicon possessing a forbidden energy gap of 1.1 eV. Aluminum is considered a conductor even though it lies in the valence III column, indicating that the number of valence electrons is not the only determining factor. The type of atomic bonding in a material also governs the electrical characteristics.

4.3.   Covalent Bonding in Semiconductors

We saw in Chapter 1 that in solid materials there are several bonding processes; however, in semiconductor work, our prime interest is in covalent bonding (Section 1.7). In semiconductor materials the bonding between atoms is covalent. Recall that covalent bonding refers to the type of interconnection or bonding between atoms in which the atoms share their valence electrons so that each atom essentially possesses eight valence electrons.

Taking germanium as an example, the covalent bonding process is illustrated in Figure 4.1. Germanium has a valence of four and shares its four valence electrons with four adjacent atoms of germanium. Each germanium atom thus appears to have a valence of eight. A covalent bond is made up of one electron from each of two atoms strongly bonded together. This type of bonding is also prevalent in silicon, a semiconductor. Silicon and germanium are elements which are semiconductors. There are semiconductor materials made up of compounds such as gallium arsenide. Gallium, which has a valence of three, bonds covalently with arsenic, which has a valence of five, to form the semiconductor compound gallium arsenide. Generally speaking, materials made up of three, four or five valence atoms in a covalent bond are semiconductors.

Structures formed by atoms bonded together covalently are called *crystal structures.* Atoms in crystals arrange themselves to share each other's valence electrons in a uniform three-dimensional pattern, depicted two dimensionally in

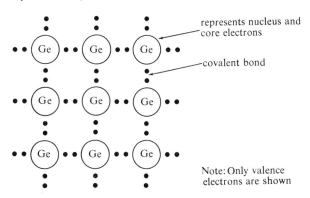

represents nucleus and core electrons

covalent bond

Note: Only valence electrons are shown

Figure 4.1.   COVALENT BONDING IN GERMANIUM STRUCTURE

Figure 4.1. Throughout the text we will refer continually to germanium and silicon crystals. In the next section we will look closely at intrinsic (pure) semiconductor crystals and their electrical characteristics.

4.4.   Intrinsic Semiconductors

Very pure semiconductor material is called *intrinsic* material. Even minute amounts of certain impurities can drastically affect the electrical properties of semiconductors. Consequently, a semiconductor would not be called truly intrinsic unless its impurity content were very small; for germanium, this is about one impurity per 10^8 germanium atoms. Silicon is called intrinsic with one impurity atom per 10^{13} silicon atoms. In practice, impurity concentrations somewhat higher than this are sometimes referred to as intrinsic.

A perfectly intrinsic germanium crystal would have a structure exactly as that shown in Figure 4.1, since it consists solely of germanium atoms covalently bonded, with no impurity atoms present. The structure in Figure 4.1 shows that *all* of the valence electrons are tightly bound to the parent atoms and to other atoms by covalent bonds. These electrons are not free to move through the crystal and therefore cannot conduct an electrical current. This is precisely the picture at a crystal temperature of absolute zero ($-273°C$, or $-460°F$). Thus, at *absolute zero*, an intrinsic semiconductor behaves like an insulator since it has no electrons available to conduct a current. The energy-band diagram of an intrinsic semiconductor crystal at absolute zero, Figure 4.2, indicates that the valence band of energies is completely filled by covalently bonded valence electrons and the conduction band is completely empty since there are no free electrons in the crystal. With no electrons in the conduction band, the semiconductor crystal cannot conduct a current and behaves essentially as an insulator.

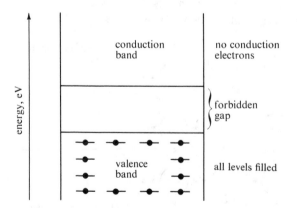

Figure 4.2.   ENERGY-BAND DIAGRAM OF AN INTRINSIC SEMICONDUCTOR AT ABSOLUTE ZERO TEMPERATURE

Let us now consider what happens as the temperature of the crystal is increased above absolute zero. External energy, such as heat, applied to an intrinsic semiconductor increases the energy of each atom in the crystal. This increase in energy may be imparted to some of the valence electrons in the structure. Through this process, a valence electron, particularly one at an energy level near the top of the valence band, may acquire sufficient additional energy to break away from its atom and become a free electron. In doing so it must break its covalent bond and acquire enough energy to jump the forbidden energy gap. Then it will exist at an energy level in the conduction band as a free conduction electron.

To illustrate this process, Figure 4.3 shows the intrinsic germanium crystal of Figure 4.1 after heat energy has been applied. In this illustration, a covalent bond has been broken by one of the valence electrons as a result of heat energy. The liberation of this valence electron has left a vacancy in the covalent structure. This vacancy is called a *hole*. The loss of a valence electron has left the parent germanium atom with a net positive charge. The broken covalent bond, or hole, is considered then to be *positively* charged with the same charge as an electron. However, a hole is not the positively charged counterpart of an electron since it has no mass. It is simply a vacancy or absence of an electron, and thus has a positive charge. A hole, by virtue of its positive charge, has a great attraction for an electron, if one should wander by.

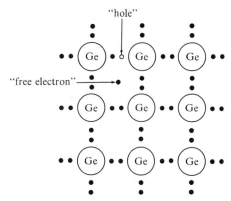

Figure 4.3.   INTRINSIC GERMANIUM STRUCTURE AFTER APPLICATION OF HEAT SHOWING A BROKEN COVALENT BOND, A FREE ELECTRON AND A HOLE

Electron-hole pairs

The excitation of a valence electron into the conduction band is always accompanied by the appearance of a hole. The freed electron and the hole it left behind are called an *electron-hole pair*. When heat is the form of energy, the generated electron-hole pairs are said to be thermally generated. The number of electron-hole pairs thermally generated depends on the temperature of the crystal. As temperature increases, more heat energy is available and more valence electrons will acquire the necessary energy to jump the forbidden gap into the conduction band. The energy-band picture of this process is shown in Figure 4.4. Notice that only the higher-level valence electrons can attain the necessary energy at room temperature (21°C or 70°F) to jump the gap.

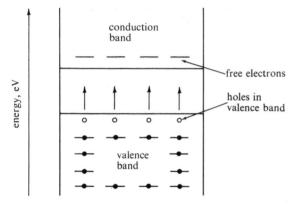

Figure 4.4.   ENERGY-BAND DESCRIPTION OF THERMAL GENERATION OF ELECTRON-HOLE PAIRS IN AN INTRINSIC SEMICONDUCTOR AT ROOM TEMPERATURE

The effect of temperature on the electrical properties of a semiconductor should now be apparent. The thermally generated electrons are free from atomic forces and are capable of producing a current if a potential difference (voltage) is applied to the semiconductor. Thus, as one would expect, the resistivity of the semiconductor decreases with increasing temperature since more electrons become available current carriers. Meanwhile, what has happened to all the thermally generated holes? The answer to this is that they, too, have become current carriers, though not in the same way as electrons. When a hole is created by an electron breaking a covalent bond, a valence electron from a neighboring atom can easily fill the hole by breaking its own covalent bond and jumping over to the first atom, leaving behind a hole in the neighboring atom. As this occurs, it appears that the hole has moved from one atom to another. This sequence of events is illustrated in Figure 4.5. Figure 4.5A shows a hole existing in the silicon atom in the lower left-hand corner. No other holes are present. In Figure 4.5B we see that a valence electron from the silicon atom in the lower right-hand corner of the figure has broken its bond and jumped over to fill the original hole, leaving behind a hole in its atom. Thus the hole has seemingly jumped from one silicon atom to another. This movement of holes, which are positively charged, constitutes a current flow. *Current flow in a semiconductor, then, is composed of electron movement and hole movement.*

If a voltage is applied to a semiconductor containing free electrons and holes, as shown in Figure 4.6, the free electrons will move from the negative terminal toward the positive terminal, and holes will move from positive to negative. From the preceding discussion of hole movement, it is clear that a

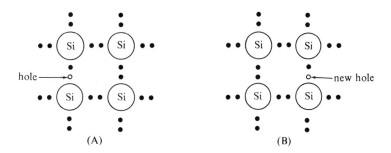

Figure 4.5.   ILLUSTRATION OF HOLE MOVEMENT CAUSED BY VALENCE
ELECTRONS JUMPING FROM HOLE TO HOLE

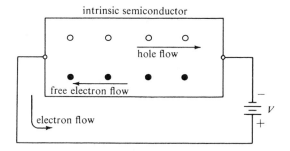

Figure 4.6.   FLOW OF CHARGES IN A SEMICONDUCTOR UNDER AN APPLIED
VOLTAGE

hole moving to the right is a valence electron moving to the left. The flow of
holes from positive to negative is therefore essentially the same as the flow of
electrons from negative to positive. Remember that holes do not actually move
but have apparent motion. Electrons are the only charges which actually move.
Thus, the current in a semiconductor can be thought of as consisting of two
parts: free electrons moving in one direction and holes moving in the opposite
direction. The total current is actually the sum of the two parts.

Recombination

It may not be apparent just what happens to the holes as they reach the edge of
the semiconductor material in Figure 4.6. What actually takes place is called
the process of *recombination*, in which some of the electrons flow from the
negative terminal of the battery and fill the holes as they enter the semiconductor. This recombination, or filling of holes, causes both the hole and electron to disappear. In reality, the process of recombination is taking place continuously throughout the semiconductor material. Free electrons wandering
through the crystal may encounter holes and recombine with them, thus an-
nihilating an electron-hole pair. What, then, keeps all the electron-hole pairs
from disappearing? The answer is that electron-hole pairs are continuously
being thermally generated. These thermally generated electron-hole pairs com-

pensate for the recombination losses, so that at a given temperature the number of electron-hole pairs in a semiconductor crystal is essentially constant.

One of the most important points to be derived from the preceding discussion is that, although hole current is actually due to the movement of electrons, it is *not* the same type of free-electron motion that makes up electron current. The electrons that cause hole current are *valence* electrons which jump from hole to hole and do not have enough energy to become free electrons. On the other hand, free electrons move freely through the crystal without being bound to any atom. Thus free electrons are able to move much more quickly through the crystal than holes, which have to move in a succession of jumps. One way to describe this comparison is to say that free electrons have a higher *mobility* than holes. This aspect will be encountered in later discussions.

We have seen that due to thermal energy an intrinsic semiconductor, which is essentially an insulator at absolute zero temperature, will contain free electrons and holes that can conduct current. At room temperature, intrinsic semiconductors such as silicon and germanium contain enough current carriers to make them fair electrical conductors, being much poorer conductors than copper or silver, but much better than an insulator such as rubber.

In summary, the resistivity and resistance of an intrinsic semiconductor decrease as its temperature increases because more valence electrons are able to break away from their covalent bonds and become free electrons. Thus more current carriers, free electrons and holes are available, increasing the semiconductor's ability to conduct current. In an intrinsic semiconductor, the total number of free electrons *equals* the total number of holes. We will see in the next section how addition of certain types of impurities will cause either free electrons or holes to be the majority current carrier.

4.5.   Extrinsic Semiconductors

Pure silicon (or germanium) is of little use as a semiconductor, except maybe as a heat- or light-sensitive resistance. Most of our modern semiconductor devices contain semiconductor materials to which certain impurities have been added to create a predominance of either free electrons or holes. Semiconductor materials containing different types of impurities are combined to produce the many useful devices presently available. The process of adding impurities to the semiconductor, called *doping*, is performed after the semiconductor material has been refined to a high degree of purity. The concentration of the added impurity, called *dopant*, is typically very minute, on the order of one part of impurity per ten million parts of pure semiconductor. The doped semiconductor is referred to as an *extrinsic* semiconductor.

N-type impurities

As was mentioned above, the effect of the impurities is to produce a predominance of either free electrons or holes. Doping impurities which add free electrons to the semiconductor material are called N-type impurities since they add *negative carriers*. Hole-producing impurities are called P-type impurities since they add *positive carriers* to the semiconductor crystal.

Let us first consider N-type impurities. The element *arsenic* is an example of this type of impurity. Referring to the periodic table of elements in Appendix I, arsenic has an atomic number of 33 and falls in the valence V column. If a small amount of arsenic is introduced into an intrinsic semiconductor crystal such as silicon, the arsenic atoms will enter into the crystalline structure and form covalent bonds with the silicon atoms. The arsenic impurities occupy positions in the crystal structure which are otherwise occupied by silicon atoms. The arsenic atoms, since they have *five* valence electrons, do not fit in exactly with the silicon crystal structure. Only *four* of their valence electrons are required in the crystal structure. The fifth valence electron does not enter a covalent bond and is thus only loosely bound to its parent arsenic atom. Only a very small amount of energy is needed to remove this electron from its atom, making it a free conduction electron. Figure 4.7 illustrates the effect of an arsenic atom in the silicon structure.

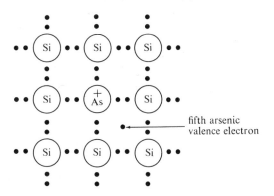

fifth arsenic
valence electron

Figure 4.7. ADDITION OF AN N-TYPE IMPURITY, ARSENIC, PRODUCES A FREE ELECTRON

For each arsenic atom present in the crystal, one virtually free electron is *donated* to the semiconductor material. For this reason, arsenic and all impurities with a valence *greater* than four are called *donor* impurities.

At absolute zero temperature, the fifth valence electron of each arsenic atom is bound to its parent atom even though it is not part of a covalent bond. However, at room temperature all these electrons have absorbed the small amount of energy needed to become conduction electrons. This point may be made clearer if we employ the energy-band concept. Figure 4.8 contains the energy-band diagrams of two silicon crystals doped with arsenic, one at absolute zero, the other at room temperature. With the addition of a donor impurity, a new energy level, the *donor* level, is introduced. As shown in Figure 4.8, it exists in the forbidden gap very close to the conduction band. At absolute zero, Figure 4.8A indicates that the valence band is filled, the conduction band is empty, and the donor level is filled, being occupied by the donated electrons from each arsenic atom which are not yet free. At room temperature this picture changes somewhat. As we know, some valence electrons will receive enough energy to jump the gap into the conduction band, leaving holes behind. This is the electron-hole pair generation we previously discussed. In addition, the donor electrons existing at the donor level easily absorb enough energy to jump into the conduction band. However, these free electrons leave no holes behind since they have broken no covalent bonds. Thus there are more free electrons in the conduction band than there are holes in the valence band. For this reason, in an N-type semiconductor, one doped with N-type donor impurities, *electrons* are the *majority* current carriers and *holes* are the *minority* carriers.

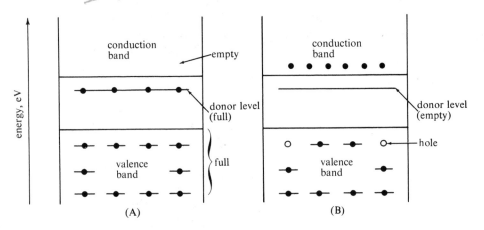

Figure 4.8.   ENERGY-BAND DIAGRAM OF SILICON DOPED WITH N-TYPE IMPURITY: (A) AT ABSOLUTE ZERO; (B) AT ROOM TEMPERATURE

Typically, the number of donated electrons is much greater than the number of thermally generated electron-hole pairs at room temperature — about one million times greater. Therefore, for an N-type semiconductor at room temperature, the number of free electrons is around one million times the number of holes. This large increase in current carriers, mainly electrons, causes the resistivity to drop well below its value for the intrinsic undoped semiconductor. Recall that in intrinsic semiconductors the number of holes and free electrons are equal, both producing current. In N-type semiconductors the majority of carriers are electrons, so current in these materials is primarily carried by electrons. This is illustrated in Figure 4.9.

A donor impurity must be an atom with a valence greater than four. The most common are phosphorous, antimony and arsenic, each with a valence of five. These atoms donate *one* free electron each. Impurity atoms with *six* valence electrons would donate *two* free electrons, and so on. It is important to note that a donor atom becomes a *positive ion* when it donates its electron. The

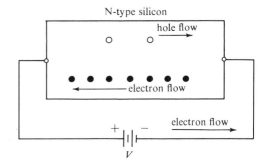

Figure 4.9.   CURRENT FLOW IN N-TYPE SEMICONDUCTOR

positive ions, however, are locked into the crystal structure and cannot move to conduct current.

P-type impurities

Let us now turn our attention to P-type impurities. The element *indium* is a common P-type impurity. Referring to the periodic table of elements in Appendix I, indium has an atomic number of 49 and a valence of *three*. If a small amount of indium is introduced into an intrinsic semiconductor crystal such as silicon, the indium atoms will enter into the crystalline structure and form covalent bonds with the silicon atoms. The indium impurities occupy positions in the crystal structure which are otherwise occupied by silicon atoms. The indium atoms, since they have only *three* valence electrons, do not fit in exactly with the silicon crystal structure. Because each indium atom has three valence electrons, it can bond covalently with only three silicon atoms. Thus one covalent bond will not be formed. This is illustrated in Figure 4.10. Figure 4.10A shows the indium atom bonding with *three* neighboring silicon atoms. The fourth bond is

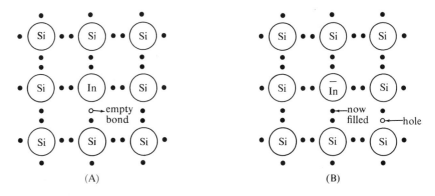

Figure 4.10.   P-TYPE IMPURITY, INDIUM, ADDED TO SILICON: (A) PRODUCES AN UNFILLED COVALENT BOND; (B) A VALENCE ELECTRON FROM NEIGHBORING SILICON ATOM JUMPS TO FILL THIS EMPTY BOND, LEAVING A HOLE BEHIND

empty since indium has only *three* valence electrons to share. This makes it very easy for an electron in a bond between two neighboring semiconductor atoms to jump in and fill this vacant bond, leaving a hole behind. Only a small amount of energy is required for this to occur. At room temperature virtually every indium atom has *accepted* a valence electron from a neighboring bond to fill its empty bond, thus producing a hole in the structure. For this reason, indium and all impurities with a valence *less* than four are called *acceptor* impurities.

At absolute zero temperature the fourth covalent bond of each indium atom is empty. However, at room temperature each of these bonds is filled by a neighboring valence electron which has absorbed the small amount of energy needed to break its own bond. This produces a hole in the semiconductor crystal. This process may become clearer if we look at the energy-band diagram of the silicon crystal when it is doped with a P-type impurity such as indium. Figure 4.11 shows this diagram at absolute zero and at room temperature. With the addition of a P-type acceptor impurity, a new energy level, the *acceptor* level, which exists in the forbidden gap very close to the valence band, is introduced. At absolute zero, Figure 4.11A indicates that the valence band is filled and the conduction band is empty. Also, the acceptor level, which is the energy level of the unfilled covalent bonds, is empty at absolute zero. At room temperature this picture changes somewhat. Thermal generation of electron-hole pairs produces some free electrons in the conduction band and holes in the valence band. In addition, electrons from the valence band have jumped up to fill the vacancies at the acceptor level. The energy required for this is much less than that required for a valence electron to jump the forbidden gap into the conduction band. Thus, virtually all the vacancies at the acceptor level are filled, leaving holes in the valence band without a corresponding conduction electron. For this reason a semiconductor doped with a P-type impurity such as indium will have, at room temperature, more holes in the valence band than electrons in

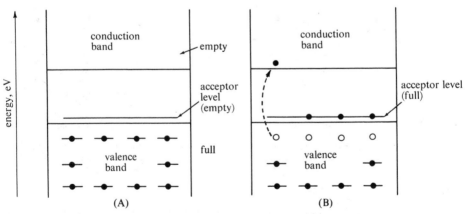

Figure 4.11.   ENERGY-BAND DIAGRAMS OF SILICON DOPED WITH A P-TYPE IMPURITY: (A) AT ABSOLUTE ZERO; (B) AT ROOM TEMPERATURE

the conduction band. For a P-type semiconductor, then, *holes* are the *majority* current carriers and *electrons* are the *minority* current carriers.

Typically, the number of holes produced by the P-type impurity atoms is about one million times greater than the number of thermally generated electron-hole pairs at room temperature. A P-type semiconductor, then, has about one million times more holes than free electrons available to act as current carriers at room temperature. Figure 4.12 illustrates current flow in a P-type semiconductor, which has a much lower resistivity than the intrinsic semiconductor at room temperature.

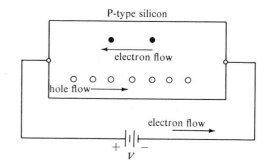

Figure 4.12.   CURRENT FLOW IN A P-TYPE SEMICONDUCTOR

An acceptor (P-type) impurity must be an atom with a valence less than four. The most common are indium, boron, aluminum and gallium, which all have a valence of three. These atoms produce one hole each. Impurity atoms with two valence electrons produce two holes each, and so on. It is important to note that an acceptor atom becomes a *negative ion* when its empty covalent bond is filled by a neighboring valence electron. These negative ions, however, are locked into the crystal structure and cannot move.

TABLE 4.1

SUMMARY OF N-TYPE AND P-TYPE MATERIAL

Characteristics	N-type	P-type
Number of valence electrons of impurity atom	Greater than 4	Less than 4
Name of impurity	Donor	Acceptor
Typical impurities	Arsenic, phosphorus, antimony, bismuth	Indium, boron, gallium, aluminum
Majority carrier	Electron	Hole
Minority carrier	Hole	Electron
Energy band in which majority carriers move	Conduction band	Valence band

The fundamentals of the two types of doping impurities have been discussed up to this point. Table 4.1 summarizes the principal characteristics of N- and P-type semiconductors.

4.6.   Effects of Temperature on Extrinsic Semiconductors

We have seen that a small quantity of impurities, N-type or P-type, can produce a large quantity of current carriers in an extrinsic semiconductor, depending on the number of donor or acceptor impurities added. This has the effect of reducing the semiconductor's resistance to current flow. For example, if we were to place an ohmmeter across a piece of N-type germanium, we might measure a resistance of about 100 ohms, whereas the same piece of intrinsic germanium would have a resistance of about 7000 ohms.

It is apparent that the ease with which current flows in an extrinsic semiconductor depends upon how many current carriers are available, and for a given amount of material this is a function of the number of impurity atoms introduced into the material. There is another important factor which contributes to the number of current carriers and thus to resistance; this important factor is the temperature of the material.

At absolute zero temperature no current carriers are available in the semiconductor material, and the material behaves essentially as an insulator with very high resistivity. As temperature is increased, two separate mechanisms produce current carriers in an N-type semiconductor: thermal generation of electron-hole pairs and impurity-produced free electrons (majority carriers). At room temperature, all the donor atoms can be presumed to have contributed to the liberation of free electrons; these electrons are in much greater abundance than the thermally generated carriers. Further increases in temperature only serve to increase the thermally generated carriers since the donor atoms have already donated their free electrons. Eventually a critical temperature is reached, 85°C for germanium and 200°C for silicon, where the number of covalent bonds that are broken is very large; the resistivity of the material is now determined by the number of temperature-produced electron-hole pairs rather than by the effects of doping. This means that at elevated temperatures the desirable effects of doping completely vanish and the material fails to perform as an extrinsic semiconductor. At these temperatures the material has approximately the same number of holes as electrons and is essentially intrinsic.

The same general effect is present in P-type material. At absolute zero, no carriers are available. At room temperature, all of the acceptor atoms have provided a hole (majority carrier) to the semiconductor material. These holes greatly outnumber the thermally generated electron-hole pairs. As the temperature is increased to the critical temperature, the thermally generated carriers are far greater in number than the acceptor-produced holes, and the semiconductor loses its extrinsic properties. The effects of temperature on both N-type and P-type semiconductors are summarized in Table 4.2.

TABLE 4.2

TEMPERATURE EFFECTS ON N-TYPE AND P-TYPE MATERIAL

Temperature	N-type	P-type
Absolute zero	No carriers — very high resistance	Same
Room temperature	Many donor-provided free electrons — few thermally generated carriers—low resistance —electrons in majority	Many acceptor-provided holes — few thermally generated carriers — low resistance—holes in majority
Critical temperature (85°C for germanium and 200°C for silicon) and above	Many thermally generated carriers—loss of extrinsic properties—holes and electrons about equal— very low resistance	Same

Doping process

To transform a specimen of intrinsic semiconductor material into either a P- or N-type requires some means of causing acceptor or donor impurities to enter the semiconductor crystal without drastically changing the crystal structure. The most common method is the process of *dopant diffusion*, in which the intrinsic specimen is heated (typically to 1200°C) while being exposed to the proper impurities in *gaseous* form. The impurity gas atoms diffuse (spread out) into the specimen and take up positions in the crystal structure. By proper control of the temperature and duration of the diffusion, any desired concentration and depth of doping can be accomplished. This doping process will be discussed in Chapter 17.

This concludes our study of semiconductor fundamentals and, up to this point, no useful semiconductor device has been mentioned. Beginning with the next chapter, we will study most of the important semiconductor devices on the market today. Practically all of these contain extrinsic semiconductor materials of both types in one crystal. The simplest combination, called a P-N junction, is the next subject to be covered.

GLOSSARY

Crystal: structure formed by atoms covalently bonded together.

Intrinsic semiconductor: semiconductor with no impurities present in its crystal structure.

Absolute zero: coldest possible temperature — 273°C or 460°F.

Thermal excitation: absorption of sufficient heat energy by a valence electron such that it can break its covalent bond and become a free electron.

Hole: vacancy caused by a valence electron breaking its covalent bond and becoming a conduction electron. It is charged positively.

Electron-hole pair: free electron and hole produced by the thermal excitation of a valence electron.

Recombination: process whereby a free electron fills a hole. This action eliminates both the electron and the hole.

Mobility: ease with which current carriers can move through the crystal.

Doping: addition of impurities to an intrinsic semiconductor.

Extrinsic semiconductor: doped semiconductor.

N-type impurities (doNor impurities): impurities with a valence of more than four which donate free electrons to the semiconductor crystal.

P-type impurities (accePtor impurities): impurities with a valence of less than four which provide holes to the semiconductor crystal.

Majority carriers: carrier (electron or hole) of greatest number in a semiconductor material.

Minority carriers: carrier of least number in a semiconductor material.

Critical temperature: temperature at which a doped crystal loses its extrinsic properties.

PROBLEMS

4.1 Which should be a better conductor: sodium or lithium?

4.2 What three factors determine the number of current carriers within a material?

4.3 Are metals such as copper considered crystals?

4.4 Sketch the symbolic crystal structure of intrinsic silicon at absolute zero. Sketch its energy-band diagram.

4.5 Sketch the same crystal structure at room temperature. Sketch its energy-band diagram.

4.6 Why is a valence electron at the top of the valence band more apt to be thermally excited than one at a lower level?

4.7 Why isn't a hole the exact positively charged counterpart of an electron?

4.8 What causes the decrease in resistivity of an intrinsic semiconductor at high temperatures?

4.9 If positive particles cannot really move, then what causes "hole current"?

4.10 What process is the opposite of thermal generation of electron-hole pairs?

4.11 Which current carrier has more mobility: a hole or an electron?

4.12 Which of the following atoms could be used as N-type impurities? Which as P-type impurities?
(a) Zinc.
(b) Gallium.
(c) Carbon.
(d) Sulfur.
(e) Beryllium.

4.13 In an N-type semiconductor, does it take more energy to thermally excite a valence electron or to liberate a donor electron? Explain.

4.14 At absolute zero, which current carriers in a semiconductor are available?

4.15 In a P-type semiconductor, does it take more energy to thermally excite a valence electron to the conduction band or to the acceptor level?

4.16 What are the majority current carriers in an N-type semiconductor? A P-type semiconductor? An intrinsic semiconductor?

4.17 Of what polarity are the dopant ions in N-type and P-type semiconductors?

4.18 Compare the relative number of electrons and holes in an N-type semiconductor at
(a) absolute zero.
(b) room temperature.
(c) critical temperature.

4.19 In a P-type semiconductor, can the electrons (minority carriers) ever outnumber the holes (majority carriers)?

REFERENCES

Branson, L. K., *Introduction to Electronics*. Englewood Cliffs, N. J.: Prentice-Hall Inc., 1967.

Foster, J. F., *Semiconductor Diodes and Transistors*, Vol. 1. Beaverton, Oregon: Programmed Instruction Group, Tetronix, Inc., 1964.

Riddle, R. L. and M. P. Ristenbatt, *Transistor Physics and Circuits*. Englewood Cliffs N. J.: Prentice-Hall, Inc., 1958.

Romanowitz, H. A. and R. E. Puckett, *Introduction to Electronics*. New York: John Wiley & Sons, Inc., 1968.

5

The P-N Junction

5.1. Introduction

By themselves, the separate P- and N-type materials are of limited practical use to us, as previously stated. If we make a junction consisting of a piece of P-type material joined to a piece of N-type material so that the crystal structure is unbroken, a *P-N junction* is formed. Many semiconductor devices contain at least one P-N junction, and for this reason it will be considered the basic building block in our study of semiconductor devices. A thorough discussion of the operation and properties of P-N junctions will be undertaken in an effort to establish a firm foundation on which to build up to devices with more than one P-N junction. The prime characteristic of the P-N junction, as we shall see, is its ability to conduct current easily in only *one* direction. The extreme usefulness of this property will readily become apparent.

5.2. The P-N Junction

Before we look at the properties of a P-N junction, let us consider, first, the condition of the separate P-type and N-type materials just before they are joined together. (Actually, we cannot just push the two pieces together; the crystal structure must remain intact at the junction of the two materials. We will not concern ourselves with the method of producing the P-N junction until later in the text.) In Figure 5.1, a small section of N- and P-type silicon is shown just prior to the formation of the P-N junction. The N-type material consists mainly of silicon atoms, a relatively small number of donor impurity atoms, free electrons (majority carriers) and a very few holes, due to thermal generation of electron-hole pairs. In the figure, the silicon atoms are not shown and should be imagined as a continuous crystal structure in the background. The donor impurity atoms are shown as positive ions, since, at room temperature, all of them have released one electron. These donor ions are fixed in the crystal structure and *cannot move*. The number of majority carriers is dependent on the number of donor atoms. The number of holes (minority carriers) depends on the temperature of the material.

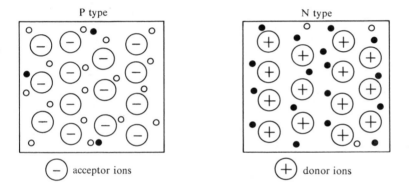

Figure 5.1.   P-TYPE AND N-TYPE SILICON PRIOR TO FORMATION OF THE
                P-N JUNCTION

Similarly, the P-type material consists mainly of silicon atoms, a relatively small number of acceptor atoms, holes (majority carriers) and a very few free electrons. The acceptor atoms are shown as negative ions, since, at room temperature, all of them will have had a valence electron from a neighboring atom jump to fill its vacant covalent bond. These acceptor ions are fixed in the crystal and *cannot move*. The number of majority carriers is dependent on the number of acceptor atoms. The number of electrons (minority carriers) depends on the temperature of the material.

Consider the situation at the moment when the P and N regions become joined together with a continuous crystal structure, as shown in Figure 5.2. In the P region there is a high concentration of holes, the majority carrier. Across the junction in the N region there is a very low concentration of holes, the minority carrier in that region. This difference in the concentration of holes in the semiconductor crystal (the joined P and N regions are now one crystal) is precisely the condition which brings about the flow of *diffusion* current as we discussed in Section 2.6. Thus we can expect that the holes in the P region will immediately begin diffusing from the region of high hole concentration across

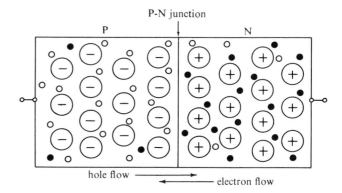

Figure 5.2.   A P-N CRYSTAL IMMEDIATELY AFTER JOINING THE P AND N
REGIONS

the junction into the region of low hole concentration. Similarly, in the N region there is a high concentration of free electrons, the majority carrier, while
across the junction in the P region there is a very low concentration of free
electrons, the minority carrier in that region. Consequently, there is a diffusion
of electrons from the N region across the junction into the P region.

Formation of the space-charge region

As the holes diffuse from the P region across the junction, they encounter a
multitude of free electrons in the N region. Due to this great abundance of free
electrons, each diffusing hole immediately recombines with a free electron and
disappears. The same thing occurs with regard to the electrons that are diffusing across the junction from the N region. They encounter a great number of
holes in the P region and in doing so immediately recombine and disappear.

It might seem that eventually all the holes from the P side would diffuse to
the N side and all the free electrons from the N side would diffuse to the P side,
but this does not occur. Only those majority carriers near the junction make it
across. The reason for this can be seen by considering Figure 5.3, which shows
the status of the P-N crystal after some of the majority carriers have diffused
across the junction and recombined.

Keep in mind that the negative acceptor ions in the P region are immobile.
When the holes on the P side near the junction diffuse into the N region, they
leave behind negative acceptor ions which have no corresponding hole. Thus
the P region is no longer electrically neutral because it has more negative
charges (ions) than positive charges (holes).

Similarly, the N region near the junction becomes lined with fixed positive
donor ions when free electrons in that region diffuse over to the P side. The N
region is no longer electrically neutral because it has more positive charges
(ions) than negative charges (electrons).

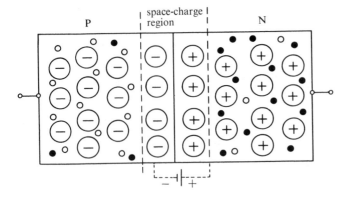

Figure 5.3.   P-N CRYSTAL AFTER MAJORITY CARRIERS HAVE DIFFUSED,
                PRODUCING SPACE-CHARGE REGION

As shown in Figure 5.3, only the fixed ions remain near the junction, negative on the P side and positive on the N side, with no charge carriers in this region. These fixed charges near the junction, which have been built up as a result of majority-carrier diffusion, tend to discourage and eventually prevent further diffusion. If a hole on the P side tries to diffuse across the junction, it is repelled by the positive donor ions lining the N side of the junction and moves back into the P region where it belongs. At the same time, electrons on the N side trying to diffuse across the junction are repelled by the negative acceptor ions lining the P side of the junction.

Summarizing what has been covered thus far, diffusion of majority carriers across the junction results in a build-up of fixed ionic charges lining the junctions. These fixed charges exert a repelling force on any further majority-carrier diffusion. This force is small at first but becomes greater as more charges diffuse across the junction. Eventually, the repelling force becomes great enough to stop further diffusion of the majority carriers. The region near the junction that contains these fixed ionic charges and no current carriers is called the *space-charge region*. See Figure 5.3. Also, because of the lack of current carriers in this region, it is referred to as the *depletion region*.

Potential barrier

The repelling force of the space-charge region is an electrical force. Actually, the fixed charges on opposite sides of the junction produce a *potential* barrier, the same as would be produced by a battery. In fact, if we could put a voltmeter across the space-charge region, we would measure a voltage equal to this potential barrier. Typically, it is about 0.3 volt for germanium and 0.7 volt for silicon. This potential barrier is indicated in Figure 5.3 by a small battery (dotted lines) across the space-charge region.

Outside the space-charge region on either side of the junction, the concentrations of charges are just as they were before the P and N regions were joined, with positive and negative charges equally distributed. The charge-concentration profile of the P-N crystal is shown in Figure 5.4. It depicts net charge density plotted against distance into the crystal. Net charge density is approximately zero everywhere, with the exception of the space-charge region. There, a large negative charge density exists on the P side of the junction and a large positive charge density exists on the N side of the junction.

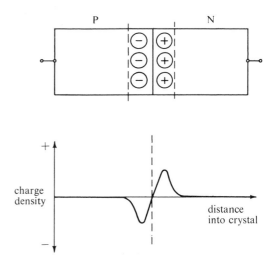

Figure 5.4.   CHARGE PROFILE OF A P-N CRYSTAL

The space-charge region in a P-N junction presents a potential barrier to any majority carriers that attempt to diffuse across the junction. This means that, in order for a free electron in the N region to move to the P region, it must gain enough energy to overcome this potential barrier. Similarly, in order for a hole on the P side to move to the N region, a valence electron in the N side must climb the potential barrier in order to fill this hole and thus have the hole appear on the N side. At room temperature, due to the heat energy, a few majority carriers will be able to overcome the potential barrier and cross the junction. A few of the holes in the P region and some of the electrons in the N region will acquire the energy to get over the barrier and diffuse across the junction. This small diffusion results in a *majority-carrier diffusion current* (the sum of the hole current and electron current) across the P-N junction. However, this diffusion current is exactly balanced by a current in the opposite direction made up of minority carriers, electrons on the P side and holes on the N side. This minority-carrier current is caused by the attraction of the space-charge region. For example, in Figure 5.3, if an electron-hole pair is produced by thermal excitation *in the space-charge region*, the electron will be accelerated toward the N side by the attraction of the positive ionic charges and the hole will be accelerated toward the P side by the negative ionic charges; this produces a *minority-carrier drift current* across the junction in a direction opposite

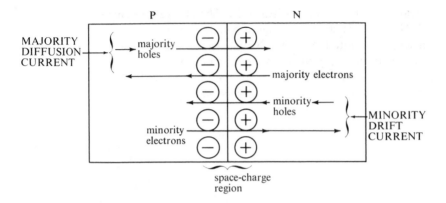

Figure 5.5.
CURRENTS ACROSS A P-N JUNCTION THAT HAS NO EXTERNAL
CONNECTIONS (UNBIASED). NET CURRENT IS ZERO.

to that of the majority-carrier diffusion current. When there are no external
connections to the P-N crystal, as in Figure 5.3, these two currents cancel each
other to produce a net junction current of zero. This is illustrated in Figure 5.5.

A P-N junction with an external voltage applied between its terminals is
said to be *biased*, and the applied voltage is called *bias voltage*. The P-N junctions in Figures 5.3 to 5.5 have no external connections and are thus *unbiased*.

Briefly summarizing the material in this section, an unbiased P-N crystal
will produce along its junction a space-charge region which acts as a potential
barrier to majority-current flow across the junction. This potential barrier
allows only a small number of majority carriers to diffuse across the junction.
It also attracts the flow of minority carriers across the junction. These two
currents are in opposite directions and cancel each other out in an unbiased
P-N crystal.

5.3.   Reverse Biasing a P-N Junction

Applying a voltage source, such as a battery, across a P-N junction with the
polarity as shown in Figure 5.6, is called *reverse biasing* the P-N junction. The
positive terminal of the voltage source, V_R, is connected to the N side and the
negative terminal of the voltage source is connected to the P side. With the P-N
junction biased in this manner, holes in the P side will be attracted to the negative terminal of the battery, and electrons in the N side will be attracted toward
the positive terminal of the battery.

Thus the majority carriers are drawn away from the junction. As they are
pulled away, more fixed donor and acceptor ions are left near the junction without a corresponding charge carrier, thus widening the space-charge region, as
shown in Figure 5.6. (Compare this with the unbiased P-N junction in Figure
5.3.) This has the effect of increasing the potential barrier. In fact, the potential
barrier is approximately equal to V_R. This increase in the potential barrier

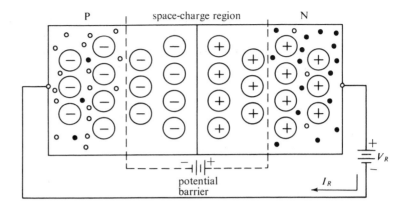

Figure 5.6.   REVERSE-BIASED P-N JUNCTION

makes it more difficult for *majority* carriers to diffuse across the junction. Actually, it takes only a very small reverse voltage, V_R, to completely halt the majority-carrier diffusion current.

The increase in the potential barrier due to V_R has the opposite effect on the *minority* carriers present in the space-charge region. They are now swept across the junction more easily, thus increasing the minority-carrier drift current. As V_R increases, the minority current increases until eventually all the available minority carriers are crossing the junction. The minority-carrier current cannot increase further, even though the reverse bias is increased. This is because the number of minority carriers is limited by temperature (thermal generation of electron-hole pairs).

Thus reverse bias applied to a P-N junction produces only the flow of minority carriers, holes from N to P and electrons from P to N across the junction. This *reverse* current, of course, also flows in the external circuit, as shown in Figure 5.7. Since it is the current flow under reverse bias, we will call it I_R. It

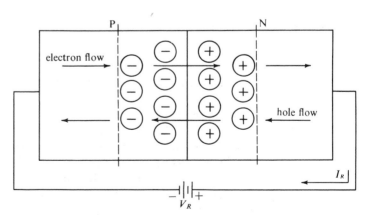

Figure 5.7.   FLOW   OF   MINORITY   CARRIERS   IN   REVERSE-BIASED   P-N
                   JUNCTION

should be pointed out that, because I_R depends on the number of minority carriers present in the P-N junction area, it will be very sensitive to temperature. As the temperature of the junction is increased, more electron-hole pairs will be generated, providing more minority carriers to the P and N regions. This, of course, will cause I_R to increase with temperature.

   In practice, this reverse current can be as low as a few nanoamperes (10^{-9} A) in silicon devices at room temperature and is typically around 100 nanoamperes. Germanium, since it has a smaller energy gap, has more minority carriers being generated at a given temperature than silicon. Consequently, it has reverse currents, typically about a few microamperes (10^{-6} A) at room temperatuure. In both silicon and germanium P-N junctions, the resistance to current in the reverse direction is very high. The charge concentration profile of a P-N junction is greatly altered by reverse bias, as can be seen by comparing Figure 5.8 with Figure 5.4. The space-charge region is much wider.

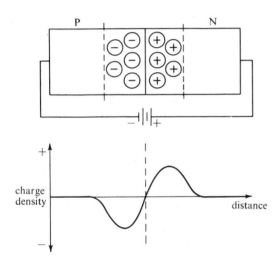

Figure 5.8.   CHARGE PROFILE OF A REVERSE-BIASED P-N JUNCTION

5.4.   Reverse Breakdown of a P-N Junction

We saw in the previous discussion that reverse biasing a P-N junction will stop all majority-carrier flow, allowing only minority carriers to cross the junction, producing a small reverse current. This reverse current is limited by the number of minority carriers present in the P-N crystal and is usually very small.

   This situation holds true as long as the reverse bias voltage, V_R, is not greater than a certain value. This certain value is called the *reverse breakdown voltage* of the P-N junction and is given the symbol V_{BD}. If the reverse bias exceeds this value the reverse current will increase very rapidly for a small increase in *reverse voltage*, producing reverse breakdown of the P-N junction.

When reverse breakdown occurs the P-N junction conducts heavily in the reverse direction, and unless the current is limited by a series resistor it may become damaged.

The phenomenon of reverse breakdown can be explained easily using the concepts we have learned thus far. Increasing the reverse voltage across a P-N junction causes the minority carriers to move across the junction at a higher speed. If this reverse voltage is high enough, the minority carriers move so rapidly that, on colliding with atoms of the crystal, they impart some of their energy to these atoms and if they have sufficient energy they can knock valence electrons out of their covalent bonds. This creates more charge carriers (free electrons and holes), which can in turn be accelerated by the reverse voltage to speeds high enough to break the bonds of other atoms. This avalanching process will produce a great number of new current carriers, causing a large increase in reverse current only when the reverse voltage is high enough to accelerate the minority carriers to the required energy. The necessary reverse voltage is V_{BD}, the reverse breakdown voltage. The process of reverse breakdown is also called *avalanche breakdown* because of the multiplying or avalanching of current carriers that takes place.

The reverse breakdown voltage, V_{BD}, of a P-N junction depends on many factors, including junction temperature and impurity concentration. Both of these effects will be discussed later.

5.5.   Forward Biasing a P-N Junction

Applying a voltage across a P-N junction with the polarity as illustrated in Figure 5.9 is called *forward biasing* the P-N junction. The positive terminal of the voltage, V_F, is connected to the P side, and the negative terminal to the N side.

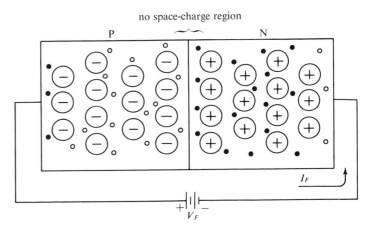

Figure 5.9.   FORWARD-BIASED P-N JUNCTION

With the P-N junction biased in this manner, holes in the P side are driven toward the junction by the positive terminal of the battery, and free electrons in the N side are repelled toward the junction by the negative terminal of the battery. The first effect of this is to neutralize some of the donor and acceptor ions in the space charge region, thus reducing the potential barrier and allowing some diffusion of majority carriers across the junction. If the forward bias, V_F, is made large enough, the potential barrier is reduced to zero and the movement of majority carriers across the junction is unimpeded, producing a relatively large *forward current, I_F*. Typical values of forward bias required to do this are 0.3 volt for germanium and 0.7 volt for silicon; these are precisely the values of potential barriers in these materials. This forward current increases very rapidly for small increases in forward bias. The nature of this forward current, illustrated in Figure 5.10, consists mainly of diffusing majority carriers.

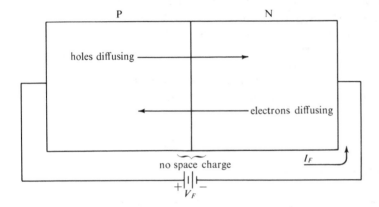

Figure 5.10.   FLOW OF MAJORITY CARRIERS IN FORWARD-BIASED P-N JUNCTION

Figures 5.9 and 5.10 show that, with sufficient forward bias, the space-charge region is completely neutralized. The charge concentration profile of the forward-biased junction shown in Figure 5.11 also illustrates this.

Let us briefly consider what happens to those majority carriers as they diffuse across the junction. The flow of holes in the P side approaching the

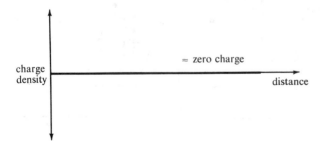

Figure 5.11.   CHARGE PROFILE OF A FORWARD-BIASED P-N JUNCTION

junction encounters electrons that have crossed over from the N side and recombination occurs, becoming more frequent as the holes proceed across the junction, until some distance into the N side all the holes recombine and disappear. Similar action involving electron flow from the N side also takes place. These diffusing majority carriers, then, cross the junction, thus becoming minority carriers (since holes are now in the N region and electrons now in the P region) and eventually recombine. The forward current in the P-N crystal, then, is caused by majority carriers diffusing across the junction as minority carriers.

Briefly summarizing, forward biasing a P-N junction with a few tenths of a volt acts to completely neutralize the P-N junction potential barrier, allowing majority carriers to diffuse freely across the junction. This produces a forward current, I_F, which increases sharply for small increases in this foward bias.

Summary of bias conditions

A summary of the various bias conditions on a P-N junction is given in Table 5.1. Note that in each case the net junction current is the combination of the majority-carrier diffusion current and the minority-carrier drift current. This table should be studied thoroughly.

TABLE 5.1

SUMMARY OF P-N JUNCTION BIAS CONDITIONS

Bias	Majority diffusion current	Minority drift current	Net junction current
Zero (unbiased)	Very small* →	Very small ←	Zero
Reverse V_R (below breakdown)	Zero	Larger. ← Depends on temperature and material. Typically 0.01 μA for silicon and 1 μA for germanium at 25°C.	← Same as minority current
Reverse breakdown ($V_R > V_{BD}$)	Zero	Very large. ← Limited by circuit's resistance.	← Same as minority current
Forward V_F	Very large. → Depends on amount of forward bias. Increases rapidly for increase in V_F. Usually limited by circuit resistance.	Zero	→ Same as majority current

*Arrows show direction of electron current.

5.6.   *I-V* Characteristic of a P-N Junction

Having thoroughly discussed the physical operation of a P-N junction, we can now look at it as a two-terminal electronic device and examine its electrical characteristics. The previous sections of this chapter have dealt with the black box portion of this device. With this background we can determine the *I-V* characteristics of the P-N junction device, which we shall call a P-N *diode*.

Figure 5.12 shows the popular electronic symbol for a P-N diode, next to its physical structure. The P region of a P-N diode is called the *anode* and the N region is called the *cathode*. Since reverse current flows from P to N in the device, it flows *with* the arrow in the symbol; forward current flows *against* the arrow in the symbol, since it flows from N to P.

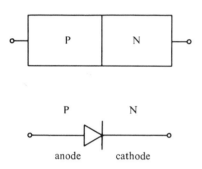

Figure 5.12.   SYMBOL FOR A P-N DIODE

The *I-V* characteristic of a typical P-N diode is shown in Figure 5.13. Let us examine it carefully. Since a P-N diode is a two-terminal device, we only need *one I-V* curve (Chapter 3). This consists of the relationship between cathode-to-anode (N side-to-P side) current, I_{CA}, and anode-to-cathode voltage, V_{AC}. When V_{AC} is positive, the anode is positive with respect to the cathode. This is forward biasing the P-N diode. Thus forward current flows from cathode to anode and I_{CA} is positive. This is the forward-bias region on the *I-V* curve in Figure 5.13. This portion of the curve shows that I_{CA} is very small until the forward voltage, V_{AC}, gets above the voltage V_{on}. V_{on}, the forward voltage needed to completely neutralize the potential barrier, is typically 0.3 volt for germanium diodes and 0.7 volt for silicon diodes. When V_{AC} is above this voltage, the forward current I_{CA} increases rapidly with increases in forward voltage. The P-N diode, then, when forward biased, conducts heavily, acting as a very low resistance.

When V_{AC} is made negative, the anode is negative with respect to the cathode. This is reverse biasing the P-N diode. Thus, reverse current flows from anode to cathode and I_{CA} is negative. This is the reverse-bias region in Figure 5.13. This portion of the *I-V* curve shows that the reverse current is very small until the reverse voltage reaches the reverse breakdown voltage, V_{BD}, after which reverse current increases rapidly. V_{BD} is usually fairly, high ranging from

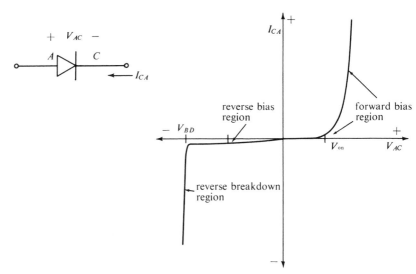

Figure 5.13.   *I-V* CHARACTERISTIC OF A P-N DIODE

about 30 volts to thousands of volts in P-N diodes. The P-N diode, then, when reverse biased, conducts very poorly (except at breakdown, which is usually avoided except in special P-N devices), acting as a very high resistance.

5.7. P-N Diode Ratings and Parameters

Ratings are the limiting values given to various diode quantities by the manufacturer. These values, if exceeded, can result in permanent damage to the diode or deterioration of its performance or life. The ratings of a semiconductor diode are based on its ability to dissipate the heat generated by power losses. When this heat raises the junction temperature above the maximum specified by the manufacturer, the device may become damaged. The most important P-N diode ratings according to symbols and definition are shown in Table 5.2.

TABLE 5.2

Rating	Symbol	Definition of rating
Peak reverse voltage	PRV	Maximum allowable *instantaneous reverse voltage* that may be applied across the diode. This rating is usually slightly below the reverse breakdown voltage of the diode.
Maximum reverse DC voltage	V_{RDC}	Maximum allowable DC reverse voltage. Usually less than PRV.
Maximum DC forward current	$I_{F(max)}$	Maximum allowable DC forward current which may flow at a stated temperature.

Various diode parameters (characteristics) are also specified by manufacturers. The two most common are given in Table 5.3.

TABLE 5.3

Parameter	Symbol	Definition of parameter
Forward voltage drop	V_F	Value of anode-to-cathode forward voltage for a specified forward current at a given temperature.
Reverse leakage current	I_R	Value of reverse current at stated temperature and reverse voltage.

A typical silicon diode *I-V* characteristic is shown in Figure 5.14. This diode has the following ratings and parameters:

Specifications (all at 25°C unless otherwise specified):
PRV: 290 volts
V_{RDC}:   250 volts
$I_{F(max)}$:   400 mA
V_F: 0.7 V @ 100 mA; 0.64 V @ 100 mA @ 55°C
I_R: 0.1 μA @ −200 V; 1 μA @ −200 V @ 55°C

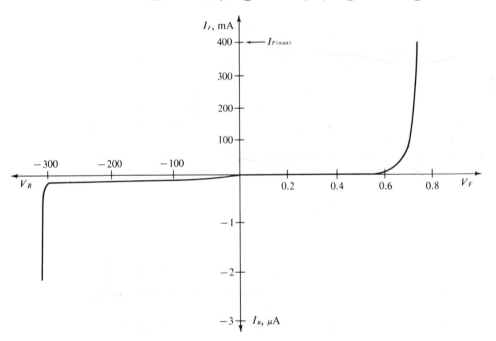

Figure 5.14.   *I-V* CHARACTERISTIC AND SPECIFICATIONS OF A TYPICAL SILICON DIODE

5.8.   Effects of Temperature on Diode Characteristics

Up to now, we have discussed the characteristics of P-N diodes at room temperature. We know semiconductor material is greatly affected by heat, so we can expect that a P-N diode will change its characteristics as its temperature changes. The primary effect of temperature on a semiconductor, as we learned in Chapter 1, is the generation of electron-hole pairs through the breaking of covalent bonds by thermally excited valence electrons. This means that at higher temperatures there are more free electrons and holes available in a semiconductor, whether it is intrinsic or extrinsic. These additional current carriers cause changes in a P-N junction's operation which are reflected in its *I-V* characteristic. We will study the principal changes closely.

In an unbiased P-N diode, the space-charge region produces a potential barrier which limits the diffusion of majority carriers across the junction to that value which exactly balances out the flow of minority carriers across the junction in the opposite direction (Section 5.2). An increase in temperature makes available more minority carriers in both regions of the P-N diode. This dictates a decrease in potential barrier so as to allow the majority-carrier diffusion to equal the minority-carrier flow, which would have increased. Thus one effect of increased minority carriers due to increased temperature is that the space-charge region (potential barrier) has decreased. This decrease in the junction potential barrier means that the forward voltage needed to cause current to flow is reduced. It also means that the reverse voltage needed to cause avalanche breakdown will increase since the minority carriers are accelerated through a narrower region and a higher voltage is needed to give them the energy needed to cause breakdown.

The second effect of increased minority carriers is that the reverse leakage current will be higher due to more available current carriers. This is very significant since P-N diodes are usually applied in circuits which depend on their ability to block current in the reverse direction.

The composite result of an increase in temperature can best be seen by comparing the *I-V* characteristics at two different temperatures. Figure 5.15 does this for a typical silicon diode. There are three main differences to be noted. First of all, in the forward-bias region it takes less voltage to produce the same current at 100°C as at 25°C. For instance, at 25°C it takes 0.85 volt to produce 100 mA of current, while at 100°C it only takes 0.75 volt to produce 100 mA. Secondly, in the reverse-bias region, the reverse leakage current is much higher at 100°C (3 μA) than at 25°C (0.1 μA). Finally, reverse breakdown occurs at 100 volts at 100°C and at 75 volts at 25°C. These changes with temperature occur for all semiconductor materials, although they occur in varying degrees. For example, the reverse leakage current in silicon doubles in value for every 11 degrees' rise in temperature, while for germanium it doubles for every 9 degrees. The decrease in forward voltage, V_F, at a given forward current is typically approximated as 2.2 mV/°C for both silicon and germanium devices, although it is actually slightly higher for germanium. The increase in reverse breakdown

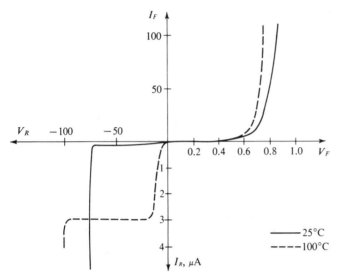

Figure 5.15.   *I-V* CHARACTERISTIC OF A SILICON DIODE AT 25°C AND 100°C

voltage at higher temperatures is different for different devices, usually increasing more for higher-voltage (higher-PRV) diodes.

Thus temperature changes cause significant changes in the *I-V* curves of P-N diodes and must be considered in the design of reliable circuits. These effects, which occur in each of the semiconductor devices which we will study, should be thoroughly understood.

5.9.   P-N Diode Circuit Analysis: Load-Line Method

The previous sections of this chapter have concentrated on the elements of the operation and characteristics of P-N diodes and their dependence upon temperature. With these elements now under our belts we can proceed to the practical aspect of employing semiconductor diodes in circuits. The basic problem consists of determining whether the diode is forward biased, reverse biased or in breakdown. There are several methods of solving this. The *load-line method*, although rarely used for P-N diodes, brings out several important points. It is used for many other devices and is thus worth our attention at this point.

The simple P-N diode circuit shown in Figure 5.16 consists of a voltage source and resistor in series with the diode. The voltage-source polarity is such that the diode is forward biased. Thus we can expect that any current that will flow as shown will be forward current. The basic problem now is to find this current, I_F, and the voltage drop across the diode, V_F. The values of I_F and V_F are called the *operating point* of the diode. In this circuit, the operating point is constant (does not change), since the source is constant. If the source were changing continuously, as it would with an AC or time-varying source, then the operating point would also change continuously.

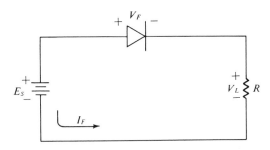

Figure 5.16.   SIMPLE DIODE CIRCUIT

We have two unknown quantities, I_F and V_F. In order to determine the values of these quantities, then, two equations or relationships between I_F and V_F are needed. The first relationship can be obtained from the circuit by using Kirchhoff's voltage law. The input voltage source, E_S, must equal the sum of the voltages across the load R and across the diode. Stated in equation form:

$$E_S = V_L + V_F \qquad \text{(5.1A)}$$

And since the voltage across R is simply $I_F \times R$, then

$$E_S = I_F \times R + V_F \qquad \text{(5.1B)}$$

Rewriting this and solving for V_F,

$$V_F = E_S - I_F \times R \qquad \text{(5.1C)}$$

The values of I_F and V_F must satisfy this equation at all times. It is a linear (straight-line) equation relating I_F to V_F for given values of E_S and R.

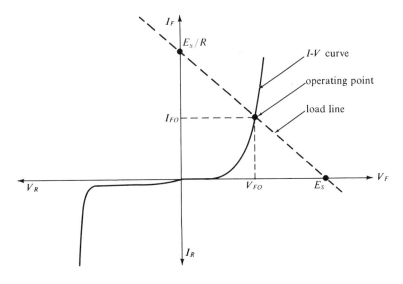

Figure 5.17.   FINDING THE OPERATING POINT OF THE CIRCUIT IN FIGURE 5.16

The second relationship can be obtained from the diode's *I-V* characteristic since it also relates I_F to V_F. This relationship is nonlinear (not a straight line) and is in the form of a curve. The values of I_F and V_F must lie on the *I-V* characteristic curve. Since the values of I_F and V_F (the operating point) must lie on both the straight line (Equation 5.1C) and on the *I-V* characteristic curve, they must lie on the intersection of the two when plotted on I_F-V_F axes. This is shown in Figure 5.17. The solid curve is the familiar *I-V* characteristic of a P-N diode. The dotted line is Equation 5.1C. In plotting this straight line, *two* points are needed. These two points are obtained by picking a value for either I_F or V_F and solving for the corresponding value of the other by using Equation 5.1C. The easiest values to use are $I_F = 0$, which gives $V_F = E_S$ according to Equation 5.1C, and $V_F = 0$, which gives $I_F = E_S/R$. These two points are indicated in Figure 5.17. The straight line is called the *load line* and the intersection of the load line and the *I-V* characteristic gives the operating point of the diode. The forward current at the operating point is I_{FO} and the voltage V_{FO}.

▶ EXAMPLE 5.1

The circuit in Figure 5.16 uses a germanium diode with the *I-V* characteristic shown in Figure 5.18, a 3-volt source and a 1-kΩ resistor. The problem is to find the operating point of the circuit.

The equation of the load line as given by Equation 5.1C is

$$V_F = 3\ \text{V} - I_F \times 1\ \text{k}\Omega \qquad (5.2)$$

The two points used to plot this line are $I_F = 0$, $V_F = 3$ V and $V_F = 0$, $I_F = 3\ \text{V}/1\ \text{k}\Omega = 3\ \text{mA}$. The load line is shown in Figure 5.18 (solid line). The intersection of the load line and *I-V* curve gives the operating

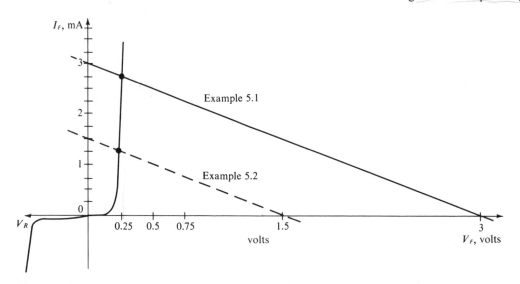

Figure 5.18.

point as $I_{FO} = $ **2.75 mA** and $V_{FO} = $ **0.25 V**. The drop across the resistor is $I_{FO} \times 1$ kΩ = 2.75 V. We can check to see that the sum of the resistor and diode voltages equals the source voltage.

$$2.75 \text{ V} + 0.25 \text{ V} = 3 \text{ V} \qquad \blacktriangleleft$$

▶ EXAMPLE 5.2

Repeating Example 5.1, using a source of 1.5 volts, the load-line equation becomes

$$V_F = 1.5 - I_F \times 1 \text{ k}\Omega \qquad\qquad (5.3)$$

and is plotted in Figure 5.18 (dotted line). The intersection of the load line and the diode curve gives an operating point of $I_{FO} = $ **1.25 mA** and $V_{FO} \approx$ **0.25 V**.

Thus, changing the voltage source from 3V to 1.5V changes the circuit current from 2.75 mA to 1.25 mA, but barely affects the diode voltage V_F. This is a result of the diode's steep forward characteristic, which signifies a small voltage change for very large current changes.   ◀

Reverse bias case

Let us now consider the simple diode circuit with the voltage source reverse biasing the diode as shown in Figure 5.19. We can expect that reverse current will flow as shown in the figure. The procedure for finding the operating point (I_R and V_R) in this case is exactly the same as in the previous circuit. First, the equation describing Kirchhoff's voltage law around the loop is written. The input voltage source, E_S, must equal the sum of the voltages across R and the diode. In equation form (noting that V_L is the load resistor voltage),

$$E_S = V_L + V_R \qquad\qquad (5.4\text{A})$$

Substituting $I_R \times R$ for voltage across R and solving for V_R, we have

$$V_R = E_S - I_R \times R \qquad\qquad (5.4\text{B})$$

The values of I_R, reverse current through the diode, and V_R, reverse voltage across the diode, must satisfy this equation at all times. They also must lie on the diode *I-V* characteristic curve. Thus by plotting Equation 5.4B and the *I-V*

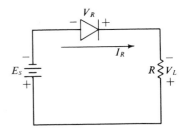

Figure 5.19.   SIMPLE DIODE CIRCUIT WITH REVERSE BIAS

characteristic on the same axes we can find their intersection, which will be the operating point. This is done in Figure 5.20. The straight-line plot of Equation 5.4B is the load line. Two points are needed to plot the load line. From Equation 5.4B we can obtain the two points shown in the figure. When $V_R = 0$, $I_R = E_S/R$ and when $I_R = 0$, $V_R = E_S$ in Equation 5.4B. The load line between these two points intersects the diode characteristic at the circuit operating point. The reverse current at the operating point is I_{RO} and the reverse voltage at the operating point is V_{RO}.

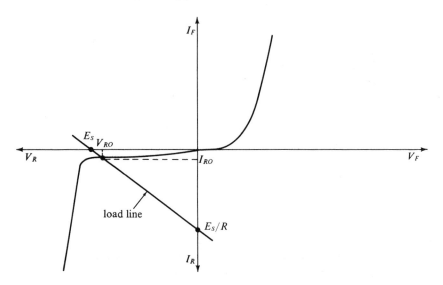

Figure 5.20.   FINDING THE OPERATING POINT OF THE CIRCUIT IN FIGURE 5.19

▶ EXAMPLE 5.3

The diode used in Examples 5.1 and 5.2 is used in the circuit of Figure 5.19 with $E_S = 50$ volts and $R = 5$ megohms (MΩ). The problem is to find the current and voltages in the circuit. The load line using $E_S = 50$ V and $R = 5$ MΩ is plotted along with the diode's characteristic curve in Figure 5.21. The intersection of the load line and characteristic curve give the operating point at $I_{RO} = $ **1 μA** and $V_{RO} = $ **45 V**. We can check these values using Kirchhoff's voltage law. The resistor voltage is 1 μA × 5 MΩ = 5V. Thus,

$$V_L + V_R = E_S$$
$$5\text{ V} + 45\text{ V} = 50\text{ V}$$

Note that most of the source voltage appears across the reverse-biased diode. This is usually the case.   ◀

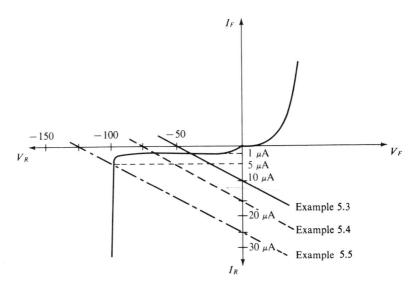

Figure 5.21.

▶ EXAMPLE 5.4

Repeat Example 5.3 for a voltage source of 75 V. The load line for this case is shown as a dashed line in Figure 5.21. The intersection of this load line and the diode curve is $I_{RO} = $ **1 μA** and $V_{RO} = $ **70 V,** the new operating point. Notice that changing the voltage source from 50 V to 75 V does not change the reverse current flowing through the diode. This is a result of the reverse current leveling off after all available minority carriers are flowing across the junction (until reverse breakdown occurs).   ◀

▶ EXAMPLE 5.5

Repeat the previous problem for $E_S = $ 125 V. The load line is plotted in Figure 5.21 (dot-dash line) and it obviously intersects the diode characteristic in its breakdown region with $V_{RO} = $ **100 V** and $I_{RO} = $ **5 μA.** Actually, any voltage source slightly above 100 V will cause the diode to break down. In reverse breakdown, the diode is capable of conducting large reverse currents and unless this current is limited, as it is in this example by the 5 MΩ resistance, the diode may burn out.   ◀

5.10.   P-N Diode Circuit Analysis: Approximate Method

The previous examples illustrate the load-line method for finding the operating point of a diode in a simple resistive circuit. This method is fairly easy to use for this simple circuit. However, it becomes virtually impossible to use in circuits which are only slightly more complicated. This makes it necessary to find a more suitable, though possibly less accurate, method of analyzing diode circuits.

We noticed in Examples 5.1 and 5.2 that with the diode forward biased, changing the source voltage from 3 to 1.5 V affected the voltage drop across the diode only slightly. This is typical of P-N diodes. Once the diode is "turned on," that is, forward biased enough so that current flows, its forward voltage changes very little for large changes in forward current. The diode is said to possess a low AC *resistance*, this being its resistance to *changes* in current, once it is turned on. For instance, in Examples 5.1 and 5.2, the germanium diode's forward current changed from 2.75 mA to 1.26 mA, while its forward voltage changed from 0.25 V to 0.24 V. Defining AC resistance as the *change* in voltage divided by the *change* in current at a certain point on the diode's characteristics, we have

$$r_f = \frac{\Delta V_F}{\Delta I_F} = \frac{0.01 \text{ V}}{1.49 \text{ mA}} = 6.7 \text{ } \Omega \tag{5.5}$$

where r_f is the diode's AC resistance in the *forward-bias* region; ΔV_F is the change in voltage (Δ is the Greek letter delta); ΔI_F is the change in current; and Ω is the Greek letter omega, representing ohms. Actually, r_f is simply the reciprocal of the *slope* of the diode curve at the point of interest. Figure 5.22 shows how one would determine r_f from a diode's characteristic. For most diodes, r_f is very small in the forward-bias region, usually only a few ohms. It changes, depending where on the characteristic it is measured. The value of r_f decreases as the forward current increases and the curve gradually gets steeper.

The relative constancy of the diodes forward voltage, once it is turned on, is a useful property for analyzing diode circuits such as the one in Figure 5.16.

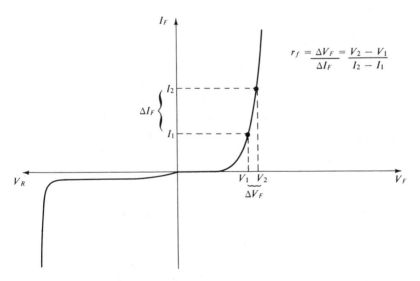

Figure 5.22.   OBTAINING r_f FROM THE DIODE CHARACTERISTIC

Since the diode voltage is essentially constant for varying current, the current in the circuit is primarily determined by the voltage source and the series resistance. The current in the circuit varies directly with the voltage source and inversely with the series resistance.

The *approximate method* of calculating the operating point of the diode in a circuit such as that in Figure 5.16 is as follows: First, calculate the approximate forward current by estimating the forward voltage of the diode and using Kirchhoff's law (Equation 5.1). Then, use this approximate current to find the forward voltage of the diode on its characteristic curve. Now, using this value of forward voltage, calculate the actual forward current. This last step is frequently unnecessary if the estimated forward voltage across the diode is close to the actual voltage.

▶ EXAMPLE 5.6

Use the approximate method to find I_{FO} and V_{FO} for the circuit in Figure 5.23 for $E_S = 3$ V.

From the diode's characteristic in Figure 5.18, we can estimate the diode's voltage at 0.25 V since it is forward biased. Thus the voltage across the resistor must be $3\ V - 0.25\ V = 2.75\ V$. The circuit current can therefore be calculated as

$$I_{FO} = \frac{2.75\ V}{1\ k\Omega} = 2.75\ mA$$

Now, looking at the diode characteristic at a current of 2.75 mA, we read a voltage of 0.25 V. Thus our original estimate of V_{FO} was a good one. These results agree exactly with those obtained by the load-line method in Example 5.1.   ◀

▶ EXAMPLE 5.7

Use the approximate method to find I_{FO} and V_{FO} for the circuit in Figure 5.23 for $E_S = 1.5$ V.

From the diode's characteristic in Figure 5.18, we can again estimate the diode's voltage at 0.25 V. Using $V_{FO} = 0.25$ V, the approximate value of the resistor voltage is $1.5\ V - 0.25\ V = 1.25\ V$. Thus,

$$I_{FO} = \frac{1.25\ V}{1\ k\Omega} = 1.25\ mA$$

Now, looking at the diode characteristic at a current of 1.25 mA, we read a voltage of 0.24 V. This is very close to our original estimate of 0.25 V. Thus we can accept these results as a very good *approximation to those calculated in Example 5.2*.   ◀

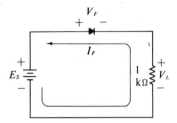

Figure 5.23.

In many cases, a circuit designer or analyst does not have the *I-V* character-istic or specifications for the diode readily available, and, therefore, must base approximations on experience. V_{FO} is usually estimated at about 0.7 V for a silicon diode and about 0.3 V for a germanium diode unless the diode's char-acteristics are known. This is usually accurate enough for most purposes; for some purposes, however, the diode's characteristics or specifications must be known.

▶ EXAMPLE 5.8

(a) Repeat Example 5.6 assuming $V_{FO} = 0.3$ V, since the diode is germa-nium. Using $V_F = 0.3$ V, the approximate value of I_{FO} is

$$I_{FO} = \frac{2.7 \text{ V}}{1 \text{ k}\Omega} = 2.7 \text{ mA}$$

Compare this with the results of Example 5.6.
(b) Change the diode in Figure 5.23 to a silicon diode and calculate the circuit current for $E_S = 8$ V.
   Since the diode is silicon, we can approximate its forward voltage as $V_{FO} = 0.7$ V. The resistor voltage is therefore approximately 8 V − 0.7 V = 7.3 V. Thus, we have

$$I_{FO} = \frac{7.3 \text{ V}}{1 \text{ k}\Omega} = 7.3 \text{ mA} \qquad ◀$$

Reverse-bias approximations

An approximate method can also be used for solving circuits such as the one in Figure 5.19, where the diode is reverse biased. The method takes advantage of the diode's characteristic in the reverse-bias region, where its reverse current is fairly constant, until reverse breakdown is reached.

The procedure is to obtain the reverse current from the diode's char-acteristic or specifications and to use it to calculate the diode voltage using Kirchhoff's voltage law. In most cases, most of the supply voltage will be dropped across the diode, since in reverse bias the current is very small and, therefore, the drop across the series resistor will be small.

▶ EXAMPLE 5.9

(a) Use the method outlined above to find the operating point of the circuit in Figure 5.24 with $E_S = 50$ V.

From the diode's characteristic in Figure 5.21, we can estimate the reverse current, I_{RO}, at 1 μA, since this is the reverse current at 50 V reverse voltage. Using this value of current, we can calculate the resistor voltage as $V_L = 1$ μA $\times$ 5 MΩ $= 5$ V. Thus, using Kirchhoff's voltage law,

$$V_{RO} = 50 \text{ V} - 5 \text{ V} = \mathbf{45 \text{ V}}$$

These values agree exactly with those in Example 5.3, obtained using the load-line method.

(b) Change R to 10 kΩ in Figure 5.24 and find the circuit operating point.

Once again we can estimate $I_{RO} = 1$ μA so that the resistor voltage is $V_L = 1$ μA $\times$ 10 kΩ $= 10$ mV. This means that the diode voltage must be

$$V_{RO} = 50 \text{ V} - 10 \text{ mV} = 49.99 \text{ V} \approx 50 \text{ V}$$

The diode voltage is essentially equal to the supply voltage. This is often the case for normal-sized load resistors (until breakdown occurs).   ◀

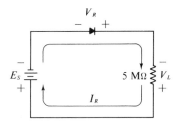

Figure 5.24.

Reverse breakdown case

The procedure is somewhat different if the supply voltage V is greater than the diode's reverse breakdown voltage. In this case, the diode will almost always be broken down and its voltage will be approximately equal to its reverse breakdown voltage. Using this value for V_{RO}, the value of I_{RO} can be calculated using Kirchhoff's law.

▶ EXAMPLE 5.10

(a) Find V_{RO} and I_{RO} for the circuit of Figure 5.24 for $E_S = 125$ V.

Since $E_S = 125$ V and the diode breaks down at 100 V, we can assume the diode is in breakdown and $V_{RO} = 100$ V. Thus the resistor voltage must be 125 V − 100 V = 25 V. The circuit current is therefore

$$I_{RO} = \frac{25 \text{ V}}{5 \text{ M}\Omega} = 5 \text{ }\mu\text{A}$$

These values agree exactly with Example 5.5.

(b) Change R to 1 kΩ and repeat. Once again, since the diode is in break-down, its voltage remains constant at V_{RO} = 100 V. The remaining 25 V of the supply voltage is across R. Thus we can calculate

$$I_{RO} = \frac{25 \text{ V}}{1 \text{ k}\Omega} = 25 \text{ mA}$$

This large value of reverse current can cause excessive diode power dissipation, with the resultant heat damaging the diode junction. For this reason, reverse breakdown is usually avoided unless R is large enough to safely limit the breakdown reverse current.   ◄

Ideal diode approximations

An *ideal* diode behaves as a perfect short circuit ($R = 0$) when forward biased, and as a perfect open circuit ($I_R = 0$) when reverse biased. In many cases in-volving diode circuits, it is possible to assume ideal diode behavior without grossly affecting the accuracy of the circuit calculations. However, in most of our subsequent work we will usually use the approximations presented above, except in cases where I_R for the diode is not known; then we will assume $I_R \approx 0$ (except for breakdown, of course), since this usually introduces in-significant error.

5.11.   Applications of P-N Diodes

A basic characteristic of P-N diodes is that they conduct current much more readily in the forward-biased state than in the reverse-biased state unless they are broken down. When forward biased, current in the circuit is limited essen-tially by the resistance in series with the diode, while in a reverse-biased circuit the diode's small reverse current is all that flows. Any device that possesses this property is called a *rectifying* device. A P-N diode, then, is often referred to as a P-N *rectifier.*

A figure of merit, M, for rectifiers is based upon its resistance to current in both its regions of operation. It is simply the ratio of the diode's reverse re-sistance to its forward resistance; that is,

$$M = \frac{R_R}{R_F} \tag{5.6}$$

where R_R and R_F are the DC *resistance* of the diode in the reverse direction and forward direction, respectively. R_R is equal to the DC reverse voltage across the diode divided by its reverse current, and R_F is equal to the forward voltage drop divided by its forward current.

For example, the diode used in the examples of the previous section had a reverse current of 1 μA at reverse voltages up to 100 V. Thus R_R at $V_R = 100$ V would be

$$R_R = \frac{100 \text{ V}}{1 \text{ μA}} = 100 \text{ M}\Omega$$

The same diode had a forward current of 2.75 mA at a forward voltage of 0.25 V. Thus R_F at $I_F = 2.75$ mA would be

$$R_F = \frac{0.25 \text{ V}}{2.75 \text{ mA}} = 91 \text{ }\Omega$$

This gives a figure of merit of

$$M = \frac{R_R}{R_F} = \frac{100 \text{ M}\Omega}{90 \text{ }\Omega} = 1.1 \times 10^6$$

This is considered a good ratio, although some P-N diodes, especially silicon types, have ratios well above ten times this value. Note that the values of R_R and R_F can vary depending on where on the diode's characteristic you take the measurements. Figure 5.25 shows the general method of finding these quantities.

The main application of rectifiers is to convert an input AC voltage to a pulsating DC voltage across a load. This is illustrated in Figure 5.26, where a silicon rectifier is placed in series with an AC voltage source and a load resistance. The AC voltage source, in this case, is simply a voltage which alternates periodically between $+50$ volts and -50 volts. When the input voltage

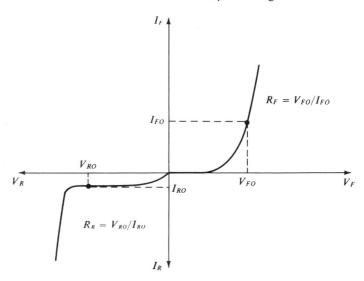

Figure 5.25.   CALCULATION OF R_R AND R_F FROM DIODE CHARACTERISTIC

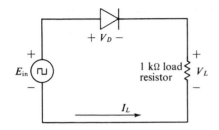

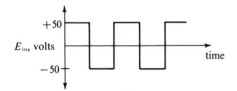

Figure 5.26.   RECTIFICATION OF AN AC VOLTAGE

E_{in} is at $+50$ volts, the diode is forward biased and when E_{in} is at -50 volts, the diode is reverse biased. When forward biased, the diode will have V_F equal to 0.7 V dropped across it. This means that the remainder of the $+50$ volts from the input must appear across the 1-kΩ load resistor. That is, 49.3 volts, which produces 49.3/1 kΩ = 49 mA of current flow, will be across the load resistor. When reverse biased, the diode will allow a small reverse leakage current to flow, which develops only a very small voltage across the load. Essentially all of the -50 volts from the input appear as reverse voltage on the diode. That is, V_D will equal -50 volts. Figure 5.27 shows the voltage wave-

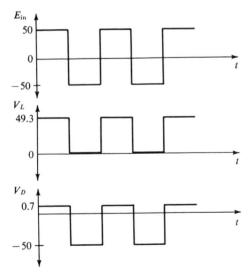

Figure 5.27.   VOLTAGE WAVEFORMS FOR CIRCUIT IN FIGURE 5.26

forms of the circuit, including E_{in}. Note that at all times $E_{in} = V_D + V_L$, satisfying Kirchhoff's voltage law. Although E_{in} alternates between a positive and a negative voltage, the voltage across the load resistor has a positive portion only. The rectifier has allowed forward current to flow and develop a positive voltage across the load, while blocking reverse current, so that no voltage appears across the load when E_{in} is negative. Thus the AC input has been rectified, producing pulsating DC* across the load.

It is important to note that in this type of circuit the rectifier must be able to withstand a reverse voltage equal to the peak negative voltage of the input, in this case 50 volts. The rectifier must have a PRV (see Section 5.7) rating of at least this amount.

▶ EXAMPLE 5.11

For the rectifier circuit in Figure 5.28, sketch the voltages V_D and V_L. Determine a safe PRV rating for this rectifier.

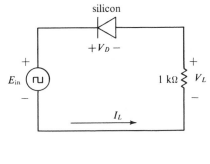

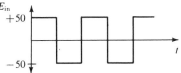

Figure 5.28.   RECTIFIER CIRCUIT WITH DIODE REVERSED

This circuit is similar to the one in Figure 5.27, except that the diode connections have been reversed. With the diode connected as such, it will become forward biased when E_{in} is at -50 volts and reverse biased when E_{in} is at $+50$ volts. Thus forward current will flow through the diode when E_{in} is at -50 volts. The voltage V_D will be about -0.7 volt, since the diode is forward biased. This leaves a voltage of -49.3 volts across the load. When E_{in} is at $+50$ volts, the diode is reverse biased, V_L will be essentially zero and V_D will be 50 volts reverse voltage on the diode. The voltage waveforms are shown in

---

Pulsating DC is simply current which flows in one direction only but is not necessarily a constant current such as pure DC.

Figure 5.29. This time the AC input has been rectified so that only negative pulsating DC appears across the load. Since the diode has 50 volts reverse voltage across it when E_{in} is positive, it must have a PRV rating of at least 50 volts.   ◄

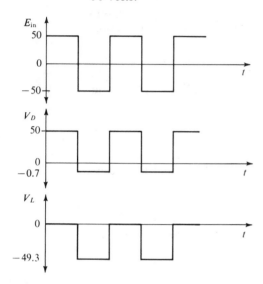

Figure 5.29.

▶ EXAMPLE 5.12

Figure 5.30 shows a rectifier circuit used to change an AC sine wave input voltage into a *half-wave* rectified sine wave. The circuit operation is the same as explained previously. The silicon diode will conduct when the input voltage exceeds +0.7 V during its positive half-cycle. The load-resistor voltage during this time will equal approximately the input-voltage waveform (less the 0.7-V diode voltage drop). When the input goes below 0.7 V and then into its negative half-cycle, the diode does not conduct; thus the resistor voltage is approximately zero and all of the input waveform is across the diode. The resistor-voltage waveform essentially contains the positive half-cycle of the input, while the diode-voltage waveform contains the negative half-cycle.   ◄

There are many applications of P-N diodes in electronic circuitry. Almost all of them, however, take advantage of the diode's rectifying property. Some other diode circuits will be left as problems at the end of the chapter.

5.12.   Power Dissipation in P-N Diodes

The power dissipated in P-N diodes as a result of current flow ($P = IV$) contributes to the temperature rise at the P-N junction in the same way as does an

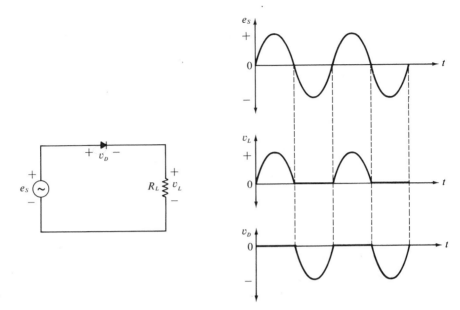

Figure 5.30.

increase in the ambient (surrounding) temperature. The power dissipation must never go above the value which causes the junction to rise above its maximum allowable temperature. If it does, the device may become permanently damaged.

The power dissipation, P_D, is simply equal to the diode's voltage multiplied by its current. The diode dissipates power in all of its operating regions: forward bias, reverse bias and reverse breakdown.

The product of power dissipation, P_D, and the *thermal resistance* of the diode, θ_{JA} (where θ is the Greek letter theta), gives the rise in junction temperature due to power dissipation. Adding this rise in temperature to the ambient temperature, T_A, gives the junction temperature, T_J. That is,

$$T_J = T_A + P_D\theta_{JA} \qquad (5.7)$$

The thermal resistance, θ_{JA}, is a physical property of the diode and its casing. It is a measure of the diode's ability to lose its heat to the surrounding air. A lower value of θ_{JA} indicates a diode which can safely dissipate more power. The units of θ_{JA} are usually °C/W or °C/mW.

▶ EXAMPLE 5.13

A silicon diode conducts a forward current of 100 mA, while its forward voltage is 0.7 V. Find the temperature at its junction if its thermal resistance, θ_{JA}, is 0.4°C/mW and the ambient temperature, T_A, is 25°C. Repeat the procedure for a T_A of 55°C.

The temperature at the junction is given by Equation 5.7. In this case, P_D is 0.7 V $\times$ 100 mA, or 70 mW. Thus at $T_A = 25°C$,

$$T_J = 25°C + (70 \text{ mW})(0.4°C/\text{mW}) = \mathbf{53°C}$$

and at $T_A = 55°C$,

$$T_J = 55°C + (70 \text{ mW})(0.4°C/\text{mW}) = \mathbf{83°C}$$

This example shows us that the junction temperature will increase as the ambient temperature increases, indicating an important point: the maximum allowable power dissipation of a diode is lower at higher ambient temperatures. ◄

▶ EXAMPLE 5.14

The silicon diode of Example 5.13 has a maximum allowable junction temperature, $T_{J(\text{max})}$, of 150°C. Find the maximum allowable power dissipation, $P_{D(\text{max})}$, at 25°C ambient temperature. Repeat for an ambient temperature of 55°C.

Stated differently, the problem is to find what value of P_D causes T_J to rise above 150°C. At 25°C, we have

$$T_J = 25°C + P_{D(\text{max})}(0.4°C/\text{mW})$$

Setting $T_J = T_{J(\text{max})} = 150°C$, we have

$$150°C = 25°C + P_{D(\text{max})}(0.4)$$

which gives

$$P_{D(\text{max})} = \mathbf{313 \text{ mW}}$$

Thus, at 25°C ambient temperature, a diode power dissipation of 313 mW causes T_J to rise to 150°C.

At 55°C ambient, we have

$$150°C = 55°C + P_{D(\text{max})}(0.4)$$

or

$$P_{D(\text{max})} = \mathbf{238 \text{ mW}}$$

Thus the diode is allowed to dissipate only 238 mW at 55°C. ◄

The power that the diode can safely dissipate decreases with ambient temperature. This is a very important consideration in the design of circuits which may be subjected to temperatures above room temperature. We will find this to be true of all semiconductor devices.

A diode may be operated above the maximum power dissipation if *heat sinks* are used. Heat sinks are large metal bases of copper and aluminum upon which the diodes may be mounted. The heat sinks may have their outside surfaces formed as fins to radiate heat more readily. The overall effect is to reduce θ_{JA} so as to allow more power dissipation without exceeding $T_{J(\text{max})}$.

Forward current in a diode is limited only by the allowable power dissipation; this depends on the type of construction, the size of the junction, the diode material and the mounting used in assembling the device. There are various processes used to manufacture P-N junctions, most of which are also employed in the production of transistors.

5.13.   Comparison of Diode Materials

The basic concepts previously discussed apply to all diodes and diode devices. There are, however, some basic differences among the different types of semiconductor materials which make one type more suitable for a particular application. For example, diode devices used for high-power rectifier applications are now primarily made of silicon as the basic material, with the doping impurities added by the process of diffusion (to be discussed later). Selenium rectifiers, germanium diodes and some copper oxide devices are sometimes used for high-power rectifiers; however, silicon is by far the most popular. The silicon diode can tolerate higher temperatures than other devices. Semiconductor diodes other than silicon have maximum operating temperatures of around 100°C, whereas silicon can typically operate at 150°C. Silicon has the added advantage of a lower reverse current, at a given temperature and voltage, than other diode devices. It also has a much higher reverse breakdown voltage (up to perhaps several thousand volts) and the ability to handle much greater forward currents. Silicon does have a disadvantage in that its forward-voltage drop is higher than some of the other devices (0.7 volt versus 0.3 volt for germanium) and, therefore, dissipates a little more power for a given forward current. Electron and hole mobilities are higher in germanium and gallium arsenide than in silicon. If speed is a consideration, the use of germanium or gallium arsenide as the basic material might be warranted. All things considered, the use of silicon devices has thus far predominated for most of the reasons mentioned above.

5.14.   P-N Junction Capacitance

A reverse-biased P-N junction can be compared to a charged capacitor. This is illustrated in Figure 5.31. The N and P regions (away from the space-charge region) are essentially low-resistance areas due to the high concentration of majority carriers, while the space-charge region, which is depleted of majority carriers, is essentially an area of effective insulation between the N and P regions. The N and P regions act as the plates of the capacitor, while the space-charge region acts as the insulating dielectric. The reverse-biased junction thus has an effective capacitance, shown in dotted lines in the figure. The value of this capacitance depends on the width of the space-charge region (distance

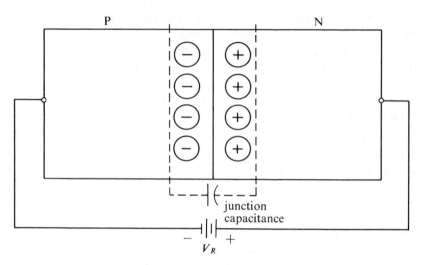

Figure 5.31.   P-N JUNCTION CAPACITANCE

between capacitor plates). Thus it depends on reverse voltage. As reverse voltage increases, the space-charge region becomes wider, effectively increasing the plate separation and decreasing the capacitance. A device known as a voltage-variable capacitor (also called a varactor) uses this property of P-N junctions in communications circuitry. This junction capacitance affects the speed of operation of P-N diodes and all P-N junction devices, as will be discussed in subsequent work.

A forward-biased P-N junction possesses junction capacitance because the space-charge region is still present, although greatly reduced in size. In fact, the narrower space-charge region results in a larger junction capacitance, since the capacitor plates are now more closely spaced. Typically, junction capacitances range from 0.1 to 4 pF in reverse bias and up to 20 pF in forward bias.

GLOSSARY

P-N junction: the joining of a P region and an N region in a single crystal structure.

Space-charge region (depletion region):   area on either side of a P-N junction which contains impurity ions, and is depleted of carriers.

Potential barrier: electrical force exerted by the space-charge region on charge carriers.

Majority-carrier diffusion current: current flow across a P-N junction carried by diffusing majority carriers.

Minority-carrier drift current: current flow across a P-N junction carried by minority carriers accelerated by the potential barrier.

Reverse bias: voltage applied to a P-N diode so as to increase the potential barrier (positive voltage on N region, negative on P region).

Forward bias: voltage applied to a P-N diode so as to neutralize the potential barrier (positive voltage on P region, negative on N region).

Reverse current: current flow across a P-N junction when it is reverse biased.

Forward current: current flow across a P-N junction when it is forward biased.

Reverse breakdown: process of minority carrier avalanching when sufficient reverse voltage is applied.

Anode: P region of a P-N diode.

Cathode: N region of a P-N diode.

PRV, V_{RDC}, $I_{F(max)}$, V_F, I_R: See Tables 5.1 and 5.2.

Load line: line representing a plot of the Kirchhoff voltage law equation in a diode circuit.

AC resistance: P-N diode's resistance to *changes* in current at a given operating point.

Rectification: process of changing an AC voltage to a DC voltage.

Rectifier: device which performs rectification.

Thermal resistance: measure of a device's ability to lose heat to its surroundings.

Junction capacitance: effective capacitance of a P-N junction.

PROBLEMS

5.1 Briefly describe the charge situation of a P-type region at room temperature. Repeat for an N-type region.

5.2 What causes majority carriers to flow at the moment a P and N region are brought together?

5.3 Why doesn't this flow continue until all carriers have recombined?

5.4 Describe the formation of the *space-charge region* in a P-N junction.

5.5 In an unbiased P-N junction, what currents are continually flowing across the junction?

5.6 Sketch the charge concentration profile of an unbiased P-N junction.

5.7 Compare the potential barriers in silicon and germanium P-N junctions.

5.8 Explain the effect of reverse bias on the space-charge region.

5.9 Which carriers conduct current when a P-N diode is reverse biased? Draw the symbol for a P-N diode showing the direction of reverse current flow.

5.10   What limits the number of reverse current carriers?

5.11   Why is the reverse current in a silicon diode much smaller than in a comparable germanium diode?

5.12   Describe the process of avalanche breakdown in a P-N diode.

5.13   What limits the reverse current in a diode which is in reverse breakdown?

5.14   Explain the effect of forward bias on the space-charge region.

5.15   Which carriers conduct forward current in a diode? Draw the symbol for a P-N diode showing the direction of forward current.

5.16   Roughly how much forward voltage is needed to cause current to flow in silicon? In germanium?

5.17   Draw the charge concentration profile for a reverse-biased P-N diode. Repeat for a forward-biased diode.

5.18   How many *I-V* curves are needed to characterize a P-N diode?

5.19   Draw the *I-V* characteristic of a typical diode. Label all significant points and regions.

5.20   In Figure 5.32, indicate the type of bias in each case.

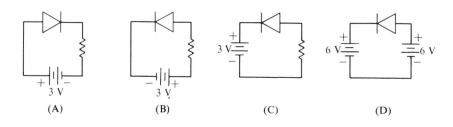

(A)	(B)	(C)	(D)

Figure 5.32.

5.21   A germanium diode has the following ratings and specifications at 25°C:

$$
\begin{aligned}
\text{PRV:} &\quad \text{200 volts} \\
V_{\text{RDC}}: &\quad \text{150 volts} \\
I_{F(\text{max})}: &\quad \text{500 mA} \\
V_F: &\quad \text{0.3V @ 100 mA} \\
I_R: &\quad \text{10 } \mu\text{A @ 100 volts}
\end{aligned}
$$

(a) What is the maximum allowable DC reverse voltage for this diode?
(b) What is the maximum allowable forward current for this diode?
(c) Could a reverse voltage of 180 volts be applied to this diode under any condition?

5.22   What is the primary effect of temperature on a semiconductor material?

5.23   What effect does an increase in temperature have on the potential barrier of a diode? How does this affect the forward characteristic of a diode? Its reverse characteristic?

5.24 A silicon diode has a reverse current of 1 μA at 25°C. What will be its approximate reverse current at 36°C? At 47°C?

5.25 Using the load-line method, find the operating point of the circuit shown in Figure 5.33.

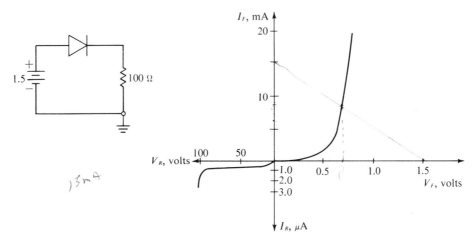

Figure 5.33.

5.26 The diode of Figure 5.33 is used in the circuit of Figure 5.34. Find the operating point of the circuit using the approximate method.

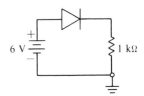

Figure 5.34.

5.27 Replace the battery in Figure 5.34 by a −50-volt battery and find the voltage across the 1-kΩ resistor. Repeat for a −150-volt battery.

5.28 Will the diode in Figure 5.35 conduct forward current?

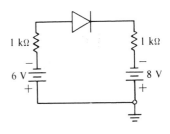

Figure 5.35.

5.29 A certain diode has an AC resistance, r_f, of 10 ohms at $I_F = 10$ mA, $V_F = 0.6$ V. What would be the value of I_F at $V_F = 0.59$ V? At 0.61 V?

5.30 Find the value of R_F, DC forward resistance, for the diode of Figure 5.33 at 10 mA. Find R_R, DC reverse resistance, at $V_R = 100$ V. What is the figure of merit, M, for this diode?

5.31 A silicon diode is used as a rectifier in the circuit of Figure 5.36. Sketch the voltage across the resistor.

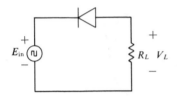

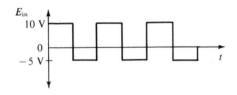

Figure 5.36.

5.32 What is the lowest PRV which the diode in Problem 5.31 can have for safe operation?

5.33 The diode in Figure 5.37 is employed as a *voltage limiter*. Its function is to limit the maximum output voltage, V_{out}, to a value of around 6.7 volts no matter how high V_{in} becomes. V_{out} can go below 6.7 volts, but not above, since the diode will become forward biased when V_{in} equals 6.7 volts; increases in V_{in} will merely serve to supply more current to the diode, whose voltage will change only slightly.
(a) Find V_{out} when $V_{in} = -2$ volts if the diode is silicon and $I_R = 1$ μA.
(b) Find V_{out} for $V_{in} = 2$ volts.
(c) Find V_{out} for $V_{in} = 4$ volts.
(d) Find V_{out} for $V_{in} = 6.7$ volts.

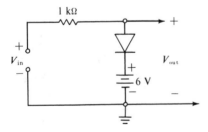

Figure 5.37.

(e) Find V_{out} for V_{in} = 10 volts.

(f) Plot V_{out} versus V_{in}.

(g) What limits the maximum *positive* value of V_{in} which can be applied to this circuit? Maximum *negative* value?

5.34 Design a diode limiter to limit the output voltage to 10.3 volts, using a germanium diode. If the input can go as high as 100 volts, what value of series resistor must be used? Assume $I_{F(max)}$ = 100 mA.

5.35 What is the power dissipation of the silicon diode in Figure 5.38?

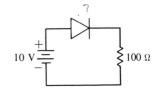

Figure 5.38.

5.36 The diode in Problem 5.35 has a thermal resistance, θ_{JA}, of 0.2°C/mW. Find the diode's junction temperature if T_A is 25°C.

5.37 If a diode has an $I_{F(max)}$ of 1 A at V_F = 1 V, what is its maximum allowable power dissipation, $P_{D(max)}$, at T_A = 25°C?

5.38 The same diode has θ_{JA} = 0.1°C/mW. What is its maximum allowable junction temperature? What is its $P_{D(max)}$ at T_A = 55°C?

5.39 Compare silicon and germanium diodes as to the following:
(a) Maximum operating temperature.
(b) Reverse breakdown voltages.
(c) Reverse leakage current.
(d) Forward voltage drop.
(e) Current handling capabilities.
(f) Operating speed.

5.40 Indicate whether junction capacitance increases or decreases with the following:
(a) Decrease in reverse bias.
(b) Increase in temperature.
(c) Increase in forward bias.

5.41 The P-N diode whose characteristic is shown in Figure 5.39 has a maximum allowable junction temperature, $T_{J(max)}$, of 125°C.
(a) Calculate the diode's thermal resistance if $P_{D(max)}$ = 500 mW at T_A = 25°C.
(b) What is $I_{F(max)}$ at a T_A of 25°C?
(c) Find the minimum value of R that can be used in the circuit shown at 25°C ambient.
(d) Calculate $P_{D(max)}$ at T_A = 40°C and repeat (b) and (c) for 40°C. (Assume that the *I-V* characteristic of the diode does not change appreciably.)

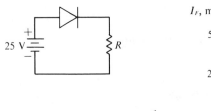
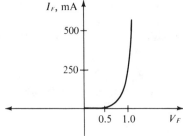

Figure 5.39.

5.42   The diode in Figure 5.40 has a reverse breakdown voltage of 100 V and a maximum power rating of 400 mW. What is the minimum value of R that can be used in this circuit?

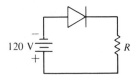

Figure 5.40.

5.43   The circuit in Figure 5.41 is supposed to function as a half-wave rectifier circuit with an output consisting of the negative half-cycles of the sine-wave input. However, when the circuit is connected, the output waveform appears as shown in the figure. What is wrong in the circuit and how would you correct the error?

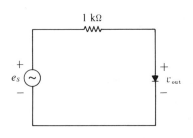
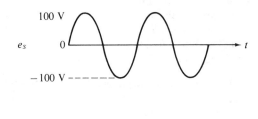

Figure 5.41.

REFERENCES

Foster, J. F., *Semiconductor Diodes and Transistors*, Vol. 2. Beaverton, Oregon: Programmed Instruction Group, Tetronix, Inc., 1964.

Romanowitz, H. A. and R. E. Puckett, *Introduction to Electronics*. New York: John Wiley & Sons, Inc., 1968.

6

Zener Diodes

6.1.   Introduction

Reverse breakdown of a P-N diode can take place due to avalanche breakdown, *zener breakdown* or a combination of both. Avalanche breakdown has already been discussed in Chapter 5; the mechanism of zener breakdown, including the effects of impurity concentration and temperature, will be explained in this chapter. Both types of reverse breakdown are employed in *zener diodes*, which are P-N diodes specifically designed to operate in the reverse breakdown region. Studying the principal applications of zener diodes will help to explain their widespread usage in electronic circuitry.

6.2.   Zener Breakdown

When a reverse-biased diode goes from a low conduction state to a state of high conduction, either zener breakdown or avalanche breakdown, or a combination of both, takes place. Avalanche breakdown, as we have seen, is a result of the ionization of covalent bonds by minority carriers accelerated across the reverse-biased junction. This results in a multiplication of carriers crossing the junction, causing reverse current to increase rapidly with reverse voltage.

When reverse breakdown in a diode occurs at a voltage greater than 5 volts, it is probably due to avalanche breakdown. When it occurs below 5 volts, it is due to zener breakdown. Zener breakdown is a result of the ionization of covalent bonds due to the high-intensity electric field that inherently exists across the narrow space-charge region. If the space-charge region is sufficiently narrow, increasing the potential difference across it will eventually develop an electric field (recall that electric field strength is inversely proportional to the distance over which a potential difference exists) that is strong enough literally to yank valence electrons out of their covalent bonds, thereby producing extra conduction electrons and holes. These additional carriers drift across the junction under the influence of reverse voltage, causing reverse current to increase rapidly and bringing about zener reverse breakdown.

In order for zener breakdown to occur, the width of the space-charge region must be made very narrow. This can be accomplished by increasing the doping in both the P and N regions, as can be seen by comparing the two silicon P-N junctions shown in Figure 6.1. In the P-N junction in part A of the figure, the space-charge region produces a potential barrier of 0.7 volt. This potential barrier is produced by a certain number of impurity ions residing near the junction. If we increase the density of the impurities in both regions, the P-N junction in part B of the figure results. Since there are now more impurity ions per unit volume, the width of the space-charge region needed to produce a 0.7-volt potential barrier is reduced. If the amount of doping is great enough, space-charge regions as narrow as a few millionths of an inch can result, making it possible for zener breakdown to occur at a few volts reverse bias.

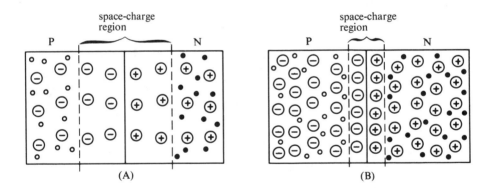

Figure 6.1.   EFFECT OF DOPING ON SPACE-CHARGE REGION

The amount of reverse voltage needed to cause zener breakdown decreases as the space-charge region becomes narrower. Thus zener breakdown voltage decreases as the amount of doping increases. The effect of an increase in temperature on zener breakdown is to cause the zener breakdown voltage to *decrease*. This situation is opposite to what happens in avalanche breakdown, where *increasing* temperature *increases* breakdown voltage.

As mentioned previously, zener breakdown occurs at voltages below 5 volts, while avalanche breakdown occurs above 5 volts. Breakdown at around 5 volts is probably a combination of both mechanisms.

6.3.   Zener Diodes

When dealing with P-N diodes as rectifiers, the entrance of the diode into its reverse breakdown region was undesirable. If a diode is purposely constructed to operate in the reverse breakdown region, the voltage across the terminals of the diode will remain fairly constant over a wide range of currents if the diode is in the breakdown condition. For a diode designed to operate in the reverse breakdown mode, the amount of doping and the junction's geometry can be varied to cause the diode to break down at a certain reverse voltage, and handle a given range of reverse currents. The name *zener diode* is given to these diodes, which may be misleading since they encompass both types of reverse breakdown mechanisms — zener and avalanche. Other names like *reference diodes* or *breakdown diodes* are sometimes used, but *zener diode* is the most widely accepted and we shall use it here. Zener diodes are designed to operate in the reverse breakdown portion of their characteristics. By careful control of impurity concentration and junction geometry, close tolerances on the reverse breakdown voltage can be obtained. Zener diodes with breakdown voltages below around 5 volts are affected by zener breakdown; zener diodes with breakdown voltages above 5 volts are affected by avalanche breakdown.

Figure 6.2 shows the *I-V* characteristic of a typical zener diode. It is not much different from that of a conventional P-N diode. However, a significant difference is the sharpness of the curve at the boundary of the breakdown region; compare this curve with the dotted curve for a conventional P-N diode. This extremely sharp curve is a desirable characteristic, as we shall see when we investigate circuit applications of zener diodes. The electronic symbol for a zener diode is illustrated in Figure 6.3 alongside that of a conventional diode.

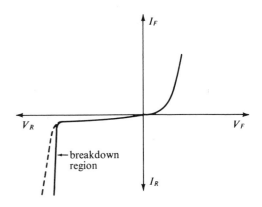

Figure 6.2.   TYPICAL ZENER DIODE *I-V* CHARACTERISTIC

Before proceeding, it should be re-emphasized that a zener diode is a P-N diode specially constructed to operate in the reverse breakdown region and whose *I-V* characteristics are essentially those of a P-N diode.

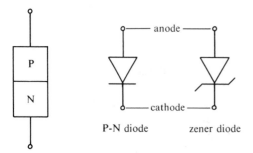

Figure 6.3.   ZENER DIODE SYMBOL

6.4.   Zener Diode Specifications and Ratings

The *I-V* characteristic of a typical zener diode is repeated in Figure 6.4, showing some of the important points on the characteristic curve. Note that reverse current is labeled I_Z, zener current; reverse voltage is labeled V_Z, zener voltage, since this is common practice. Zener diodes are meant to operate in their breakdown region, and, as such, this is the most important region on the zener curve. The portion of the reverse characteristic where the device just begins going into breakdown is called the *zener knee*. The zener current at the zener knee is labeled I_{ZK}. This is the minimum value of zener current which must be supplied in order to assure that the zener diode is operating in its breakdown region.

Once the zener diode reaches breakdown (below the knee), the zener current increases sharply with voltage. In this region (the zener region), the voltage across the zener is fairly constant over a wide variation in current. Manufacturers stipulate a test point on the zener breakdown curve somewhere below the knee. The current at this point is labeled I_{ZT}, *zener test current*. The zener voltage at this point is labeled V_{ZT}. V_{ZT} is called the *nominal zener voltage* and is measured at the zener test current. For a given type of zener diode, manufacturers usually place a tolerance of 5, 10 or 20 percent on the value of V_{ZT}, similar to the tolerances on resistors. For example, if a zener diode has a nominal value V_{ZT} of 10 volts with a 10 percent tolerance, we can expect zener diodes of this type to have values V_{ZT} of 10 volts $\pm$ 10 percent; that is, anywhere from 9 to 11 volts.

The maximum allowable zener current is labeled I_{ZM} and is the maximum zener current which can be safely conducted by the zener diode. I_{ZM} is determined by the maximum power dissipation rating of the zener diode, $P_{Z(\max)}$. As with ordinary P-N diodes, this maximum power dissipation is determined by

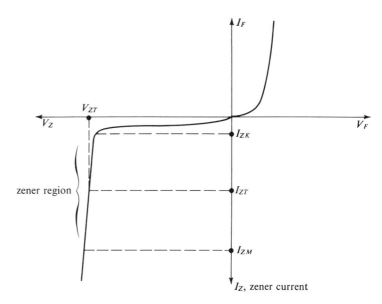

Figure 6.4.   ZENER DIODE CHARACTERISTIC CURVE

the maximum allowable junction temperature, $T_{J(max)}$. Exceeding this rating can lead to destruction of the zener diode.

Zener impedance

Another important parameter of zener diodes is called *zener impedance*, Z_Z. The zener impedance is essentially the AC resistance (impedance) of the zener diode (similar to AC resistance of an ordinary P-N diode; see Section 5.10). It is simply the reciprocal of the slope of the zener curve. Zener impedance is usually measured just above the knee in the zener region, Z_{ZK}, and at the zener test point, Z_{ZT}. Zener impedance decreases with increases in zener current, since the slope of the zener curve gets steeper as current increases. Thus Z_{ZT} is always much smaller than Z_{ZK}. Zener impedance is important since it is a measure of the steepness of the zener region and, as such, indicates how much the reverse voltage across the zener diode will change for a change in zener current. That is,

$$Z_Z = \frac{\Delta V_Z}{\Delta I_Z} \text{ (ohms)} \qquad (6.1)$$

and is obtained from the zener curve in the same manner as was the AC resistance of a P-N diode in Section 5.10.

A small value of Z_Z is usually desirable since it indicates a steep curve where relatively large changes in current (ΔI_Z) result in very small changes in voltage (ΔV_Z). As we will see later, this is an important characteristic of zener diodes when operated in the breakdown region.

Temperature effects

The effect of temperature on the breakdown voltage of zener diodes was quali-
tatively discussed in Section 6.2. It is usually important in circuit design and
analysis to know quantitatively what this effect will be. For this reason, manufacturers often supply a typical *zener voltage temperature coefficient* as part of
the zener diode specifications. The zener voltage temperature coefficient, denoted K_T, indicates the *percentage* change in nominal zener voltage, V_{ZT}, for
each degree centigrade of change in junction temperature. For a zener diode
with a K_T of 0.05 percent per °C, each degree of *increase* in junction temperature will *increase* the value of V_{ZT} by 0.05 percent of its nominal value. A K_T of
-0.05 percent per °C means that each degree of *increase* in temperature will
decrease the value of V_{ZT} by 0.05 percent of its nominal value. Zener diodes
which break down due to zener breakdown (see Section 6.2) have a *negative*
temperature coefficient, while those which breakdown due to avalanche breakdown have a *positive* K_T value. The value of K_T increases as the nominal zener
voltage increases.

A typical zener diode specification which contains values of all these parameters is shown in Figure 6.5. Much valuable information about this zener diode
can be ascertained from these specifications. In fact, with this information, one
could make a fairly accurate sketch of the *I-V* curve of this zener. Let us take a
look at some of the information which can be derived from these specs.

The particular zener diode has a maximum power rating, $P_{Z(\text{max})}$, of 1 watt
at an ambient temperature of 25°C and has a nominal zener voltage, V_{ZT}, of 20
volts with a 10 percent tolerance. Thus a zener diode of this type can have a
value of V_{ZT} anywhere in the range of 18 to 22 volts at 25°C. Its temperature coefficient, K_T, is 0.075 percent per °C and can be used to determine the value of
V_{ZT} at elevated junction temperatures.

ELECTRICAL CHARACTERISTICS (@ 25°C AMBIENT)

V_{ZT}:	20 V $\pm$ 10% (measured at I_{ZT})
I_{ZT}:	12.5 mA @ V_{ZT} = 20 V
I_{ZK}:	0.25 mA @ V_{ZK} = 4 V
I_{ZM}:	32 mA
Z_{ZT}:	22 ohms max (measured at I_{ZT})
Z_{ZK}:	750 ohms max (measured at I_{ZK})
K_T:	0.075%/°C
$P_{Z(\text{max})}$:	1 watt: θ_{JA} − 100°C/watt
I_R:	1 μA @ V_R = 6 V

Figure 6.5.   TYPICAL ZENER DIODE SPECS

▶ EXAMPLE 6.1

If a zener diode with the specs given by Figure 6.5 has its junction temperature
increased from 25°C to 75°C, find the value of V_{ZT} at the higher temperature.

Since K_T gives the percentage change per degree, it is advisable first to use K_T to calculate the voltage change per degree and then to multiply by the number of degrees *change* in temperature. Thus the voltage change per degree is given by

$$\frac{\Delta V_{ZT}}{°C} = \frac{K_T}{100\%} \times V_{ZT} \tag{6.2}$$

which for this zener diode becomes

$$\frac{0.075\%/°C}{100\%} \times 20 \text{ V} = 0.015 \text{ V}/°C$$

Multiplying this by the change in temperature, in this case $75° - 25° = 50°$, we have the total voltage change, ΔV_{ZT}, given by

$$\Delta V_{ZT} = (0.015 \text{ V}/°C) \times (50°C)$$
$$= 0.75 \text{ V}$$

Adding this change in voltage to the nominal V_{ZT} gives the value of V_{ZT} at 75°C. That is,

$$V_{ZT} \text{ (at 75°C)} = V_{ZT} \text{ (at 25°C)} + \Delta V_{ZT}$$
$$= 20 \text{ V} + 0.75 \text{ V}$$
$$= 20.75 \text{ V}$$

The general formula for calculating the zener voltage at temperatures other than 25°C is given by

$$V_{ZT} \text{ (at } T_1) = V_{ZT} \text{ (at 25°C)} + \frac{K_T}{100} \times V_{ZT} \text{ (at 25°C)} \times (T_1 - 25) \tag{6.3}$$

where T_1 is the temperature of interest. Note that if T_1 is *less* than 25°C, then V_{ZT} at T_1 will be *less* than V_{ZT} at 25°C, since K_T is positive. The opposite is true if K_T is negative.   ◄

Zener diode junction temperature depends on ambient temperature and power dissipation, exactly as discussed in Section 5.11 and given by Equation 5.6 for the P-N diode. Essentially all the material in Section 5.11 can be applied to the zener diode as well, since it is, after all, a special type of P-N diode.

▶ EXAMPLE 6.2

For the zener diode with specs given by Figure 6.5, determine the maximum allowable junction temperature.

The maximum junction temperature occurs when the maximum rated power is being dissipated. Thus

$$T_{J(\text{max})} = T_A + P_{Z(\text{max})} \times \theta_{JA} \tag{6.4}$$

For this zener diode,

$$T_{J(\text{max})} = 25°C + (1 \text{ W}) \times (100°C/W)$$
$$= 125°C \qquad \blacktriangleleft$$

The knee current, I_{ZK}, for the zener diode of Figure 6.5 is given as 0.25 mA, indicating that at least this amount of zener current must be flowing to assure that the zener is operating on the steep portion of its characteristic curve.

The maximum zener current, I_{ZM}, is given as 32 mA for this zener diode. The manufacturer cannot reliably guarantee operation of this zener diode at currents above this value.

Also given as part of the zener diode specs is the value of reverse current, I_R, at some value of reverse voltage much less than V_{ZT}. This is 1 μA at 6 volts for the zener of Figure 6.5, indicating a reverse leakage current of 1 μA before the zener reaches breakdown.

More detailed zener specifications than those in Figure 6.5 are usually the case. Information on the physical characteristics and markings, as well as additional electrical specifications, are included. Appendix II contains a typical manufacturer's zener diode spec sheet.

6.5.   The Zener Diode as a Circuit Element

The zener diode is a special type of P-N diode and, as such, the methods of analyzing its operation in circuits are the same as those covered in Chapter 5. Both the load-line method and the approximate method can be used for zener diodes; however, the latter is usually employed because of its simplicity. The load-line method will be used in some circuits to help explain operation of the zener diode.

Before turning to specific zener diode applications, let us consider the simple circuit of Figure 6.6. The zener diode in this circuit has the *I-V* characteristic depicted in the figure. This circuit is similar to one studied in Section 5.9. The polarity of the voltage, E_{in}, is such that the zener diode is reverse biased. Looking at the *I-V* characteristic, the zener breakdown voltage is around 10 volts. Thus, for values of E_{in} below 10 volts, only a small reverse current will flow. The voltage across the zener diode will be equal to E_{in}, and across the load resistor, it will be essentially zero (Example 5.12). For values of E_{in} greater than 10 volts, the zener diode will be in its breakdown region if the zener current, I_Z, is above the knee current, I_{ZK}, which is 0.1 mA for this zener.

Consider an input voltage of 12.5 volts and a load resistor of 5 kΩ. The procedure for finding the exact operating point of the zener diode starts with drawing the circuit load line (Section 5.9), which is shown in Figure 6.6. It is a straight line connected between $V_Z = 12.5$ V and $I_Z = 12.5$ V$/5$ k$\Omega = 2.5$ mA. The point at which this load line intersects the zener characteristic is the operating point of the zener, which in this case is $V_Z \approx 10$ V and $I_Z \approx 0.5$ mA. The

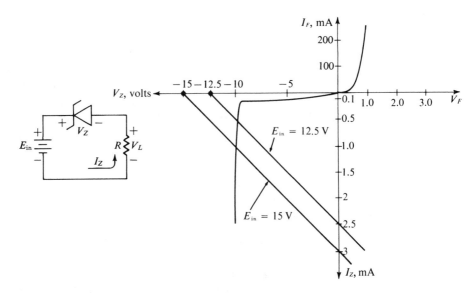

Figure 6.6.   SIMPLE ZENER DIODE CIRCUIT AND ZENER CHARACTERISTIC
CURVE

12.5-volt input is sufficient to cause the zener to operate in its breakdown region with $V_Z = 10$ V. The voltage V_L across the 5-kΩ load resistor accounts
for the other 2.5 volts of the input, since

$$E_{in} = V_Z + V_L \tag{6.5}$$

As a check, we have $I_Z = 0.5$ mA at the operating point and this current flows
through the 5-kΩ load so that

$$V_L = I_Z \times 5 \text{ k}\Omega = 2.5 \text{ V}$$

If we change E_{in} to 15 volts and repeat the above procedure, we find the
operating point to be the intersection of the 15-volt load line and the zener
characteristic. The zener is still operating in its breakdown region but at a
greater zener current. From the figure, it can be seen that $V_Z \approx 10.1$ V and
$I_Z \approx 1.0$ mA at the operating point. Also, $V_L \approx 5.0$ V.

Let us now compare these two cases, summarized in Table 6.1. The main
point to be noticed in this comparison is the relative constancy of the voltage
across the zener diode as the zener current increases. The increase in input

TABLE 6.1

E_{in}	V_Z	I_Z	V_L
12.5 V	10 V	0.5 mA	2.5 V
15 V	10.1 V	1.0 mA	5.0 V

voltage from 12.5 volts to 15 volts results in a change in current from 0.5 mA to 1.0 mA (almost double). This change in current produces only a 0.1-volt increase in the voltage across the zener (a 1 percent change). Most of the 2.5-volt change in the input voltage appears across the load resistor. The ability of a zener diode to maintain a nearly constant voltage for varying currents arises from the shape of its characteristic in the breakdown region. The more nearly this portion of the characteristic approaches a perfectly vertical line, the more constant the zener voltage will be for varying zener currents. A measure of this constant voltage property of zener diodes is the zener impedance, Z_Z. A small value of Z_Z indicates that a change in zener current will result in a small change in zener voltage. From Table 6.1 we can calculate the value of Z_Z for this zener diode in the region of its operation. The current, I_Z, changes 0.5 mA, resulting in a 0.1-volt change in V_Z. Thus

$$Z_Z = \frac{\Delta V_Z}{\Delta I_Z} = \frac{0.1 \text{ V}}{0.5 \text{ mA}} = 200 \, \Omega$$

This value is somewhat high since some zener diodes operate with zener impedances of less than 100 ohms and as low as a fraction of an ohm. Actually, the value of Z_Z decreases for larger values of I_Z. For this reason, zeners should be operated at currents much greater than the zener knee current if a better constant-voltage property is desired.

Most applications of zener diodes make use of their constant-voltage characteristic in the breakdown region. This will be pointed out in the following sections, where we will examine some of the standard zener diode circuits.

6.6.   The Zener Diode as a Voltage Regulator

Probably the most popular application of zener diodes is in producing a *regulated* output voltage from a practical DC power supply; that is, from a source having internal resistance.

To help understand how a zener diode achieves *voltage regulation*, we will briefly study a practical DC voltage supply. Figure 6.7 shows a practical DC source represented by an ideal source, E, and a series internal resistance, R_{int}. The output terminals of the source are labeled xx, and these are the terminals accessible to the user. In other words, these terminals are connected to the load to which DC power is being applied. The internal resistance, R_{int}, is internal to the power supply and generally cannot be modified by the user.

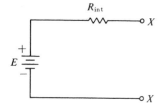

Figure 6.7.   PRACTICAL VOLTAGE SOURCE

Consider the case where $E = 10$ volts and $R_{int} = 100\ \Omega$. If we measure the voltage at the terminals of the voltage source with no load connected, we will measure 10 volts. This no-load voltage is represented by E_{NL}. Thus $E_{NL} = 10$ volts. Consider now a load resistor, R_L, connected to the voltage source terminals as in Figure 6.8. The load resistor will draw current, I_L, from the voltage source, developing an output voltage, E_L. This voltage can be easily calculated using Ohm's law in Equation 6.6. The load current, I_L, is given by

$$I_L = \frac{E}{R_{int} + R_L} \tag{6.6}$$

The load voltage, E_L, is then given by

$$E_L = I_L R_L = E\left(\frac{R_L}{R_L + R_{int}}\right) \tag{6.7}$$

It can be seen from Equation 6.7 that the voltage across the load is dependent on the load resistance. For example, with $R_L = 400\ \Omega$, we have, from Equation 6.7,

$$E_L = 10\left(\frac{400}{400 + 100}\right) = 8 \text{ volts}$$

Thus the output voltage has dropped from 10 volts at no-load to 8 volts at a 400-Ω load. The drop in output voltage due to loading is caused by R_{int}; that is, the load current develops a voltage drop across it equal to $I_L \times R_{int}$. This voltage drop leaves less of the supply voltage available to the load, causing the load voltage to decrease.

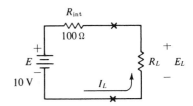

Figure 6.8.   LOADED PRACTICAL VOLTAGE SOURCE

The percentage change in output voltage from a no-load condition to a loaded condition on the power supply is called the *percentage regulation* of the power supply and is determined by applying

$$\text{percentage regulation} = \left(\frac{E_{NL} - E_L}{E_L}\right) \times 100\% \tag{6.8}$$

For the case under consideration, we can use this equation to calculate

$$\text{percentage regulation} = \left(\frac{10 - 8}{10}\right) \times 100 = 20\%$$

This indicates a 20 percent change in output voltage going from a no-load to a loaded condition. This is typical for an unregulated power supply.

The task of a regulated power supply is to maintain a nearly constant output voltage over the range of load current (no-load to full-load current). The zener diode accomplishes the necessary regulation in the circuit shown in Figure 6.9. In this circuit, a series resistor, R_S, and a zener diode are connected to the terminals of the unregulated supply. The output is taken across the zener diode and the terminals labeled "oo" are the output terminals of the regulated supply. The function of the zener diode is to keep the output voltage fairly constant over a wide variation in load current. This is accomplished by operating the zener in its breakdown region, where its voltage varies only slightly with changes in zener current. The zener breakdown voltage must, of course, be less than E in order to insure operation of the zener in its breakdown region. The resistor R_S is necessary to keep the zener current limited to a value below I_{ZM} (see Section 6.4).

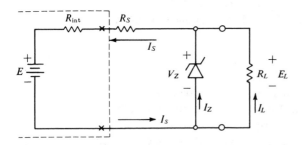

Figure 6.9.   ZENER REGULATED SUPPLY

Circuit operation

Circuit operation may best be described by assuming some circuit values. Let $E = 10$ V, $R_{int} = 100 \,\Omega$, $R_S = 100 \,\Omega$, $V_{ZT} = 6$ V. With these values for the circuit, consider the case with no load; that is, $R_L = \infty$. In this case, $I_L = 0$ and the current supplied by the voltage source all flows through the zener; that is, $I_Z = I_S$. To calculate I_S, we can use Kirchhoff's voltage law:

$$E = I_S \times (R_{int} + R_S) + V_Z \qquad (6.9)$$

If we assume that the zener diode is in breakdown (it should be if $I_S > I_{ZK}$), then V_Z will be approximately equal to $V_{ZT} = 6$ V. The output voltage, which is equal to V_Z, is $E_{NL} = 6$ V. Substituting for R_{int}, R_S, E and V_Z, we have, using Equation 6.9,

$$10 \text{ V} = I_S \times (200 \,\Omega) + 6 \text{ V}$$

Solving for I_S,

$$I_S = \frac{4 \text{ V}}{200 \,\Omega} = 20 \text{ mA}$$

Thus the current flowing through the zener is 20 mA. This is the maximum current which will flow through the zener in this regulator circuit, since there is no load resistor to shunt some of the current away from the zener. The I_{ZM} rating of this zener should then be greater than 20 mA to insure safe operation.

If we now connect a load resistor across the output terminals of the regulated supply, it will draw current depending on its resistance. This current is subtracted from the 20 mA being supplied by the voltage source, leaving less available for the zener current. However, V_Z will change very slightly with changes in I_Z as long as I_Z is sufficiently above the value I_{ZK}. If we assume a value of $I_{ZK} = 1$ mA for this zener, then we can consider that V_Z will remain at approximately 6 V as long as I_Z is kept greater than 1 mA. Consider a load resistor $R_L = 600\ \Omega$ in the circuit which is redrawn in Figure 6.10. With 6 volts across the zener, the load voltage is also 6 volts, as indicated. The load current is therefore 6 V/600 Ω = 10 mA. The current supplied by the voltage source is still 20 mA, as calculated previously. This means that 10 mA of this current is being supplied to the load, leaving $I_Z = 10$ mA. This is a result of applying Kirchhoff's current law at the output terminals. That is,

$$I_S = I_L + I_Z \tag{6.10}$$

The 10 mA of zener current is still well above the knee current, so that the zener is definitely in its breakdown region where $V_Z = 6$ volts.

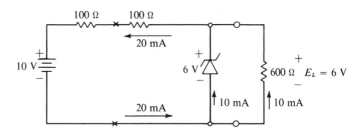

Figure 6.10.

If we calculated the percentage regulation of this circuit using Equation 6.8, the result would be zero, indicating perfect regulation, since $E_L = E_{NL} = 6$ volts. In practice, however, the output voltage does drop slightly when a load is connected since V_Z is not perfectly constant as I_Z changes. Just how much V_Z changes depends on the slope of the zener breakdown curve; that is, the change depends upon the zener impedance. A low value of zener impedance will result in a small variation in V_Z as the regulated supply is loaded, giving a percentage regulation generally close to zero percent.

There is a limit to how much load current can be supplied before the circuit of Figures 6.9 and 6.10 begins to lose regulation. As mentioned previously, at least 1 mA of current must be supplied to the zener to keep it operating in its breakdown region. This means that, of the 20 mA supplied by the voltage source, no more than 19 mA can be drawn by the load if good regulation is to be maintained. This stipulation, of course, determines the minimum value of R_L that can be used. That is,

$$R_{L(min)} = \frac{V_Z}{I_{L(max)}} \qquad (6.11)$$

which in this case becomes

$$R_{L(min)} = \frac{6 \text{ V}}{19 \text{ mA}} = 315 \text{ } \Omega$$

Any value of R_L below 315 Ω would draw more than 19 mA of current, and thus pull the zener out of its breakdown region.

▶ EXAMPLE 6.3

A zener diode with the following specifications is used in the voltage regulator circuit of Figure 6.9, with $E = 20$ volts and $R_{int} = 100$ Ω:

$$V_{ZT} = 10 \text{ V @ } I_{ZT} = 24 \text{ mA}$$
$$I_{ZK} = 1 \text{ mA}; I_{ZM} = 80 \text{ mA}$$

(a) Calculate the value of R_S required to insure that I_Z stays below I_{ZM}. The no-load case gives the most current flow through the zener; that is, $I_Z = I_S$. If we use Equation 6.9 and set $I_S = I_{ZM} = 80$ mA, we can solve for the value of R_S which will produce 80 mA of current flow. Substituting our values in this equation, we have

$$20 \text{ V} = 80 \text{ mA} \times (100 \text{ } \Omega + R_S) + 10 \text{ V}$$

Solving for R_S, we obtain

$$R_S = \frac{10 \text{ V}}{80 \text{ mA}} - 100 \text{ } \Omega = 25 \text{ } \Omega$$

Any value of R_S below 25 Ω would cause more than 80 mA of current to flow through the zener in the no-load case.

(b) Using $R_S = 25$ Ω, calculate the minimum load resistor, $R_{L(min)}$, which can be used and still have voltage regulation maintained. Since $I_{ZK} = 1$ mA, the maximum load current will be 80 mA − 1 mA = 79 mA. Using Equation 6.11, $R_{L(min)}$ is given by

$$R_{L(min)} = \frac{10 \text{ V}}{79 \text{ mA}} = 126.6 \text{ } \Omega \qquad ◀$$

To summarize this discussion, it can be stated that the addition of a zener diode to an unregulated power supply will generally improve the output voltage regulation. This improvement is a direct result of the zener diode's ability to maintain a relatively constant voltage for a wide range of zener current when operated in its breakdown region.

6.7.   The Zener Diode as a Reference Element

There are many applications in which it is desirable to maintain a constant voltage between two points in a circuit and use this voltage as a reference to which another circuit voltage may be compared. The difference between the compared voltage and the reference voltage is usually amplified and used to perform some control function. Elaborate power supply voltage regulator circuits, measurement circuits and servomechanism circuits use this type of arrangement, which is shown in Figure 6.11.

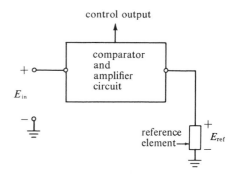

Figure 6.11.   COMPARISON CIRCUIT

A zener diode of an appropriate voltage rating can be advantageously used as the reference element. Its constant-voltage characteristic in the breakdown region makes it desirable for this application. The circuit of Figure 6.11 is redrawn in Figure 6.12 using a zener diode as the reference element. The reference voltage, E_{ref}, in this circuit is equal to the zener breakdown voltage. Current is supplied to the zener by the main circuit power supply, E_S, through a series resistor, R_S. Obviously E_{ref} must be less than E_S. The value of R_S is chosen to insure that the zener is operating well below the knee on the good part of its characteristic.

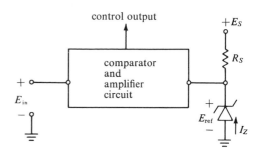

Figure 6.12.   ZENER DIODE AS THE REFERENCE ELEMENT

▶ EXAMPLE 6.4

The circuit of Figure 6.12 is to use the zener diode specified in Example 6.3. The supply voltage, E_S, is $+24$ volts. Calculate a value for R_S to use in this circuit. The value of R_S must be small enough to assure that I_Z is greater than $I_{ZK} = 1$ mA and yet not so small that I_Z is greater than $I_{ZM} = 80$ mA. In this circuit, since $E_S = 24$ V and $E_{ref} = 10$ V, the voltage drop across R_S is 14 V. Thus we have

$$I_Z \times R_S = 14 \text{ V}$$

For $I_Z = I_{ZK} = 1$ mA, we can calculate $R_{S(max)}$, the maximum value of R_S we can use:

$$R_{S(max)} = \frac{14 \text{ V}}{1 \text{ mA}} = 14 \text{ k}\Omega$$

For $I_Z = I_{ZM} = 80$ mA, we can calculate $R_{S(min)}$, the minimum value of R_S we can use.

$$R_{S(min)} = \frac{14 \text{ V}}{80 \text{ mA}} = 175 \text{ }\Omega$$

In practice, a value of R_S somewhere between these limits would be used.   ◀

There are, of course, many more applications for the zener diode in today's vast complexity of electronic circuits. Most of these applications take advantage either of the zener's constant-voltage characteristic or of its ability to block current in the reverse direction until the applied voltage is greater than the breakdown voltage. Use of the zener diode in some of these other applications will be introduced in the questions at the end of this chapter.

GLOSSARY

Zener breakdown: onset of heavy current conduction in the reverse direction due to the high-intensity electric field across a narrow space-charge region.

Zener diode: a P-N diode designed to operate in the reverse breakdown region (also called *reference diode* or *breakdown diode*).

Zener knee: region of the zener diode characteristic curve at point where breakdown just begins.

I_Z, V_Z: zener current and voltage.

I_{ZK}: zener knee current.

I_{ZT}: zener test current.

I_{ZM}: maximum allowable zener current.

V_{ZT}: nominal zener voltage measured at I_{ZT}.

$P_{Z(max)}$: maximum allowable zener power dissipation.

Z_Z: zener impedance.

K_T: zener voltage temperature coefficient.

Voltage regulation: ability to maintain a nearly constant voltage over a wide range of current levels.

Percentage regulation: percentage decrease in output voltage of a power supply from no-load to loaded conditions.

Reference voltage: constant voltage used as a reference to which another voltage is compared.

PROBLEMS

6.1 Name the *two* types of reverse breakdown which can occur in P-N diodes. Which occurs at lower voltages?

6.2 How does increased doping of both sides of a P-N junction affect the space-charge region? How does it affect reverse breakdown voltage?

6.3 Would you expect the breakdown voltage of a 4-volt zener diode to increase with an increase in temperature? Why?

6.4 What is the principal difference between the *I-V* characteristics of a zener diode and those of an ordinary P-N diode?

6.5 Consider the zener diode *I-V* characteristic in Figure 6.13. From this characteristic,
(a) estimate the value of I_{ZK}.
(b) determine V_{ZT}, nominal zener voltage, if I_{ZT} = 30 mA.

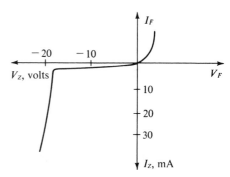

Figure 6.13.

6.6 Data taken on a certain zener diode is given in Figure 6.14. From this data, determine
(a) approximate zener impedance, Z_{ZK}, at I_{ZK} = 1 mA.
(b) approximate zener impedance, Z_{ZT}, at I_{ZT} = 10 mA.

6.7 The zener diode of Problem 6.6 is specified to have a nominal zener voltage of 9.1 V ± 5 percent. Does the zener of Figure 6.14 fall within specifications?

I_Z, mA	V_Z, volts
0.0	0.00
0.5	7.50
0.9	8.99
1.0	9.00
1.1	9.01
5.0	9.20
9	9.24
10	9.25
11	9.26
12	9.27
15	9.30
20	9.33
40	9.38

Figure 6.14.

6.8 Refer to the silicon zener diode specification sheet in Appendix II. From this spec sheet, determine the following:
(a) Which zener diode of this series is least dependent on temperature?
(b) Which is most dependent on temperature?
(c) What is the maximum junction temperature, $T_{J(\text{max})}$?
(d) What is the maximum power dissipation, $P_{Z(\text{max})}$?

6.9 From the zener spec sheet in Appendix II, determine the value of V_{ZT} for the 1N3827 at the following temperatures:
(a) $T_J = 25°C$.
(b) $T_J = 100°C$.
(c) $T_J = 175°C$.

6.10 Repeat Problem 6.9 for the 1N3824 diode.

6.11 The zener diodes specified in Appendix II have a $P_{Z(\text{max})}$ of 1 watt. The spec also indicates that this value should be derated by 6.67 mW for each degree that ambient temperature is above 25°C. Calculate
(a) $P_{Z(\text{max})}$ at $T_A = 50°C$.
(b) $P_{Z(\text{max})}$ at $T_A = 100°C$.

6.12 A 1N3826 zener diode (Appendix II) is used in the circuit below. Determine the maximum value of E_{in} which should be used for safe operation.

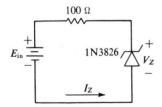

Figure 6.15.

6.13   In the circuit of Figure 6.15 calculate the voltage across the resistor for $E_{in} =$ 15 volts; $E_{in} = 3$ volts.

6.14   A particular unregulated 30-volt DC source has its output voltage drop to 27 volts when loaded with 100 Ω. Calculate the percentage regulation for this source.

6.15   A zener regulated voltage supply employs a 1N3830 zener diode (Appendix II) and a 15-volt unregulated supply with an internal resistance of 50 Ω. The regulated voltage must supply current to a minimum load resistance of 100 Ω. Calculate the largest value for series resistance, R_S, which could be used.

6.16   Why is it advantageous to operate a zener diode at high current levels (somewhat higher than necessary to keep it operating in its breakdown region) if it is to be used as a voltage regulator?

6.17   It is desired to convert a 15-volt unregulated DC power supply into a zener regulated power supply with *two* separate output voltages of 7.5 volts and 12.6 volts. Design a circuit which will accomplish this using the zener diode series in Appendix II.

6.18   The 1N3821 zener diode is to be used as a reference voltage for a circuit such as that in Figure 6.12. Calculate the limits on series resistor R_S. What value of R_S should be used to keep $E_{ref} = V_{ZT}$?

6.19   A 24-V zener diode is used in the circuit shown in Figure 6.16. This circuit is called a *clipper*, or *limiter*, circuit. The input voltage e_S is a 100-V peak-to-peak sine wave.
(a) Sketch the waveform of output voltage.
(b) The zener diode in this circuit has $I_{ZK} = 3$ mA, $I_{ZT} = 20$ mA and $I_{ZM} = 40$ mA. What is the *smallest* value of R which should be used in this circuit?

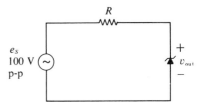

Figure 6.16.

6.20   Reverse the zener diode in Figure 6.16 and repeat the preceding problem.

6.21   The circuit in Figure 6.17 is supposed to convert the 12-V input supply to a *regulated* 6-V supply by using a zener diode with $V_{ZT} = 6$ V $\pm$ 5 percent. A technician constructs the circuit and makes the following sequence of measurements using a VOM as a voltmeter:

$$V_{ab} = 12 \text{ V} \qquad V_{cd} = 4 \text{ V} \qquad V_{ac} = 6 \text{ V}$$

From these measurements, the technician sees that Kirchhoff's voltage law is not satisfied and that the zener diode voltage is only 4 V. He/she concludes that the zener diode is faulty and replaces it, but the results are the same. Which of the following could be probable causes of these results?

(a) Both zener diodes are faulty.

(b) R_S is too large in value.

(c) R_S is burned out (open circuited).

(d) The resistance of the VOM is loading the circuit.

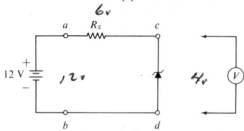

Figure 6.17.

REFERENCES

Foster, J. F., *Semiconductor Diodes and Transistors*, Vol. 2. Beaverton, Oregon: Programmed Instruction Group, Tetronix, Inc., 1964.

Silicon Zener Diode and Rectifier Handbook. Phoenix: Motorola, Inc., 1961.

Sowa, B. A. and M. M. Toole, *Special Semiconductor Devices*. New York: Holt, Rinehart and Winston, Inc., 1968.

7

Tunnel Diodes

7.1.   Introduction

A revolutionary, relatively new semiconductor device developed from the P-N junction is the *tunnel diode** invented in 1958 by Dr. Leo Esaki. The principles of operation of the tunnel diode are remarkably different from those of other semiconductor devices. A detailed explanation of these principles will not be presented here, as they are not necessary to understand the action of a tunnel diode in a circuit. The tunnel diode will be used mainly as a vehicle for introducing the student to the concept of *negative resistance*, which will be encountered in later circuit work. Some of the important applications of this device require more background than the student is expected to have at this time and will be passed over in favor of simpler circuits which illustrate the use of negative resistance. For a more detailed and higher-level discussion of tunnel diodes, the student is advised to consult the references listed at the end of this chapter.

7.2.   The Tunnel Diode

As was discussed in the previous chapter, increasing the amount of doping impurities added to the P and N regions of a P-N junction reduces the voltage at

---

*Sometimes called an "Esaki diode."

which the diode enters a reverse breakdown condition. This is illustrated in
Figure 7.1. It might be thought that there would be a limiting case when the
reverse breakdown voltage is reduced to zero volts. However, experiments
have shown that it is possible to dope many semiconductor materials heavily
enough to cause reverse breakdown at a slight forward bias. When a larger forward bias is applied, the device goes out of reverse breakdown and the current
falls to a small value until forward bias turn-on occurs.

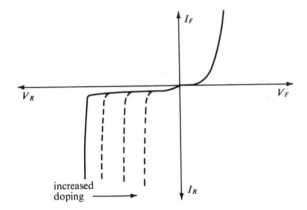

Figure 7.1.   EFFECT OF DOPING ON REVERSE BREAKDOWN VOLTAGE

The breakdown mechanism in these devices is known as the *tunnel effect;*
this is a process whereby electrons, with apparently insufficient energy to do so,
can surmount a potential barrier. In this case, the barrier is the space-charge
potential barrier. Because of the heavy doping, the space-charge region in a
tunnel diode is very narrow (less than a millionth of an inch), which is why
electrons can tunnel through the barrier. This tunnelling gives rise to an additional current across the P-N junction at very small forward bias, which disappears as forward bias is increased. Figure 7.2 is an illustration of electrons
tunnelling through the space-charge barrier.

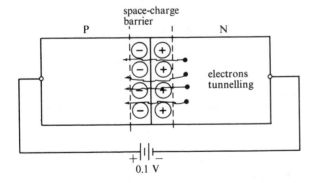

Figure 7.2.   ELECTRONS TUNNELLING

A typical tunnel diode *I-V* characteristic curve (solid curve) is shown in Figure 7.3 along with that of a conventional P-N diode (dotted curve). Notice that with no applied voltage there is no current flow through the tunnel diode (point *A*). With the application of reverse voltage, reverse current begins to flow immediately, since the device is very heavily doped. This is completely different from what occurs in the conventional P-N diode where reverse current is very small for values of reverse voltage below the relatively high reverse breakdown voltage. As forward voltage is applied to the tunnel diode, forward current immediately begins to flow until a point is reached where the current starts to decrease as forward voltage is increased (point *B*). This region of decreasing current (point *B* to point *C*) is called a *negative resistance region* since the AC resistance (r_F in Chapter 5) is negative; that is, a positive ΔV_F produces a negative ΔI_F. Further increase in voltage past point *C* causes the tunnel diode to enter the normal forward conduction mode of the conventional P-N diode. The entire region of the tunnel diode curve from point *A* to point *C* is due to the tunnelling effect which occurs as a result of heavy impurity doping. This effect is not present to the right of point *C*, where normal P-N diode operation takes place.

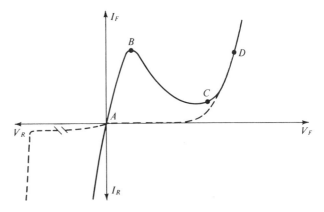

Figure 7.3.   TUNNEL DIODE *I-V* CURVE

A detailed discussion on the nature of the tunnel current (point *A* to point *C*) would require some understanding of quantum mechanics and will not be presented here. It should be noted, however, that the speed of the electrons which tunnel through the space-charge barrier is much faster than the speed of electrons which diffuse from the N side to the P side (normal forward current in a P-N diode). In fact, the tunnelling electrons travel very close to the speed of light across the P-N junction. For this reason, tunnel diodes are particularly suited for use at very high frequencies (theoretically as high as 10^7 MHz). The principal applications of the tunnel diode take advantage of its fast operating speed. Figure 7.4 shows some symbols that indicate a tunnel diode. The first symbol is the one which will be used from this point on.

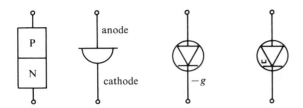

Figure 7.4.   TUNNEL DIODE SYMBOLS

7.3.   Tunnel Diode Specifications

The *I-V* curve of a typical germanium tunnel diode is drawn in Figure 7.5 showing the important tunnel diode parameters (I_P, I_V, V_P, V_V and V_{FP}). The *peak current*, I_P, is measured at the point where the current starts decreasing with voltage (point *B*) and *peak voltage*, V_P, is the corresponding voltage at that point. The *valley current*, I_V, is measured at the point where current begins increasing again (point *C*) and *valley voltage*, V_V, is the corresponding voltage at

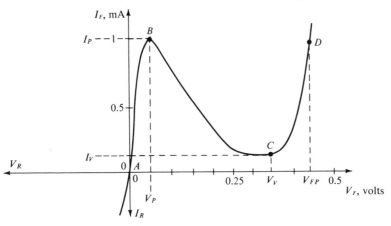

Figure 7.5.   *I-V* CURVE FOR GERMANIUM TUNNEL DIODE

that point. The *forward-point voltage*, V_{FP}, is measured in the region of normal P-N diode conduction at a forward current equal to I_P (point *D*). In the area of negative resistance between points *B* and *C*, the value of AC resistance or of AC *conductance* can be calculated. Most often AC conductance is used for the tunnel diode and is given the symbol *g*. It is given by

$$g = \frac{\Delta I_F}{\Delta V_F} \text{ mhos (ohms}^{-1}) \tag{7.1}$$

and is calculated in the negative resistance region at the point where the curve is the steepest. Thus *g* is always a negative quantity. An approximate value of *g* can be obtained using

$$g \approx \frac{-2(I_P - I_V)}{(V_V - V_P)} \tag{7.2}$$

The values of peak voltage, valley voltage and forward-point voltage are determined primarily by the semiconductor material used in construction of the tunnel diode. Peak current is determined by the geometry of the junction. Valley current and the AC conductance depend on a number of factors. Typical values of these parameters for germanium tunnel diodes are $V_P = 55$ mV, $V_V = 350$ mV, and $V_{FP} = 500$ mV. For gallium arsenide tunnel diodes, $V_P = 150$ mV, $V_V = 500$ mV and $V_{FP} = 1100$ mV. These voltages are typical at room temperature. Typical values of I_P, I_V and g can be obtained from Figure 7.5 as $I_P = 1$ mA, $I_V = 0.125$ mA and $g = -0.01$ mhos.

Temperature affects each of these parameters. Peak voltage, valley voltage, AC conductance and forward voltage all decrease with increasing temperatures, while the valley current increases with increasing temperature. The peak current may increase or decrease with temperature depending on the doping level and the semiconductor material. For a typical germanium tunnel diode, V_P decreases 60 μV for each degree increase in temperature; V_V and V_{FP} decrease 1 mV per degree. Note that this is less than for a conventional diode (Section 5.8), whose forward voltage decreases typically 2.2 mV per degree. In general, tunnel diodes are less sensitive to temperature than P-N diodes and other semiconductor devices. Tunnel diodes may operate at temperatures as high as 340°C, whereas conventional silicon diodes stop working at 200°C and germanium at about 100°C.

There are other important tunnel diode parameters which will not be covered here but are thoroughly discussed in this chapter's references.

7.4.   The Tunnel Diode as a Circuit Element

The many features of a tunnel diode make it a very useful circuit element. The negative resistance characteristic makes it possible to use the tunnel diode as an amplifier, an oscillator and an extremely rapid switching device. It can operate at speeds and temperatures at which other semiconductor devices cannot function. Besides these features, it also is relatively insensitive to the damaging effects of nuclear radiation environments which are very harmful to other semiconductors. This gives it a priority in space vehicles and missile circuitry and in applications related to nuclear power production and control.

As the first step in understanding the tunnel diode's operation in a circuit, consider the circuit shown in Figure 7.6 along with the *I-V* characteristic for the tunnel diode used in the circuit. The 0.5-volt battery in series with the tunnel diode and 100-Ω resistor forward biases the tunnel diode. In order to find I_F, tunnel diode forward current, and V_F, tunnel diode voltage, the load-line procedure is used as in the two previous chapters. The equation of the load line in this case is

$$0.5 \text{ V} = I_F \times 100 + V_F \qquad (7.3)$$

Superimposing this load line (solid line) on the *I-V* curve should give us the operating point of the tunnel diode. However, a remarkable situation has

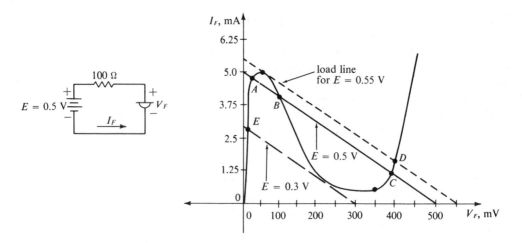

Figure 7.6.   SIMPLE TUNNEL DIODE CIRCUIT

occurred. The load line intersects the tunnel diode curve at *three* separate
points. The question now becomes: If all three points, *A*, *B* and *C*, lie on both
the load line and the *I-V* curve, which point is the actual operating point of the
circuit? It certainly cannot be all three points at the same time. Consider point
B. It lies on the negative resistance portion of the tunnel diode *I-V* curve. This
circuit cannot operate at point *B* since it is an *unstable* point. That is, even if the
circuit were somehow made to operate at this point, any little circuit fluctua-
tion or spurious noise would cause the circuit to immediately move to point *A*
or point *C* and remain there. This instability is analogous to a ball resting on
the peak of a hill as in Figure 7.7A. With the ball in this position, any small dis-
turbance will cause it to roll down either side of the hill and come to rest at the
bottom. Point *A*, on the other hand, is a *stable* point of operation. That is,
small circuit fluctuations will not cause the operating point to shift. It will always return to point *A*. Point *C* is also a *stable* operating point. These stable
points are analogous to a ball resting in a bowl as in Figure 7.7B. Small dis-
turbances may cause the ball to move up the side of the bowl, but it will always
come back to rest at the bottom.

Points which lie on positive resistance portions of the tunnel diode curve,
such as points *A* and *C*, are always stable points. Points which lie on negative

(A)   (B)

Figure 7.7.   (A) UNSTABLE POINT; (B) STABLE POINT

resistance portions of the curve, such as point *B*, *are unstable if the load line also intersects the curve at stable points.*

In the circuit under consideration, point *A* is at $I_F = 4.75$ mA and $V_F = 25$ mV; point *C* is at $I_F = 1.0$ mA and $V_F = 400$ mV. The circuit can operate stably at either point. Before the circuit is connected both V_F and I_F are zero; after the circuit is connected the circuit will move up the *I-V* curve to the stable operating point at point *A*. In order for the circuit to operate at point *C*, the load line must be momentarily moved *above* the *peak point*. This can be done by momentarily increasing the battery voltage to 0.55 V, causing the load line to move to a new position (shown by a dotted line) where it intersects the *I-V* curve only at point *D*, a stable point. The circuit will remain at point *D* as long as the input voltage is kept at 0.55 V. However, if the battery voltage is reduced back to 0.5 V, the circuit will not return to point *A*, but rather will move along the *I-V* curve to point *C* and remain there. Operation will return to point *A* only if the load line is momentarily moved *below* the valley point by reducing the input voltage to, say, 0.3 V (shown by a dashed line). This will cause the operating point to shift to point *E*, a stable point. When the battery is increased back to 0.5 V, the operating point will move up the *I-V* curve to point *A*. Thus it can be seen that this circuit does not have a unique operating point, but will operate at either *A* or *C* depending on what has occurred previously. Since this circuit has two stable operating points, it is considered to be a *bistable* circuit. This property, unlike any circuit or device we have yet studied, results from negative resistance of the tunnel diode.

Figure 7.8 summarizes the variation of tunnel diode voltage in this circuit as the battery voltage is varied from zero to 0.5 V, up to 0.55 V, back to 0.5 V, down to 0.3 V and back up to 0.5 V.

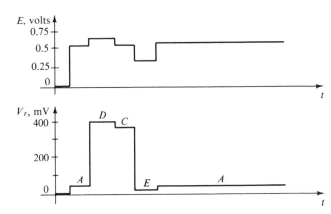

Figure 7.8.   VARIATION OF V_F AS BATTERY VOLTAGE CHANGES

It is possible for the circuit of Figure 7.6 to have a load line which intersects the tunnel diode curve in its negative resistance region and nowhere else. For example, a battery voltage of 0.2 volt and a resistance of 20 ohms would produce such a load line. This is exactly the manner in which tunnel diodes are biased when they are used in amplifier or oscillator circuits. The operating point in this situation would be on the negative resistance portion of the tunnel-diode curve. Such an operating point is *conditionally stable:* that is, it will be stable if care is taken to keep circuit inductance to a very low value; otherwise, undesirable circuit oscillations will take place and the circuit will never stabilize. For this reason, design of tunnel-diode amplifiers is critically dependent on the physical circuit layout.

7.5.   Tunnel Diode Switching Circuits

In the previous section we saw that with the proper load line the tunnel diode circuit is bistable and the circuit can be made to switch from one stable operating point to another. For purposes of our discussion we will refer to the two possible operating points as the *low-voltage point* and the *high-voltage point*. In Figure 7.6, point A would be the low-voltage point and point C would be the high-voltage point. The low-voltage point is always at a value of V_F which is less than V_P and the high-voltage point is always at a value of V_F which is greater than V_V.

In the circuit of Figure 7.6 we switched the circuit operating point from A to C by increasing the input voltage. A more commonly used procedure is to increase the input current until it is at a value greater than the peak-point current. This causes the tunnel diode to switch to its high-voltage region (to the right of the valley point), since in the low-voltage region (to the left of the peak point) the tunnel diode current can only go as high as I_P. Consider the circuit of Figure 7.9, which employs the tunnel diode used in Figure 7.6. Assume, to

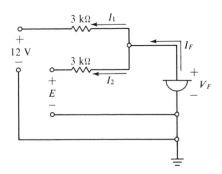

Figure 7.9.   TUNNEL DIODE SWITCHING CIRCUIT

start, that $E = 0$ volts and that the tunnel diode is in its low-voltage state with $V_F \approx 0$. We now can calculate I_F to see whether the tunnel diode is actually in its low-voltage state. We have

$$I_F = I_1 + I_2 \tag{7.4}$$

The current I_1 is approximately

$$\frac{12 \text{ V}}{3 \text{ k}\Omega} = 4 \text{ mA}$$

since $V_F \approx 0$. The current I_2 is essentially zero since $E = 0$. Thus, from Equation 7.4,

$$I_F = 4 \text{ mA} + 0 \text{ mA} = 4 \text{ mA}$$

This is less than the peak current for this tunnel diode which is given as 5 mA in Figure 7.6. Thus the tunnel diode actually is in its low-voltage state with $V_F \approx 0$ (to be exact, it is a few millivolts, but for our purposes we can call it zero). In order to get the tunnel diode into its high-voltage state, the total input current, $I_1 + I_2$, must, at least momentarily, exceed I_P. If E is increased to 3.3 volts, then I_2 will be approximately

$$\frac{3.3 \text{ V}}{3 \text{ k}\Omega} = 1.1 \text{ mA}$$

and I_F will then be 5.1 mA. This is greater than I_P, which means that the tunnel diode must be in its high-voltage state. In this state we can usually approximate V_F by V_{FP} volts, which is 0.425 volt for this diode. Since V_F is now approximately 0.425 volt, we must recalculate I_1 using this value. I_1 is given by

$$I_1 = \frac{(12 - 0.425) \text{ V}}{3 \text{ k}\Omega} = 3.86 \text{ mA}$$

and I_2 is given by

$$I_2 = \frac{(3.3 - 0.425) \text{ V}}{3 \text{ k}\Omega} = 0.96 \text{ mA}$$

resulting in $I_F = 4.82$ mA. Notice that this is less than I_P, but the diode has already switched to the high-voltage state. It will remain in this state with $V_F \approx 0.425$ volt until I_F is reduced below the valley current (0.5 mA for this diode). This is because the tunnel diode high-voltage state exists only for currents greater than I_V (see *I-V* curve). When I_F goes below I_V the tunnel diode switches back to its low-voltage state with $V_F \approx 0$.

▶ EXAMPLE 7.1

For the circuit of Figure 7.9, determine the variation of V_F as the voltage, E, is varied according to the waveform in Figure 7.10. One procedure is to calculate I_F each time the value of E changes. If the tunnel diode is initially in the low-voltage state and I_F is increased above I_P, the tunnel diode then switches to its high-voltage state. If the tunnel diode is initially in the high-voltage state and I_F is decreased below I_V, the tunnel diode then switches to its low-voltage state. A simpler procedure would be to determine what value of E is needed to cause the tunnel diode to switch to the high-voltage state and what value of E is needed to get the tunnel diode back to the low-voltage state.

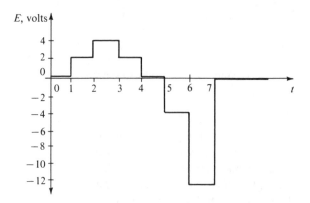

Figure 7.10.

If the diode is initially in the low state, then $I_1 = 4$ mA as calculated previously. Thus I_2 must increase to just above 1 mA for I_F to go above $I_P = 5$ mA and cause switching. This necessitates a value of E greater than 1 mA $\times$ 3 kΩ = 3 volts to cause the tunnel diode to switch from its low state to its high state. If the diode is initially in the high state, then $V_F = V_{FP} = 0.425$ volt and $I_1 = 3.86$ mA as calculated previously. Since I_F must drop below $I_V = 0.5$ mA for switching to occur, I_2 must be made to flow opposite the direction shown in Figure 7.9 so as to subtract from I_1. That is,

$$I_1 + I_2 = 0.5 \text{ mA}$$
$$3.86 \text{ mA} + I_2 = 0.5 \text{ mA}$$

or

$$I_2 = -3.36 \text{ mA}$$

Thus E must be negative and is given by

$$E = 3 \text{ k}\Omega \times (-3.36 \text{ mA}) + 0.425 \text{ V} = -9.675 \text{ V}$$

The value of E must be more negative than this value in order to switch the diode from its high state to its low state. Looking at the waveform above, $E = 0$ initially and the tunnel is initially in its low state with $V_F = 0$. It will stay there until E jumps to 4 volts, whereupon it goes to its high state with $V_F \approx 0.425$ volt. The diode remains in its high state until E drops to -12 volts, whereupon it switches to $V_F \approx 0$ and remains there. The waveform of V_F corresponding to the input voltage (E) waveform is shown in Figure 7.11. ◀

The chief advantage of tunnel diodes in switching circuits lies in their high operating speed. This property, inherent in the tunnelling process, makes them likely candidates for increased application in high-speed computers and communication systems.

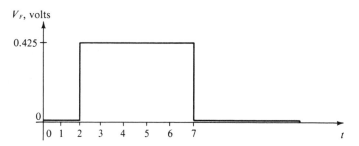

Figure 7.11.

GLOSSARY

Tunnel effect: process whereby electrons, with apparently insufficient energy to do so, can surmount a potential barrier (tunnel through the barrier).

Negative resistance region: that portion of the tunnel-diode *I-V* curve where increases in V_F cause I_F to decrease.

I_P: peak-point current.

V_P: peak-point voltage.

I_V: valley current.

V_V: valley voltage.

g: AC conductance.

V_{FP}: forward-point voltage.

Unstable operating point: possible circuit operating point at which the circuit cannot stably exist.

Bistable circuit: circuit with two possible stable operating points.

Low-voltage point or region: that region of the tunnel diode characteristic for values of V_F less than V_P.

High-voltage point or region: that region of the tunnel diode characteristic for values of V_F greater than V_V.

PROBLEMS

7.1 What condition is necessary for the tunnel effect to occur in a P-N junction?

7.2 Sketch the *I-V* characteristics of a tunnel diode and a P-N diode on the same set of axes.

7.3 Why can't a tunnel diode be used as a rectifier?

7.4   Why are tunnel diodes particularly useful at high frequencies?

7.5   From the tunnel diode *I-V* curve in Figure 7.12 determine (a) I_P, (b) V_P, (c) I_V, (d) V_F, (e) V_{FP} and (f) g (use Equation 7.2).

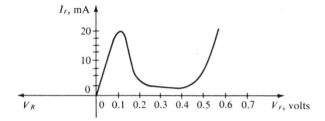

Figure 7.12.

7.6   How is a tunnel diode affected by temperature in comparison to a conventional P-N diode?

7.7   Can a tunnel diode be biased so that the load line intersects the *I-V* curve in only *two* points, both of which are stable?

7.8   A 1N3150 tunnel diode has the specifications listed in Appendix II. In Figure 7.13, what will be the value of I_F and V_F if $E = 6$ volts? If $E = 24$ volts?

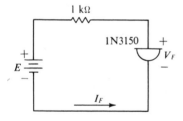

Figure 7.13.

7.9   In the circuit of Figure 7.13, sketch the waveform of V_F if E has the waveform shown in Figure 7.14.

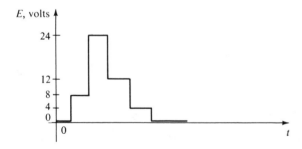

Figure 7.14.

7.10 The circuit in Figure 7.15 uses a 1N3150 tunnel diode. Inputs E_1 and E_2 can be either 6 volts or 0 volts and are controlled by switches S_1 and S_2.

(a) Choose values of R_1 and R_2 so that the tunnel diode will go to the high-voltage state only if E_1 *and* E_2 are both at $+6$ volts. This is called an "and" circuit.

(b) Choose values of R_1 and R_2 so that the tunnel diode will switch to the high-voltage state if *either E_1 or E_2 or both* are at 6 volts. This is an "or" circuit.

(c) Add another branch to this circuit with input E_3. Choose resistor values so that the diode switches when two out of three inputs are 6 volts. This is a "majority" circuit.

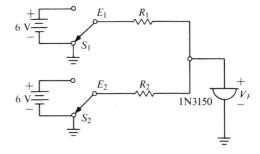

Figure 7.15.

7.11 The 1N3150 tunnel diode is used in Figure 7.16. The current I_1, which is supplied by the 6-volt source, serves to keep the tunnel-diode current above the valley current level at all times. The current I_2 is supplied by the input E_{in}. This input voltage is monitoring a certain manufacturing process. It is normally at zero volts. However, when a momentary irregularity occurs in the process, E_{in} becomes 6 volts for the duration of the malfunction. The duration may be very short and, unless someone monitors E_{in} continuously, a malfunction may go undetected. This tunnel-diode circuit is designed to *detect* when E_{in} goes to 6 volts and *store* this information. It does this by remaining in its low-voltage state while E_{in} is zero and then switching to its high-voltage state when E_{in} becomes 6 volts. If E_{in} returns to zero volts, the tunnel diode

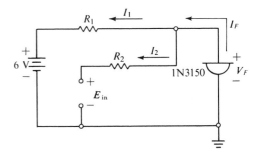

Figure 7.16.

will remain in its high-voltage state, since I_1 will supply enough current to keep $I_F > I_V$. Thus the tunnel diode remains in its high state even if E goes to 6 volts only momentarily. The tunnel diode essentially *remembers* that the input was 6 volts. It is said to have a *memory*. This memory comes about due to its negative resistance characteristics.

Choose values of R_1 and R_2 to insure that the circuit operates as explained above.

REFERENCES

Chow, W. F., *Principles of Tunnel Diode Circuits*. New York: John Wiley & Sons, Inc., 1964.

Sowa, W. A. and J. M. Toole, *Special Semiconductor Devices*. New York: Holt, Rinehart and Winston, Inc., 1968.

Tunnel Diode Manual. Syracuse, New York: General Electric Company, 1964.

8

Optoelectronic Devices

8.1. Introduction

In the last few years the fields of optics and electronics have combined to form a new technology called *optoelectronics*. In our study we will be interested in optoelectronic devices; that is, devices whose electronic properties are affected by *light* energy. The term *light* will be used in its broadest sense to include the visible, infrared and ultraviolet regions of the frequency spectrum.

The role of optoelectronic devices (also referred to as photoelectric devices) in today's technology is everincreasing. Missile and satellite systems, television, computer systems and even oil burners and clothes dryers use light-sensitive devices. In most of these applications the input is a beam of light energy, and before any study of optoelectronic devices can be made, a brief discussion of light and its properties is necessary.

8.2. The Nature of Light

It is commonly known that radio, television and other communications systems are made possible by the transmission of electromagnetic energy through space.

This electromagnetic energy is associated with high-frequency alternating current; a typical FM radio transmission frequency is 100 MHz (10^8 Hz).

Light energy also moves through space, but at frequencies in the range of 10^{15} Hz. To get an overall picture of the range of light frequencies, a graph, or *frequency spectrum*, is displayed in Figure 8.1. Note that frequency is shown increasing from right to left so that infrared radiation is the *lowest* frequency of light and ultraviolet is the *highest* frequency. In the visible portion of the spectrum, the frequency of the light energy increases in the order of the colors of the rainbow.

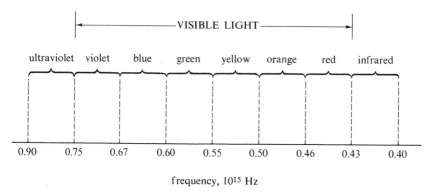

frequency, 10^{15} Hz

Figure 8.1.   FREQUENCY SPECTRUM OF LIGHT ENERGY

Wavelength

Light energy, like all electromagnetic energy, propagates through space at the same speed. This speed is about 186,000 miles per hour, or 3×10^8 meters per second. This leads to a basic mathematical relationship that holds true for all electromagnetic waves:

$$\lambda = \frac{c}{f} \qquad \textbf{(8.1)}$$

where c is the speed of light in meters/s, f is the frequency in cycles/s and λ is the *wavelength* in meters. Wavelength, λ, represents the distance that an electromagnetic (or light) wave will travel in one cycle of its frequency. Stated another way, wavelength is the distance between each wave crest (peak) in space.

The meter is a very impractical unit for wavelength since λ will usually be a very small fraction of a meter. A better unit is the *angstrom*, whose symbol is Å.

$$1 \text{ Å} = 10^{-10} \text{ meters}$$

▶ EXAMPLE 8.1

Calculate the wavelength in angstroms of a green light whose frequency is 0.58×10^{15} Hz.

$$\lambda = \frac{c}{f} = \frac{3 \times 10^8}{0.58 \times 10^{15}} \text{ meters}$$
$$= 5.17 \times 10^{-7} \text{ meters}$$
$$= 5170 \times 10^{-10} \text{ meters}$$
$$= \mathbf{5170 \text{ Å}} \qquad \blacktriangleleft$$

The reason for introducing wavelength is due to the fact that scientists and engineers usually use wavelength rather than frequency to specify the various colors of light. Figure 8.2 shows a wavelength spectrum of light energy. Note that wavelength increases from left to right, with ultraviolet light having the shortest wavelength and infrared having the longest.

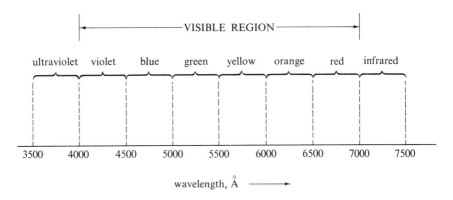

Figure 8.2.   WAVELENGTH SPECTRUM OF LIGHT

Measurements have shown that the human eye can respond to wavelengths of 3800 Å to 7600 Å, but it does not respond equally to all wavelengths. Eyes are more sensitive to yellow and yellow-green light (5200–5800 Å) than they are to blue light (4500 Å) or red light (6500 Å).

There are many sources of light, such as the sun, fluorescent lights, incandescent (tungsten) lights and burning substances like candles or oil. Each of these has its own color or shading. Fluorescent lights tend to be bluish in color so that most of their visible light is in the wavelength range of 4500–5000 Å. A tungsten lamp generates electromagnetic energy over a very wide range of wavelengths, from around 4000 Å to 30,000 Å. Only a small portion of this is perceived by the eye (3800–7600 Å). Sunlight consists of all the visible wavelengths in relatively equal amounts, giving the appearance of being "white light."

Quantum theory of light

Most of the phenomena involved in optoelectronic devices are best explained by the quantum theory of light. According to this theory, light consists of discrete quanta, or packets of energy, called *photons*. Photons are uncharged particles, each possessing an amount of energy that depends on the frequency of the light waves as given by

$$E = h \times f \tag{8.2}$$

where E is energy in electronvolts (eV), f is frequency in Hz and h is a constant of proportionality (called Planck's constant) equal to 4.137×10^{-13}.

▶ EXAMPLE 8.2

How much energy is possessed by the photons of red light whose wavelength is 7000 Å?

$$\lambda = \frac{c}{f}$$

so that

$$f = \frac{c}{\lambda} = \frac{3 \times 10^8}{7000 \times 10^{-10}} = 0.43 \times 10^{15} \text{ Hz}$$

Thus the photon energy (using Equation 8.2) is

$$E = 4.137 \times 10^{-13} \times 0.43 \times 10^{15} \text{ eV}$$
$$\simeq 178 \text{ eV} \qquad ◀$$

Equation 8.2 shows that photon energy increases with frequency. Therefore, light at higher frequencies (shorter wavelengths) will possess greater amounts of energy. The important thing to understand is that a given frequency of light possesses a given amount of energy (Equation 8.2) which is carried by the photons.

Before proceeding further, we should become familiar with the measure of illumination; that is, the amount of light present. Illumination is measured in units of *foot-candles* (fc). A foot-candle is a standard unit of illumination kept by the U.S. Bureau of Standards. Typical average solar illumination at noon on a sunny day is 8000 foot-candles. Moonlight provides around 0.02 foot-candles of illumination.

8.3. Classification of Optoelectronic Devices

Optoelectronic devices can generally be categorized as *light emitters* or *photodetectors*. Light emitters convert electrical energy into light energy. Junction lasers and light-emitting diodes (LEDs) are examples of this type. Photodetectors convert incident light energy into electrical energy; that is, the light energy carried by the incident photons is absorbed and transferred to electrons in the photodetector material.

Photodetectors can be further divided into the following subgroups:

(a) *Bulk photoconductive devices*, where the conductivity of the material changes as a function of the amount of incident light.

(b) *Photojunction devices* which contain a P-N junction that is exposed to light energy. Photodiodes, solar cells, phototransistors, photo-FETs and photothyristors are included in this group.

8.4.   Bulk Photoconductive Cells

Devices whose resistance to current changes as a function of light are called *photoconductive cells* (also called photoresistive cells, or simply photoresistors). Photoconductive cells are fabricated from semiconductor materials such as germanium, silicon, selenium and cadmium compounds. They generally consist of a thick film of the semiconductor material deposited on an insulating substrate with metallic leads attached to each end of the semiconductor. A glass cover is used to allow light to fall on the semiconductor. This structure is illustrated in Figure 8.3.

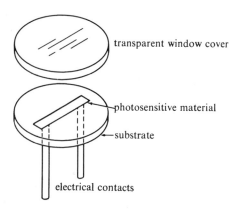

Figure 8.3.   STRUCTURE OF A TYPICAL PHOTOCONDUCTIVE CELL

Photoconductive process

When light strikes the semiconductor material, the incident photons may impart their energy to valence electrons through collisions. If the energy of the photons is greater than the inherent energy gap of the material, these valence electrons can be excited into the conduction band to become free conduction electrons. These liberated electrons leave corresponding holes in the crystal structure. Thus the effect of the incident light is to produce electron-hole pairs in addition to those present in the intrinsic semiconductor material due to heat. This increase in available current carriers produces a *decrease* in the material's resistivity and, therefore, also in its resistance.

The most sensitive photoconductor materials are those which are essentially insulators in the dark, so that the additional carriers generated by the absorption of light energy cause the greatest possible change in resistivity of the cell. For this reason, semiconductor materials that have practically no thermally generated current carriers at room temperature are used as photoconductors. These materials, which include cadmium sulfide (CdS), cadmium selenide (CdSe) and cadmium telluride (CdTe), have wider energy gaps than silicon and germanium. Thus they do not contain very many thermally generated electron-hole pairs. Because of their different energy gaps (2.45 eV for CdS, 1.74 eV for CdSe, 1.45 eV for CdTe), the response of each of these materials to the different colors of light will be different since the photons of each color possess different energies (recall Equation 8.2). CdS is most sensitive to green-yellow light (5500 Å), CdSe to red (7000 Å) and CdTe to infrared light (8000 Å).

Spectral response

The spectral response of a photoconductive cell is a plot of the relative response in percent versus light wavelength in angstroms. Each photoconductive cell will be most sensitive to one wavelength of light. At this one wavelength, the cell is said to have its peak response (100 percent). At other wavelengths of light, the cell's response will be less than 100 percent. Figure 8.4 shows the

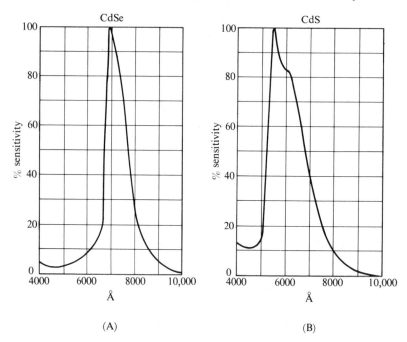

(A)   (B)

Figure 8.4.   SPECTRAL RESPONSES OF TYPICAL CdSe AND CdS PHOTO-
CONDUCTIVE CELLS

spectral responses for typical CdSe and CdS photoconductive cells. Note that the CdSe cell has its peak response at about 6900 Å (red), while the CdS cell is most sensitive to 5500 Å (yellow-green).

CdS photocells currently enjoy the widest application among photoconductive devices mainly due to their high sensitivity. The resistance of a CdS crystal in the dark may be 10,000 to 100,000 times greater than its resistance when illuminated with 100 foot-candles of light. However, the protective enclosure of commercial photocells absorbs some light, thus lowering this range to about 1 to 600. The smallest CdS cells respond to light intensities as low as 0.0001 foot-candle. The CdS crystal may be pure, but more often it contains a small amount of impurity, called an activator, which adds to the sensitivity of the device. The activator employed is usually silver, antimony or indium.

One of the most important parameters of the photoconductive cell is its resistance at different levels of illumination. Cell resistance as a function of illumination for a typical CdS cell is shown in Figure 8.5, along with the electronic symbol for photoconductive or photoresistive cells. Note that the scales used in the plot are log scales.

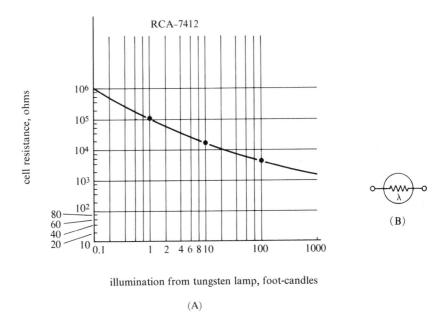

illumination from tungsten lamp, foot-candles

(A)

Figure 8.5.   (A) RESISTANCE VS ILLUMINATION FOR TYPICAL CdS CELL; (B) SYMBOL FOR PHOTOCONDUCTIVE CELL

▶ EXAMPLE 8.2

Using the photocell described in Figure 8.5 in the circuit of Figure 8.6, calculate the circuit current for $E = 10$ volts and $E = -10$ volts if the cell is subjected to 10 foot-candles of illumination.

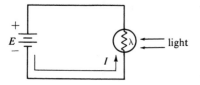

Figure 8.6.

From the curve in Figure 8.5, the cell resistance at 10 foot-candles is 20 kΩ. With $E = 10$ volts and using Ohm's law, I is easily calculated as 0.5 mA flowing in the direction shown. If the 10-volt battery is reversed so that $E = -10$ volts, then I is simply -0.5 mA. That is, the current is the same only in the opposite direction. The photoconductive cell conducts the same in both directions. After all, it is simply a resistor whose resistance value depends on illumination. This is one of the few semiconductor components a technician will encounter that does not contain a P-N junction.   ◀

The resistance of a photoconductive cell is often indirectly expressed in terms of the current drawn through the cell at a given voltage across the cell and at a given light level. For example, the cell of Example 8.2 would be rated at 0.5 mA at 10 volts under 10 foot-candles of illumination.

Speed of response

The resistance of CdS cells does not change rapidly with rapid changes in incident illumination but requires some time to reach its steady-state value.* That is, in part, due to the activator impurity, and is the price paid for added sensitivity. This effect is illustrated in the circuit of Figure 8.7 using the photocell

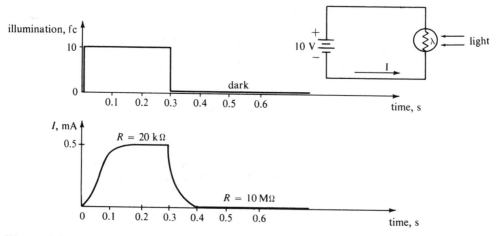

Figure 8.7.   ILLUSTRATION OF TYPICAL RESPONSE TIME OF CdS PHOTOCELL

---

*Steady-state value is the value the resistance eventually reaches under a constant illumination.

with the characteristics of Figure 8.5. If 10 foot-candles are applied instantaneously at time = 0 seconds (s), the photocell will not respond immediately but will gradually drop in resistance until its resistance is 20 kΩ. When the cell is dark, its resistance is above 10 megohms, or 10 million ohms; from Figure 8.7 we see that it takes approximately 0.1 s for the current in the circuit to increase from zero to 3.3 mA after the light is applied. On the other hand, when the illumination is instantaneously removed, it takes some time for the photocell to return to its high resistance state. The current in the circuit gradually decays to zero about 0.1 s after light is removed.

This relatively slow buildup and decay of current upon the application or removal of illumination limits the use of CdS photocells to applications where the light levels are not changing any faster than around 100 times per second (100 Hz). For applications at higher speeds, CdS photocells give way to other types, most popular of which is CdSe. CdSe photocells can operate at speeds 100 times greater than CdS photocells but are roughly 100 times less sensitive. Thus the CdS cell is used where a large resistance change for a given variation of light is important, and the CdSe cell is used where fast action is necessary.

Semiconductor photocells are adversely affected by increases in temperature which produce electron-hole pairs. These thermally generated current carriers cause the cell resistance to decrease. This lowers the cell's sensitivity to light energy. A CdS cell is not affected as much as other types because of its wide energy gap. CdSe, with a smaller energy gap, is more temperature sensitive.

8.5. Photoconductive Cell Ratings and Specifications

In addition to the resistance-versus-illumination curve of Figure 8.5, manufacturers normally supply certain ratings and specifications. Figure 8.8 contains some typical ratings and specifications for a CdS photocell. The ratings are maximum values of cell voltage, current and power dissipation which should not be exceeded. One specification given is the *photocurrent* at a given voltage (12 volts) under a given illumination (1 foot-candle). It ranges between 65 and 275 μA at 25°C for this particular cell. Also given is the value of *decay current*,

CdS PHOTOCONDUCTIVE CELL

Type: RCA–7412
Maximum voltage: 200 volts
Maximum power dissipation: 50 mW
Maximum current: 5 mA
Photocurrent at 12 volts and 1 foot-candle: 65 to 275 μA @ 25° C
Decay current: 1 μA (10 seconds after removal of above illumination) @ 25° C

Figure 8.8.   TYPICAL RATINGS AND SPECIFICATIONS FOR A CdS PHOTOCELL

which is the current flowing through the photocell a certain time (10 seconds) after the illumination (1 foot-candle) is removed. The decay current is essentially the DC current through the cell when it has been in the dark for greater than 10 seconds. It is also called the *dark current*.

8.6.   Applications of Photoconductive Cells

The photoconductive cell can be used in many applications that require the control of a certain function or event according to the absence, presence, color or intensity of light. Its numerous applications include door openers, burglar alarms, flame detectors, smoke detectors, and lighting control for street lamps in residential and industrial areas. It has been used in automatic business machines to read holes on cards or punched tape, in X-ray measurements and in photographic equipment. Many of the applications require the use of the photocell in conjunction with a relay. The circuit of Figure 8.9 is a very useful circuit of this type. In this circuit the relay will not be energized as long as the photocell resistance is high enough to keep the current, I, below the pull-in current of the relay. As the cell is illuminated its resistance decreases, causing I to increase, thus energizing the relay. The relay contacts control the desired function. In a garage door opener the relay would control the power applied to the motor mechanism. In this circuit, the photocell must have a current rating greater than the value needed to energize the relay and a voltage rating greater than the supply voltage.

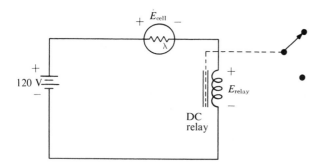

Figure 8.9.   PHOTOCELL-RELAY CIRCUIT

▶ EXAMPLE 8.3

A relay with a pull-in current of 2 mA and a DC resistance of 10 kΩ is used in the circuit of Figure 8.9. The cell is a type 7412 (Figures 8.5 and 8.8).

    (a) Find the current, I, when the cell is dark.

    (b) Determine the illumination needed on the cell in order to barely energize the relay.

When this cell is in the dark it has a resistance equal to 12 megohms. This is determined from the cell specifications in Figure 8.8, which say that the cell conducts 1 μA of current in the dark with 12 V applied to it. Thus

$$R_{dark} = \frac{12 \text{ V}}{1 \text{ } \mu A} = 12 \text{ megohms}$$

To find the circuit I in this case using Ohm's law, we have

$$I = \frac{120 \text{ V}}{R_{dark} + R_{relay}} \approx \frac{120 \text{ V}}{12 \text{ megohms}} = 10 \text{ } \mu A$$

This is not enough to energize the relay.

The relay requires 2 mA to be energized. The supply voltage must supply this amount of current to just energize the relay. The current, I, is given by

$$I = \frac{120 \text{ V}}{R_{cell} + R_{relay}} = \frac{120 \text{ V}}{R_{cell} + 10 \text{ k}\Omega} = 2 \text{ mA}$$

In order to get $I = 2$ mA, the total resistance of relay and cell must be 60 kΩ. Thus R_{cell} must be 50 kΩ. The amount of light needed to produce a cell resistance of 50 kΩ is obtained from Figure 8.5 as approximately **3 foot-candles.** ◀

▶ EXAMPLE 8.4

Repeat the previous example for a relay with pull-in current of 10 mA and a DC resistance of 1 kΩ.

According to the specs on the 7412 in Figure 8.8, it can handle a maximum current of only 5 mA and thus cannot be used in this circuit with this particular relay. A higher-current photocell must be used.   ◀

▶ EXAMPLE 8.5

Repeat Example 8.3 using a 240-volt battery.

According to the specs on the 7412, it can handle a maximum voltage of 200 volts. Thus it cannot operate in this circuit, since in the dark condition the total supply voltage of 240 volts would appear across its high resistance (12 MΩ).   ◀

The circuit of Figure 8.7 can also operate on AC voltage, using an AC relay, since the photocell acts as a bilateral resistor (it conducts equally in both directions). Much more sophisticated circuits utilize photocells in conjunction with other semiconductor devices. A number of these circuits will be brought out in the problems at the end of this chapter and in subsequent chapters.

8.7.   Photovoltaic Cells

Photovoltaic devices are devices that develop an electromotive force when illuminated. That is, they convert light energy directly to electrical energy. No outside source of electrical energy is required to produce current flow as in photoconductive devices. Figure 8.10 represents the basic construction of a *junction photovoltaic cell*. It consists of a silicon P-N junction as in a normal P-N diode. With no light applied to the junction, the circuit shown will produce no current flow (no bias on the P-N diode). However, if the space-charge region near the junction is illuminated, an interesting phenomenon can be observed. In the circuit, there will be a flow of current caused by a voltage appearing at the terminals of the photovoltaic cell. This voltage is labelled V_{PH}, photo-voltage, and the resulting current, I_{PH}, photocurrent. The photovoltage arises from the generation of electron-hole pairs in the space-charge region due to light energy. The electrons are swept into the N side by the positive space charge and the holes are swept into the P side by the negative space charge. This movement charges the N side negatively and the P side positively, with the resultant potential difference between the two sides being V_{PH}. The illuminated P-N junction acts like a source of electrical energy, similar to a battery. It can supply electrical power to a load. As with any source of electrical energy, the output voltage is a maximum when the load is an open circuit ($R_L = \infty$). This is the open-circuit voltage, V_{OC}. The output current is a maximum when the load is a short circuit ($R_L = 0$). This is the short-circuit current, I_{SC}.

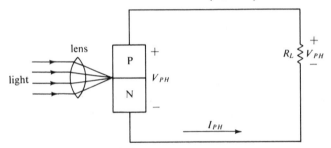

Figure 8.10.   JUNCTION PHOTOVOLTAIC CELL

In Figure 8.11, the *I-V* characteristic of a typical silicon photovoltaic cell is drawn, along with the symbol for photovoltaic devices, for various values of illumination. If we consider the curve for 100 foot-candles, we see that under no-load (with $I_{PH} = 0$) the output voltage of the cell is 0.42 volt. This is the open-circuit voltage, V_{OC}, and is the maximum output voltage that the cell can supply at 100 foot-candles. If the cell is supplying current to a load, its output voltage will drop, as can be seen from this *I-V* curve, until, with the cell short-circuited ($V_{PH} = 0$), I_{PH} is 1 mA. This is the short-circuit current, I_{SC}, and is the maximum current the cell can supply at 100 foot-candles. Increasing the illumination on the cell to 1000 foot-candles causes the *I-V* curve for the cell to move upward. V_{OC} increases to 0.55 volts and I_{SC} increases to 10 mA. At 10,000 foot-candles, V_{OC} increases to 0.58 volts and I_{SC} to 100 mA. Thus the amount of current and voltage that the cell can supply increases as the illumina-tion increases. For most photovoltaic cells the short-circuit current, I_{SC}, is directly proportional to illumination. That is,

$$I_{SC} = K \times L \qquad \textbf{(8.3)}$$

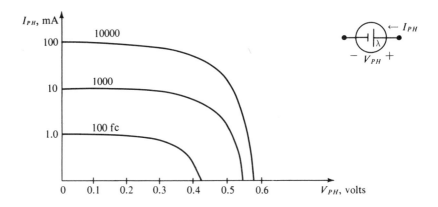

Figure 8.11.   TYPICAL PHOTOVOLTAIC CELL $I\text{-}V$ CHARACTERISTIC AT $25°$ C

where L is the illumination in foot-candles and K is the constant of proportionality. For the cell of Figure 8.11,

$$I_{SC} = 10^{-5} \times L$$

with I_{SC} given in amperes. For example, at $L = 100$ foot-candles, this equation gives

$$I_{SC} = 10^{-5} \times 100 = 10^{-3} \text{ A} = 1 \text{ mA}$$

The open-circuit voltage, V_{OC}, also increases with illumination but it eventually levels off at high values of illumination, typically to 0.5–0.7 V for silicon cells.

A photovoltaic cell acts as a *power converter*. It can supply electrical power to a load when light power is applied to the cell. The efficiency with which it transforms light power to electrical power is given by:

$$\% \text{ efficiency} = \frac{\text{electrical power output}}{\text{light power input}} \times 100 \qquad (8.4)$$

A high efficiency is, of course, desirable, since it indicates a higher electrical power output for a given light power input. The most efficient type of photovoltaic cell on the market today is the silicon P-N junction photovoltaic cell. It can have efficiencies of 10 to 15 percent and, for this reason, is widely used in space applications where sunlight is the only source of light. When used in this application, these cells are called *solar cells*, or *solar batteries*.

Despite the high efficiency of silicon cells, many commercial photovoltaic cells use selenium as the semiconductor, because it gives more stable characteristics with temperature and age. Selenium cells have an efficiency usually around 1 percent. Its *I-V* characteristics are similar to those in Figure 8.11.

The speed of response of photovoltaic cells is considerably greater than for CdS photoconductive cells. Typically, they can respond to light levels which are changing 10,000 to 50,000 times per second (10 kHz to 50 kHz).

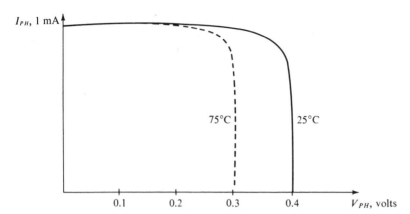

Figure 8.12.   EFFECT OF TEMPERATURE ON PHOTOVOLTAIC CELL CHAR-
ACTERISTIC

The effect of temperature on the characteristics of photovoltaic devices is illustrated in Figure 8.12. The open-circuit voltage decreases as temperature increases from 25°C to 75°C. Thus the cell cannot supply as much voltage, and consequently as much power, at 75°C as it can at 25°C. Note that the short-circuit current is practically independent of temperature.

8.8.   Photovoltaic Cells as Circuit Elements

Figure 8.13 is a typical photovoltaic cell circuit. The cell is acting as the electrical power source supplying voltage and current to the load. Obviously, in this circuit the voltage across the load resistor is equal to the output voltage of the cell. That is,

$$V_L = V_{PH} \tag{8.5}$$

And since $V_L = I_{PH} \times R_L$, we have

$$I_{PH} \times R_L = V_{PH} \tag{8.6}$$

This equation is a straight-line equation relating I_{PH} to V_{PH}. If we plot this line on the I_{PH}-V_{PH} coordinates, its intersection with the characteristic curve of the

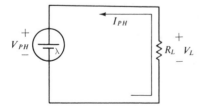

Figure 8.13.   PHOTOVOLTAIC CELL SUPPLYING POWER TO A LOAD

cell will give us the operating point of the circuit. This is illustrated in Figure 8.14 for a 1000-Ω resistor. The operating point becomes $V_{PH} = 0.4$ V and $I_{PH} = 0.4$ mA. The power being supplied by the cell is given by the product $V_{PH} \times I_{PH}$. In this case, it is 0.16 mW of power that is being supplied to the 1000-Ω load.

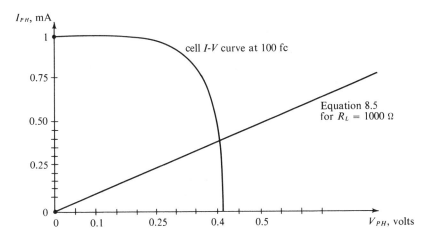

Figure 8.14.   DETERMINING OPERATING POINT OF FIGURE 8.13

8.9.   Applications of Photovoltaic Cells

The principal applications of photovoltaic cells are in the fields of (1) photographic exposure meters, (2) foot-candle meters, (3) lighting control and (4) automatic iris control on cameras. All of these take advantage of using no electrical power source other than the cell itself. Typical of these is the foot-candle meter circuit shown in Figure 8.15. In this circuit, the photovoltaic cell is in series with a current meter. The current meter is usually a very low resistance and essentially the cell is operating into a short circuit. Thus the photocurrent flowing in the circuit will be the cell's short-circuit current, I_{SC}. The current, we know, will vary proportionately with illumination, as will the reading of the

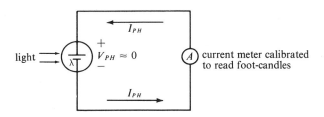

Figure 8.15.   FOOT-CANDLE METER CIRCUIT

current meter. The current meter can be calibrated to read foot-candles, thus serving as an extremely simple, yet very practical and popular light meter.

As far as space and missile systems are concerned, *solar cells* find their greatest application here. Solar cells are usually made up of many silicon photo-voltaic cells set in series in order to obtain high enough voltages; each cell contributes only a few tenths of a volt. In space vehicles, solar cells are used as secondary power sources which can supply electrical power when the vehicle is illuminated by sunlight. When the vehicle is in the dark, the solar cells do not operate and storage batteries provide the vehicle power. The solar cells serve to recharge the storage batteries during the light hours. Solar cells can be constructed to supply large amounts of power by increasing the area exposed to sunlight.

8.10.   Photodiodes

A photodiode, like the photovoltaic cell, is a P-N junction device. However, a photodiode is designed to operate with *reverse* bias applied to its P-N junction. We know from our previous work with P-N diodes that a very small reverse leakage current flows when the junction is reverse biased. This reverse leakage current is caused by thermally generated electron-hole pairs produced in the depletion region and swept across the junction by the intense electric field (reverse bias). We have seen that increases in temperature produce more electron-hole pairs and therefore more leakage current. The same effect occurs if light is allowed to enter the P-N junction area, where photons can be absorbed and their energy used to free electron-hole pairs. Increasing light intensity will cause the diode reverse current to increase.

A photodiode is designed to have a very low *thermal* leakage current, which would be the only reverse current to flow in the dark. It also has the characteristic that its reverse current increases almost proportionately with increases in light intensity. Figure 8.16 shows a graph of reverse current versus illumination for a typical silicon photodiode biased at -20 V. Also shown is the photodiode symbol. Note that at zero illumination the photodiode current is essentially zero and increases linearly with illumination.

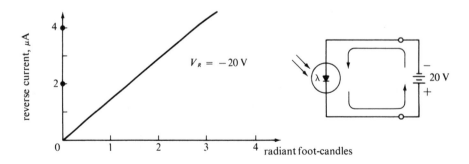

Figure 8.16.   TYPICAL PHOTODIODE CURRENT-ILLUMINATION RELATIONSHIP

A typical photodiode application is shown in Figure 8.17. The load R_L will have a current flowing through it dependent on the amount of illumination on the photodiode. As shown, the circuit is of limited usefulness because the photodiode current is very low (μA) even at high light intensities. Putting it simply, a current of a photodiode cannot energize a typical relay. As such, its current would have to be amplified before it could be useful. However, the response time of photodiodes is relatively fast (1–10 MHz) and makes them useful in high-speed counting and in soundtrack pickups.

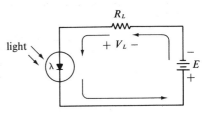

Figure 8.17.

▶ EXAMPLE 8.6

Determine the load current and voltage in the circuit of Figure 8.17 using the photodiode of Figure 8.16, $E = 20$ V, $R_L = 100$ kΩ and a 2 foot-candle (fc) illumination.

At 2 fc the photodiode current from Figure 8.16 is approximately **2.8 μA.** Thus the load voltage is

$$V_L = I \times R_L = 2.8 \ \mu A \times 100 \ k\Omega$$
$$= \textbf{0.28 volts}$$

This leaves 20 V − 0.28 V = 19.72 V across the diode. Even though the photodiode voltage is less than 20 V, the characteristic of Figure 8.16 can be used because reverse current of the diode is not affected greatly by decreases in reverse bias down to a few volts.   ◀

Photodiode *I-V* curves

Figure 8.18 shows a family of *I-V* curves for a typical silicon photodiode. Each curve represents a different amount of incident illumination. The curve for zero illumination is simply the normal P-N diode *I-V* curve showing very little current in the reverse-bias region and significant current when sufficiently forward biased. However, under illumination the curves show a rather interesting effect. Even with *zero* voltage across the diode (short circuit) there will be a considerable amount of reverse current. This effect is due to the internal space-charge potential across the P-N junction (Chapter 5), which is enough to accelerate the photon-produced electron-hole pairs across the junction. In fact, this reverse

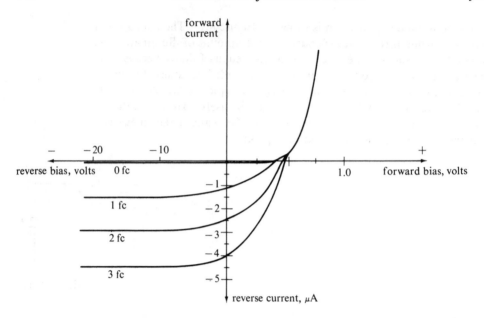

Figure 8.18.   PHOTODIODE *I-V* CURVES

current will flow until a sufficient *forward* bias is applied to neutralize the space-charge region. At that point, the diode conducts in the forward direction like a normal P-N diode.

8.11.   Light-Emitting Diodes (LEDs)

We have seen that P-N junctions can absorb light and produce an electrical current. The opposite is also possible; a junction diode can also emit light. The emission of light occurs under forward bias due to the *recombination* of electrons and holes.

In a P-N diode, free electrons from the N-type material diffuse into the P region when forward biased (Chapter 5). Once in the P region these free electrons encounter a preponderance of holes and eventually recombine; that is, a free conduction electron fills a vacancy in the covalent structure and thus becomes a valence electron. In doing so, the electron loses a certain amount of energy as it jumps from the conduction band down to the valence band. In silicon or germanium diodes, the energy that the recombining electrons lose is dissipated in the form of heat. If other semiconductor materials such as gallium arsenide and gallium phosphide are used to form P-N diodes, the energy lost by recombining electrons is given off in the form of *light* energy.

The frequency of the emitted light is directly related to the energy lost by the recombining electrons. The greater the energy gap of the semiconductor material, the more energy a recombining electron loses and the greater the frequency of emitted light radiation (the lower the wavelength). Gallium arsenide

(GaAs) diodes emit light energy in the infrared region (8800 Å). Gallium phosphide diodes emit light in the red region (6800 Å). By combining various materials and by varying the doping levels, light-emitting diodes can be made to emit many different wavelengths of light.

Figure 8.19 displays some of the data available for a typical GaAs LED, along with the circuit symbol for an LED. Part A of the figure shows the diode's forward-bias *I-V* characteristic with its typical shape. This particular diode has a maximum current rating of 100 mA. Part B shows a plot of radiant power output versus forward current. As can be seen, the light power output varies almost linearly with forward current. This is a characteristic of most LEDs.

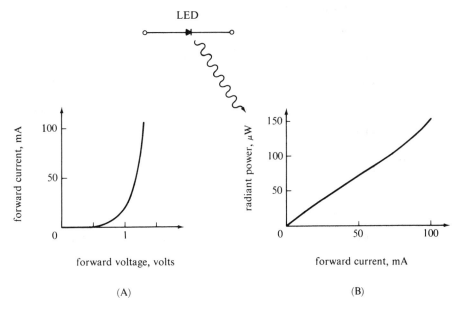

Figure 8.19.   TYPICAL GaAs LED CHARACTERISTICS

Advantages and disadvantages

LEDs have recently come into general use in applications that previously employed other types of light sources. LEDs possess extremely fast response times (typically a few nanoseconds), allowing them to be used in high-frequency pulsed applications. They also possess a long lifetime compared to lamps and can operate from very low voltages, making them compatible with integrated circuits. In addition, the frequency spectrum of some LEDs has a sharp peak, which is near the peak response of many silicon photodetectors, making it easy to match light source and detector.

On the other hand, LEDs are easily damaged by overvoltage or overcurrent, which makes extra care necessary in circuit design. LEDs, like all semiconductor devices, are affected by temperature. In most LEDs, an increase in temperature reduces the amount of radiant output power and also causes the radiant frequency to shift downward.

This concludes our present study of optoelectronic devices. Additional light-sensitive devices will be introduced at appropriate points in subsequent chapters.

GLOSSARY

Optoelectronics: technology employing devices that operate both optically and electronically.

Wavelength: distance that an electromagnetic (or light) wave will travel in one cycle.

Photons: particles of light energy.

Foot-candle: quantity of illumination.

Light emitters: devices which convert electrical energy into light energy.

Photodetectors: devices which convert light energy into electrical energy.

Photoconductive (photoresistive) cells: devices whose resistance varies with illumination.

Decay current (dark current): current through an optoelectronic device under zero illumination.

Photovoltaic (solar) cells: devices that can supply electrical power when illuminated.

Photodiodes: P-N diodes whose reverse current varies with light intensity.

LEDs: P-N diodes which emit light when forward biased.

PROBLEMS

8.1 A certain photodetector is most sensitive to radiation at 4800 Å. To what color of light is this photodetector most reponsive?

8.2 Calculate the wavelength of light which is at a frequency of 0.62×10^{15} Hz.

8.3 How much energy is possessed by the photons of blue light?

8.4 Which color of light has the greater energy content, red or green?

8.5 Explain the principle of operation of photoconductive cells.

8.6 Why are photoconductive cells of CdS, CdSe and CdTe used more widely than those of Si or Ge?

8.7   An RCA–7412 photoconductive cell is used in Figure 8.20. Determine the curcuit current at an illumination of 100 foot-candles.

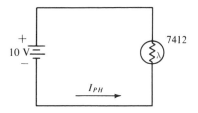

Figure 8.20.

8.8   Why are CdS cells unsuitable for applications where light levels change rapidly?

8.9   How does heat affect the sensitivity of semiconductor photocells?

8.10   Refer to the specifications for the 7412 photocell (Figure 8.8). What is the maximum illumination that should be applied to the cell in the circuit of Figure 8.20?

8.11   Change the source to 90 V in Figure 8.20. Calculate the maximum allowable illumination.

8.12   The *dark resistance* of a photoconductive cell is obtained from its specifications by dividing applied voltage by the decay (dark) current. What is the dark resistance of the RCA–4413 (Appendix II)?

8.13   For the circuit in Figure 8.21, calculate V_{out} for the following values of illumination: 0, 1, 10, 100 foot-candles. Plot V_{out} versus illumination.

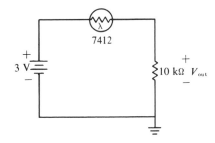

Figure 8.21.

8.14   Design a simple circuit, similar to Figure 8.21, in which V_{out} *decreases* as illumination increases.

8.15   The circuit in Figure 8.22 energizes the relay when the cell is in the dark and de-energizes it when the cell is illuminated. The relay pulls in at 3 mA, drops out at 2 mA and has a resistance of 6 kΩ. Determine the value of R needed to insure that the relay drops out at exactly 10 foot-candles.

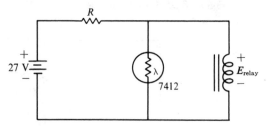

Figure 8.22.

8.16 The circuit in Figure 8.23 utilizes *two* 7412 cells. Output voltage will be high only if both cell *A and* cell *B* are illuminated at the same time. Calculate E_{out} for the following conditions:
(a) Cell *A* dark; cell *B* dark.
(b) Cell *A* dark; cell *B* at 10 foot-candles.
(c) Cell *B* dark; cell *A* at 10 foot-candles.
(d) Cell *B* and cell *A* both at 10 foot-candles. This is a photocell *and* circuit.

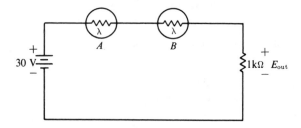

Figure 8.23.

8.17 Design a simple *or* circuit, using *two* 7412 cells, in which the output voltage is high if *either* cell *A or* cell *B or both* are illuminated (10 foot-candles).

8.18 The circuit in Figure 8.24 uses a 7412 cell and 1N3150 tunnel diode. The tunnel diode voltage is low when the cell is dark. When the cell is illuminated the tunnel diode switches to its high state. Calculate the value of R such that the tunnel diode switches to its high state when the illumination is 10 foot-candles or above.

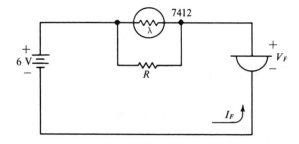

Figure 8.24.

8.19 Briefly explain the photovoltaic effect in a junction photovoltaic cell.

8.20 A particular photovoltaic cell delivers 12 mW of power to a load when it is supplied with 98 mW of light power. What is this cell's efficiency?

8.21 For the circuit of Figure 8.13 determine the following, using the cell characteristic in Figure 8.14:
(a) The operating point using R_L = 500 Ω.
(b) The power supplied to the load.
(c) Repeat (a) and (b) for R_L = 250 Ω.
(d) Plot power output versus R_L for R_L = 250 Ω, 500 Ω and 1000 Ω.

8.22 The photodiode of Figure 8.16 is used in the circuit of Figure 8.25. Determine the output voltage for illuminations of 1, 2 and 3 foot-candles.

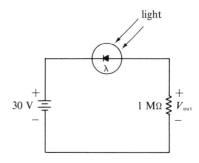

Figure 8.25.

8.23 What is a major disadvantage of photodiodes?

8.24 Explain the basic operation of LEDs.

8.25 The LED of Figure 8.19 is used in the circuit of Figure 8.26. Determine the approximate amount of light output power.

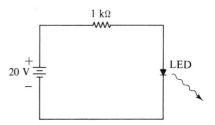

Figure 8.26.

REFERENCES

Brophy, J. J., *Semiconductor Devices*. New York: McGraw-Hill Book Company, 1964.

De Boo, G. J. and C. N. Burrows, *Integrated Circuits and Semiconductor Devices*. New York: McGraw-Hill Book Company, 1971.

Phototubes and Photocells. Lancaster, Pa.: Radio Corporation of America, 1963.

Sowa, W. A. and J. M. Toole, *Special Semiconductor Devices*. New York: Holt, Rinehart and Winston, Inc., 1968.

9

Basic Junction Transistor Operation

9.1.   Introduction

The junction transistor is a *two-junction*, three-terminal device whose operation can be understood from our previous study of the P-N junction diode. Since the invention of the transistor in 1948 there has been a rapidly expanding effort to utilize and develop many types of semiconductor devices. Transistors themselves have replaced bulky vacuum and gas tubes in performing many jobs. Transistors offer several advantages over tubes. Among them are: (1) they have a much smaller size; (2) no heater or filament is required; (3) very low operating voltages can be used; (4) they consume low power, resulting in greater circuit efficiency; (5) they have long life with essentially no aging effects; (6) they are essentially shockproof.

The material in this chapter describes the physical structure, theory of operation and electrical characteristics of junction transistors. The basic behavior of the transistor in DC circuits will be emphasized. In addition, the transistor's behavior as a switch will be studied.

9.2.   Junction Transistor Structure

Figure 9.1 shows the symbolic structure of an NPN *junction transistor*. It contains *three* separate extrinsic regions present in one crystal structure. The two outside regions are doped so they are N type and the center region is doped so it is P type. For reasons which will become clear later, the center P region is called the *base*, the first N region is called the *emitter* and the second N region is called the *collector.* There are two P-N junctions present. The P-N junction formed by the emitter and base regions is called the *emitter-base (E-B) junction*. The P-N junction formed by the collector and base regions is called the *collector-base (C-B) junction*. A terminal or lead is connected to each region, making the transistor a three-terminal device.

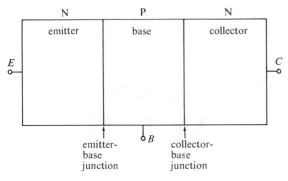

Figure 9.1.   NPN TRANSISTOR STRUCTURE

Figure 9.2 shows a similar structure for the *PNP junction transistor*. In a PNP transistor the doping is just the opposite of that in an NPN type. We shall study both types of transistor since each is widely used. The operation of each type is essentially the same except for differences in voltage and current polarities. For this reason, the discussion will concentrate on the NPN type, pausing periodically to relate the discussion to the PNP type.

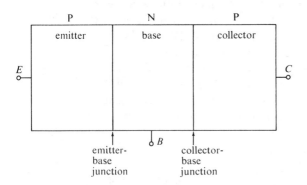

Figure 9.2.   PNP TRANSISTOR STRUCTURE

9.3.   Junction Transistor Operation

Consideration of Figure 9.1 and 9.2 will reveal that there are four possible ways of biasing the two transistor junctions. These are enumerated in Table 9.1.

TABLE 9.1

Condition	E-B junction	C-B junction	Region of operation
I	Forward biased	Reverse (or un-) biased	Active
II	Forward biased	Forward biased	Saturation
III	Reverse biased	Reverse (or un-) biased	Cutoff
IV	Reverse biased	Forward biased	Inverted

Of these four possible combinations, only one interests us at the moment: condition I, where the *E-B* junction is forward biased and the *C-B* junction is reverse (or un-) biased. Let us consider this condition of operation, which is referred to as *active operation*, by modifying the drawing of Figure 9.1 to look like Figure 9.3. In this circuit arrangement the battery V_{BE} acts to forward bias the *E-B* junction and the battery V_{CB} acts to reverse bias the *C-B* junction. The currents flowing through each terminal have been labeled I_E (emitter current), I_B (base current) and I_C (collector current). The directions indicated are the assumed positive directions for an NPN transistor.

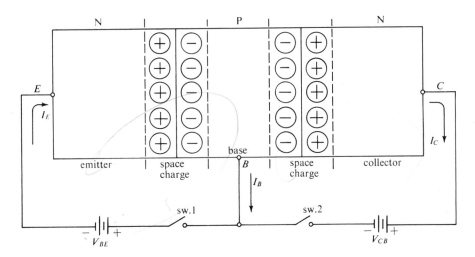

Figure 9.3.   BIASING THE NPN TRANSISTOR FOR ACTIVE OPERATION

Switch 1 closed

If only switch 1 is closed, the *E-B* junction will be forward biased. Since the emitter and base regions are essentially just like a P-N diode, we can expect that a relatively large current will flow across the *E-B* junction. This current will consist of majority carriers from the emitter and base diffusing across the junction. This case is summarized in Figure 9.4. The total current flow across the

E-B junction is the sum of the electron diffusion current and hole diffusion current. In junction transistors the base region is deliberately doped *very lightly* compared to the emitter region. Because of this, the electrons diffusing from the emitter usually make up over 99 percent of the total current. Another effect of the light base doping is that many of the electrons which have entered from the emitter will move through the base region toward the positive terminal of the battery without finding a hole with which to recombine. Only a very small number of the diffusing electrons will find a hole with which to recombine in the base region. We should note in Figure 9.4 that large equal currents flow into the emitter lead and out of the base lead ($I_E = I_B$) and no collector current flows ($I_C = 0$).

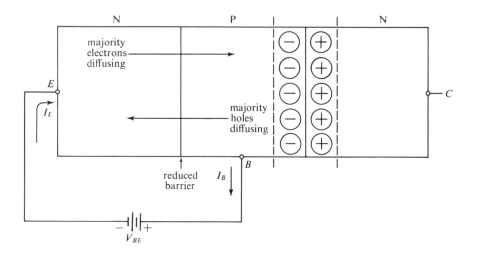

Figure 9.4.   SWITCH 1 CLOSED, SWITCH 2 OPEN

Switch 2 closed

Referring again to Figure 9.3, if only switch 2 is closed the C-B junction will become reverse biased. This case is summarized in Figure 9.5. With reverse bias, the only current flow across the C-B junction will be the reverse leakage current made up of thermally generated minority carriers which are accelerated by the potential barrier. It will be relatively small and, of course, dependent on temperature. We should note in Figure 9.5 that the current will actually flow into the base terminal. Thus I_B will be negative since it actually flows opposite to the assumed direction. In this case very small currents flow out of the collector lead and into the base lead ($I_C = -I_B$) and no emitter current flows ($I_E = 0$). The small collector current in this case is given a special symbol, I_{CBO}. It represents the *collector leakage current* which flows when $I_E = 0$ and the C-B junction is reverse biased.

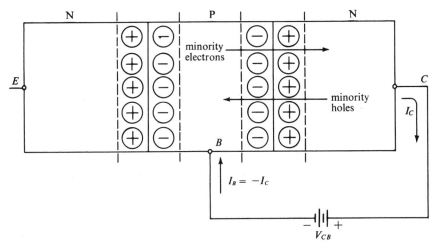

Figure 9.5.   SWITCH 2 CLOSED, SWITCH 1 OPEN

Both switches closed

Referring again to Figure 9.3, what should we expect to occur if both switch 1 and switch 2 are closed? It would seem from the previous two discussions that both I_E and I_B would be rather large currents and I_C would be a very small current. However, the results of closing both switches are almost entirely unexpected. I_E is a large current, as expected, but I_B turns out to be a very small current and I_C turns out to be a large current. The reason for I_C being large,

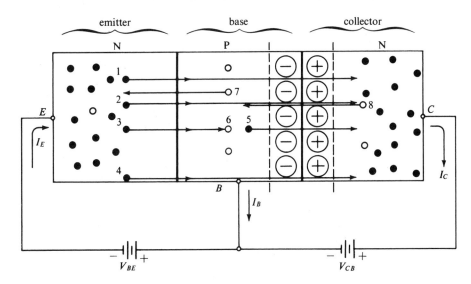

Figure 9.6.   SWITCH 1 AND SWITCH 2 CLOSED; NPN TRANSISTOR OPERATING IN THE ACTIVE REGION

while I_B is small, must be investigated. The drawing in Figure 9.6 depicts this situation. Some of the holes and electrons have been numbered to facilitate the following description of transistor action.

Consider first the *E-B* junction. It is, of course, still forward biased by the voltage V_{BE}. As such, majority carriers will diffuse across the junction, as exemplified by electrons nos. 1, 2, 3 and 4 from the emitter to base and hole no. 7 from base to emitter. Since the base is very lightly doped, most of this majority diffusion current is carried by the electrons and very little by the holes. Now let us consider the *C-B* junction. It is still reverse biased by the voltage V_{CB}. As such, only minority carriers will cross the junction, as exemplified by electron no. 5 from base to collector and hole no. 8 from collector to base. These minority carriers are thermally generated and their current is essentially the I_{CBO} discussed previously. However, there are minority electrons in the base region which are not a result of heat but have diffused from the emitter. These are electrons nos. 1–4 which become minority carriers once they enter the base. These electrons diffuse through the base region and toward the *C-B* junction. Many of them diffuse as far as the space-charge region where they are then accelerated by the potential barrier into the collector in the same way as the thermally generated electrons. In this diagram electrons nos. 1, 2 and 4 make it through to the collector. These additional electrons crossing the *C-B* junction add to the normal reverse leakage current I_{CBO} to give a larger collector current. Electron no. 3, however, does not cross through the base to the collector since it meets hole no. 6 in the base region and recombines with it. In general, it is desirable to have most of the electrons coming from the emitter reach the collector. For this reason, the base region is deliberately made *very narrow* (0.001 in.) so that the diffusing electrons will reach the collector before encountering a hole with which to recombine. The light doping in the base also contributes to the difficulty of recombination in the base region.

It should now be apparent how the terms *emitter* and *collector* come about. With forward bias applied to the *E-B* junction, the emitter *emits* or injects its majority carriers into the base region. Many electrons are injected, many more than can find a hole in the base with which to recombine. The electrons that cannot find a hole diffuse toward the *C-B* junction and upon reaching it are accelerated by the potential barrier there and *collected* by the collector region. In a good transistor only a small percentage of the injected electrons recombine in the base. The greater part of them reach the collector. This is accomplished by making the base very narrow and doping it very lightly. For each electron that reaches the collector region an electron will leave the other end of the collector and go to the positive end of the V_{CB} source. For each electron that recombines in the base an electron leaves the base region and goes to the positive end of the V_{BE} source. Since most of the emitter-injected electrons reach the collector, the collector current will be large; normally it is approximately the same as the emitter current. The base current, then, will be very small.

The amount of emitter current which will flow in the transistor is essentially determined by the magnitude of forward bias, V_{BE}. The base thickness and its degree of doping effectively determine how much of the emitter current will reach the collector and become collector current. The value of reverse bias, V_{CB}, on the *C-B* junction has a slight effect on collector current. There has to be a space-charge region surrounding the *C-B* junction in order to attract and accelerate the diffusing electrons into the collector region. Increasing this space-charge region will cause more emitter electrons to reach the collector, since the diffusing electrons will not have as far to go. Increasing V_{CB} essentially narrows the effective base width which the electrons must traverse before reaching the space-charge region. Thus we can expect for a given value of I_E that I_C will increase slightly and I_B will decrease slightly as V_{CB} is increased.

Summary of active region operation

Figure 9.7 summarizes what has been said thus far about transistor operation in the *active* region. Symbolically it shows the relative sizes of the three transistor currents. I_E is the largest, I_C is almost equal to I_E and I_B is much smaller than either I_E or I_C.

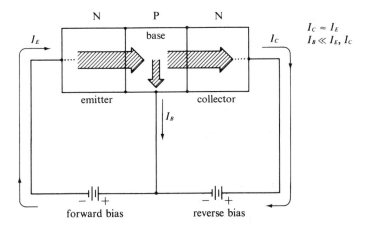

Figure 9.7.   SUMMARY OF ACTIVE REGION OPERATION

Basic transistor relationships

At this point we can begin writing down some of the relationships among the various transistor currents. First of all, invoking Kirchhoff's current law we can state that

$$I_E = I_B + I_C \qquad (9.1)$$

This equation is a simple statement of what has been said up to now about the emitter current distributing itself as collector current and base current. Secondly, from the discussions above we can state that the collector current is made up of two parts: the fraction of emitter current which reaches the collector and the C-B junction reverse leakage current, I_{CBO}. In equation form,

$$I_C = \alpha_{DC}I_E + I_{CBO}* \tag{9.2}$$

where α_{DC} is the fraction of emitter current, I_E, which reaches the collector. Solving for α_{DC} in the above equation we obtain

$$\alpha_{DC} = \frac{I_C - I_{CBO}}{I_E} \tag{9.3A}$$

In cases where I_{CBO} is very small compared to total collector current, it can be deleted from the above equation to give

$$\alpha_{DC} \approx \frac{I_C}{I_E} \tag{9.3B}$$

Here α_{DC} is given as simply the ratio of DC collector current to DC emitter current in the transistor. For typically good transistors α_{DC} is usually in the range of 0.900 to 0.999, indicating that most of the emitter current becomes collector current.

If we now use Equations 9.1 and 9.2 to solve for I_B in terms of I_E we will obtain

$$I_B = (1 - \alpha_{DC})I_E - I_{CBO} \tag{9.4}$$

This equation along with Equation 9.2 allows us to determine I_C and I_B for given values of I_E.

▶ EXAMPLE 9.1

A certain transistor has a value of α_{DC} of 0.98 and a collector leakage current I_{CBO} of 1 μA. Calculate base and collector currents when I_E = 1 mA. Repeat for I_E = 0 μA.

(a) With I_E = 1 mA, and using Equation 9.2, we can solve for I_C:

$$I_C = 0.98 \times 1 \text{ mA} + 1 \text{ μA}$$
$$= 0.98 \text{ mA} + 0.001 \text{ mA} = \textbf{0.981 mA} = \textbf{981 μA}$$

Using Equation 9.4 for I_B we obtain

$$I_B = (1 - 0.98) \times 1 \text{ mA} - 1 \text{ μA}$$
$$= 0.02 \text{ mA} - 0.001 \text{ mA}$$
$$= 0.019 \text{ mA} = \textbf{19 μA}$$

The sum of I_B and I_C should equal I_E. Checking this we have

$$I_E = I_B + I_C$$
$$1 \text{ mA} = 0.019 \text{ mA} + 0.981 \text{ mA} = 1 \text{ mA}$$

---

*I_{CBO} is usually very small and can often be neglected, except at high temperatures.

We could have used Equation 9.1 to calculate I_B once we had I_C and obtained the same result. Note that I_C and I_E are almost equal and I_B is very small. This is usually the case when the transistor is operating in the *active region*.

(b) With $I_E = 0$ and using Equation 9.2 we obtain for I_C

$$I_C = 0.95 \times 0 + 1 \ \mu A$$
$$= 1 \ \mu A = I_{CBO}$$

And using Equation 9.1 to find I_B we have

$$I_B = -1 \ \mu A$$

These results should be expected since, with no emitter current flowing, the only current present across the *C-B* junction is leakage current. (See Figure 9.5.) This is operation in the *cutoff* region.   ◄

▶ EXAMPLE 9.2

Measurements on a certain transistor produce the following data:

I_E	I_C	I_B
0	0.01 μA	−0.01 μA
10 μA	9.91 μA	0.09 μA

From this data determine (a) I_{CBO}, (b) α_{DC} and (c) I_C when $I_E = 20 \ \mu A$.
   (a) From the measurements at $I_E = 0$ the value of I_C is seen to be 0.01 μA. Thus $I_{CBO} = $ **0.01 μA.**
   (b) Using Equation 9.3B and substituting the values $I_E = 10 \ \mu A$ and $I_C = 9.91 \ \mu A$ we obtain

$$\alpha_{DC} \approx \frac{9.91 \ \mu A}{10 \ \mu A} = \textbf{0.991}$$

   (c) Using the values of α_{DC} obtained above we can calculate I_C using Equation 9.2 with $I_E = 20 \ \mu A$ and neglecting I_{CBO}.

$$I_C = 0.991 \times 20 \ \mu A = \textbf{19.82 μA} \qquad I_C = \alpha \ I_E$$

Again note that I_C is almost equal to I_E.   ◄

Everything that has been said thus far is also applicable to PNP transistor operation. The only differences are the following for the PNP type:

(a) The emitter injects *holes* into the base region since the emitter is P type.
(b) The bias voltages have polarity opposite those for the NPN transistor.
(c) The current directions for I_B, I_C and I_E are also the reverse of those for the NPN type.

The operation of the PNP transistor in the active region is summarized in Figure 9.8. Compare it with Figure 9.6 for the NPN transistor. Note that the holes and electrons have interchanged. In the NPN transistor electrons are the principal carrier, while in the PNP type the holes have this role. Equations 9.1–9.4, developed previously for the NPN transistor, are equally valid for the PNP transistor. Thus Examples 9.1 and 9.2 could just as easily have used PNP transistors.

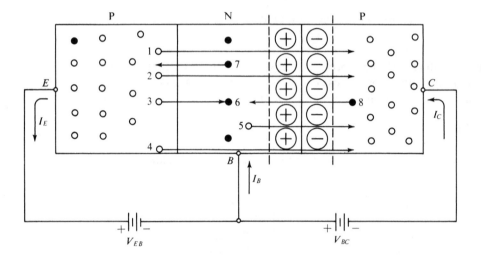

Figure 9.8.   PNP TRANSISTOR OPERATING IN THE ACTIVE REGION

To summarize what has been said thus far concerning transistor operation in the *active* region, the following points should be re-emphasized:

(a) When the *E-B* junction is forward biased the emitter injects its majority carriers (electrons for NPN, holes for PNP) into the base region.

(b) Since the base region is lightly doped and very narrow, most of these injected carriers diffuse to the *C-B* junction without recombining in the base region.

(c) If the *C-B* junction is reverse biased, the space-charge region at the junction serves to accelerate these carriers into the collector region. These carriers which are added to the normal reverse leakage current, I_{CBO}, at the *C-B* junction contribute to a large collector current.

(d) The base current is typically very small since most of the injected carriers reach the collector.

At this point, a brief description of the other conditions of operation found in Table 9.1 should be given. Conditions II and III, *saturation* and *cutoff*, are extremely important and must be completely understood if intelligent use of transistors is to be accomplished. Condition IV, *inverted operation*, is used only in certain special circuits and, as such, will not be stressed here.

Referring to Table 9.1, it is seen that operation in the *saturation region* occurs when both transistor junctions are forward biased. In this condition, the C-B junction space-charge region has almost completely diminished and the ability of the collector to collect the emitter-injected carriers is greatly reduced. In fact, if the forward bias on the C-B junction is sufficient, the collector will stop collecting carriers and will begin emitting carriers into the base in exactly the same way as the emitter. Much more will be said about saturation in the following sections.

Operation in the *cutoff region* occurs when both transistor junctions are reverse biased. In this condition, the emitter will *not* inject majority carriers into the base region. The only emitter current flowing will be a reverse leakage current and, consequently, the collector current will also consist of a reverse leakage current. The cutoff region also includes the case where the emitter circuit is open; that is, $I_E = 0$. In either case, with no injected emitter current, the transistor is said to be cut off and only very small leakage currents flow. More will be said about cutoff in the following sections.

Inverted operation is simply using the transistor's collector as the emitter and vice versa. This operation is quite different from the normal operation (Condition I, active) since the emitter and collector are not doped exactly the same. The emitter is the most heavily doped region of the three. The base is the most lightly doped and the collector is doped at some intermediate level. Interchangeability of collector and emitter is therefore not possible. However, some transistor applications utilize this type of operation for reasons which need not be discussed here.

9.4. The Common-Base Configuration

The electronic symbols for both the NPN and PNP transistor are shown in Figure 9.9. The arrow on the emitter of each type indicates the direction of emitter current flow when the E-B junction is *reverse* biased. In other words, it is opposite to the direction of forward-bias current through the emitter.

Recall from our discussion in Chapter 3 that, when analyzing a three-terminal device, such as a junction transistor, one of the terminals is chosen as the *common* terminal, one as the *input* and one as the *output*. Let us first consider the *base* as the *common* terminal, the *emitter* as the *input* terminal and the *collector* as the *output* terminal. This configuration is called the *common-base* configuration. An NPN transistor in the common-base configuration is shown in Figure 9.10, with the bias voltages applied for operation in the active region. The E-B junction is forward biased, producing I_E and I_B in the direction shown. The C-B junction is reverse biased, producing I_C in the direction shown. In the common-base (abbreviated Com.-B) configuration, the emitter terminal is considered the input. The emitter current is the input current and the collector current is the output current. For a particular I_E input there will be a corresponding I_C output.

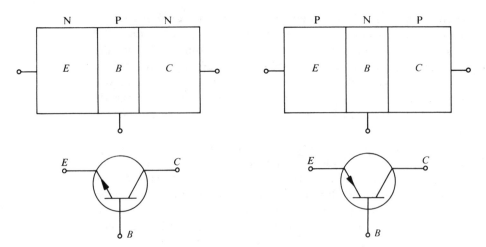

Figure 9.9.   ELECTRONIC SYMBOLS FOR NPN AND PNP TRANSISTORS. ARROW ON THE EMITTER INDICATES THE DIRECTION OF CURRENT FLOW WHEN *E-B* JUNCTION IS REVERSE BIASED.

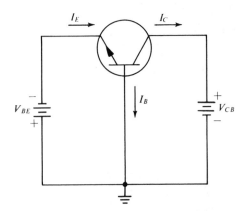

Figure 9.10.   NPN TRANSISTOR IN COMMON-BASE CONFIGURATION AND BIASED IN THE ACTIVE REGION

▶ EXAMPLE 9.3

The *E-B* junction of the transistor in Figure 9.10 is forward biased so that an emitter current of 10 mA flows. Determine I_C and I_B under these conditions if $\alpha_{DC} = 0.99$.

(a) Using Equation 9.2 and neglecting I_{CBO}. $I_C = 0.99 \times 10$ mA $= 9.90$ mA.

(b) Using Equation 9.1, we have $I_B = I_E - I_C = 10$ mA $- 9.90$ mA $= 0.10$ mA $= 100$ μA.   ◀

In the Com.-B configuration, the parameter α_{DC} is called the *common base DC current gain* since it relates the DC output current, I_C, to input current, I_E. The term *gain* seems to imply an increase in current from input to output. However, since α_{DC} is always less than one, the output current in Com. B is *always less* than the input current. That is, the *current gain* is always less than one. In Example 9.3 above, the output current was 99 percent of the input current since α_{DC} was 0.99.

A summary of the basic transistor equations for the Com.-B configuration is presented in Figure 9.11.

$$I_C = \alpha_{DC} I_E + \boxed{I_{CBO}}$$

$$I_B = (1 - \alpha_{DC}) I_E - \boxed{I_{CBO}} \qquad (I_{CBO} \text{ can usually be neglected})$$

$$\alpha_{DC} = \frac{I_C - \boxed{I_{CBO}}}{I_E}$$

Figure 9.11.   COMMON-BASE TRANSISTOR EQUATIONS

9.5.   Common-Base Characteristic Curves

The preceding equations do not, by any means, completely describe the transistor behavior. For one thing, the value of α_{DC} is not constant, but varies with I_C and V_{CB}. Up to a point, α_{DC} *increases* as I_C *increases;* it then *decreases* for further increases in I_C. It also *increases* as V_{CB} *increases*. Secondly, these equations assume that I_E is known. To find I_E, the variation of I_E with V_{BE} must be available. For these reasons, the transistor characteristic curves are often utilized in analysis and design.

Recall from Chapter 3 that for a three-terminal device *two* sets of *I-V* characteristics — the *input characteristics* and the *output characteristics* — are needed to describe completely its static operation. For the NPN transistor in the Com-B configuration, typical input characteristics are as shown in Figure 9.12. These are the *common-base input characteristics* and they relate the input current, I_E, to the voltage between the emitter and base, V_{BE}, for different values of collector-base reverse voltage, V_{CB}. For a given value of V_{CB}, the curve relating I_E to V_{BE} is essentially that of a P-N junction diode, which is, of course, the *E-B* junction. The *E-B* junction becomes a better diode as V_{CB} increases. That is, there will be a greater I_E for a given V_{BE} as V_{CB} increases. The effect of V_{CB} comes about as the result of the increased space-charge region at the *C-B* junction as V_{CB} is increased. This effectively increases the ease with which majority carriers diffuse from the emitter to base by exerting a greater force on them once they enter the base. It can be seen from Figure 9.12 that this effect is slight except for small values of V_{CB} and, as will be shown, can be neglected in many cases.

Note that for negative values of V_{BE} the NPN transistor *E-B* junction is reverse biased and only a small leakage current flows through the emitter. This is the condition for *cutoff* when $I_E = 0$.

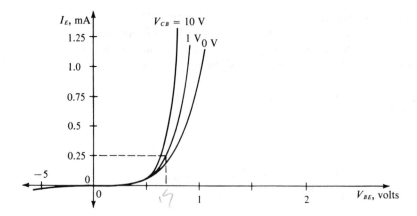

Figure 9.12.   COMMON-BASE INPUT CHARACTERISTICS FOR A TYPICAL NPN SILICON TRANSISTOR

For the same NPN transistor in the Com.-B configuration, a set of output characteristics are shown in Figure 9.13. These are *the common-base output characteristics* and they relate the output current, I_C, to the voltage between collector and base, V_{CB}, for various values of input current, I_E.

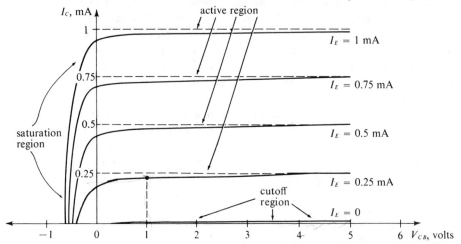

Figure 9.13.   COMMON-BASE OUTPUT CHARACTERISTICS FOR NPN SILICON TRANSISTOR

A study of these output characteristics reveals several interesting points:

(a) The collector current is approximately equal to the emitter current for positive values of V_{CB} (*C-B* junction reverse biased). This is the *active* region.

(b) All the curves have a slight slope for positive values of V_{CB}, indicating that I_C increases slightly as V_{CB} increases for a given I_E. This indicates that α_{DC} increases slowly with V_{CB}.

(c) The collector current is not zero when I_E is zero, but has a value of I_{CBO}. That is, when the transistor is in *cutoff*, only a very small value of I_C flows.

(d) As V_{CB} becomes negative (*C-B* forward biased), I_C begins to decrease for a given I_E. This is the *saturation* region.

► EXAMPLE 9.4

Determine the collector current and emitter current for the transistor with the characteristics of Figures 9.12 and 9.13 when $V_{BE} = 0.7$ volts and $V_{CB} = 1$ volt. Calculate α_{DC} at this point.

(a) First, from the input characteristics, the value of I_E is obtained at $V_{BE} = 0.7$ V using the curve for $V_{CB} = 1$ V. It is $I_E = 0.25$ mA. Now, using the output characteristics, the value of I_C is obtained at $V_{CB} = 1$ V using the curve for $I_E = 0.25$ mA. It is $I_C = 0.225$ mA.

(b) The value of α_{DC} at this point is 0.225 mA/0.25 mA = 0.9.   ◄

The Com.-B output characteristics, often called the *collector characteristics*, or *collector family of curves*, do not vary radically from transistor to transistor since α_{DC} is very close to *one* for most transistors for values of V_{CB} greater than one volt. This is an important property which will be utilized in certain applications.

The Com.-B characteristics of a PNP transistor are essentially the same as those in Figures 9.12 and 9.13, except for the polarities on the V_{BE} and V_{CB}, plus the fact that I_C flows in the opposite direction. Simply change the polarities on the V_{BE} and V_{CB} values given in these figures and the curves become PNP characteristics.

9.6.   Analyzing Common-Base Circuits: Load-Line Method

In analyzing a transistor Com.-B circuit, it is first necessary to determine input current, I_E. Once I_E is known, the output current, I_C, can be determined. Again, two methods of analysis will be presented. The *load-line method* is essentially a graphical technique and is used mainly in the analysis of transistor amplifiers. The *approximate method* is used for quick, easy calculations; though not as accurate as the load-line method, it is often used in determining the DC operating point when a great deal of accuracy is not needed.

In either method, the value of I_E must be determined first. In determining I_E, the diodelike characteristics of the *E-B* junction are utilized. Consider the circuit of Figure 9.14. The emitter current is supplied by the voltage V_{EE} through the series resistor R_E. We can write the Kirchhoff's voltage equation for the input portion of the circuit. We have

$$V_{EE} = I_E \times R_E + V_{BE} \qquad\qquad \textbf{(9.5)}$$

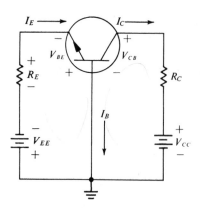

Figure 9.14.   TYPICAL DC COMMON-BASE CIRCUIT

If we rearrange this equation to solve for I_E, it becomes

$$I_E = \frac{V_{EE} - V_{BE}}{R_E} \tag{9.6}$$

The voltage V_{BE} is the voltage drop across the *E-B* junction when it is forward biased. As with P-N diodes, this voltage is typically about 0.7 volt for silicon and 0.3 volt for germanium units. Using this fact in Equation 9.6 makes the value of I_E readily obtainable.

▶ EXAMPLE 9.5

A silicon transistor is used in the circuit of Figure 9.14 with $V_{EE} = 10.7$ V and $R_E = 500$ Ω. Determine I_E. Repeat for a germanium transistor.

(a) Using $V_{BE} = 0.7$ V in Equation 9.6, we have

$$I_E = \frac{10.7 \text{ V} - 0.7 \text{ V}}{500 \text{ }\Omega} = 20.0 \text{ mA}$$

(b) Using $V_{BE} = 0.3$ V in Equation 9.6, we have

$$I_E = \frac{10.7 \text{ V} - 0.3 \text{ V}}{500 \text{ }\Omega} = 20.8 \text{ mA} \qquad ◀$$

This procedure for calculating the value of I_E is essentially the same for both the load-line and approximate methods of analysis.

In using the load-line method to find the value of I_C, the procedure is to write the Kirchhoff's voltage equation for the output portion of the circuit and plot this equation (load line) on the output characteristic curves (I_C and V_C co-ordinate axes). The intersection of this load line with the output curve corresponding to the value of I_E previously calculated gives the circuit operating point. Referring to Figure 9.14, the Kirchhoff's voltage equation for the output portion of the circuit is

$$V_{CC} = I_C \times R_C + V_{CB} \tag{9.7}$$

This equation relates I_C to V_{CB} and is called the load-line equation. Plotting this load line on the output characteristics is done by choosing two points on the line. The easiest points to choose are $(I_C = 0, V_{CB} = V_{CC})$ and $(I_C = V_{CC}/R_C, V_{CB} = 0)$.

▶ EXAMPLE 9.6

The silicon transistor of Figure 9.14 has the output characteristic curves drawn in Figure 9.15. Find I_C and V_{CB} for $I_E = 20$ mA if $V_{CC} = 40$ volts and $R_C = 1.6$ kΩ.

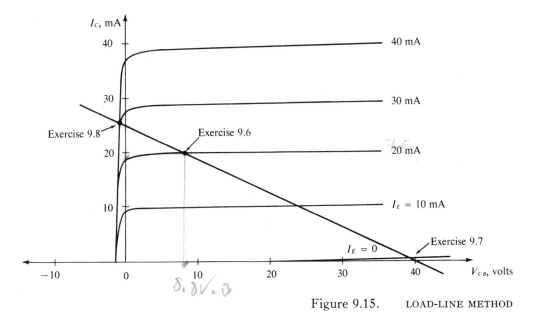

Figure 9.15.   LOAD-LINE METHOD

The load-line equation is 40 V $= I_C \times 1.6$ kΩ $+ V_{CB}$. The plot of this line on the characteristic curves is shown in the figure. Note that the two points used to plot the load line are $(I_C = 0, V_{CB} = 40$ V$)$ and $(I_C = 40$ V$/1.6$ kΩ $= 25$ mA, $V_{CB} = 0)$.

The intersection of this load line and the output curve for $I_E = 20$ mA is the operating point. It is seen to be $I_C = 19.5$ mA and $V_{CB} = 8.8$ volts. The voltage across the 1.6-kΩ resistor is 19.5 mA $\times 1.6$ kΩ $= 31.2$ V. This can be checked by substituting these values into the load-line equation. In this case, the operating point lies in the active region of the transistor characteristic. ◀

▶ EXAMPLE 9.7

Repeat the previous example using $I_E = 0$.

The intersection of the load line and the $I_E = 0$ curve gives the new operating point as $I_C \approx 0$ and $V_{CB} = 40$ V. When the transistor is in cutoff, all of the supply voltage, V_{CC}, is dropped across the *C-B* junction; none is dropped across the resistor.   ◀

▶ EXAMPLE 9.8

Repeat Example 9.6 using (a) $I_E = 30$ mA and (b) $I_E = 40$ mA.
   (a) The intersection of the load line and the $I_E = 30$ mA curve gives the new operating point as $I_C \approx 25.0$ mA, $V_{CB} \approx -0.5$ V. This operating point lies in the *saturation* region since V_{CB} is negative.
   (b) The operating point when $I_E = 40$ mA is seen to be $I_C \approx 25.0$ mA, $V_{CB} \approx -0.5$ V. This operating point also lies in the saturation region. The important point to notice is that an increase in I_E from 30 mA to 40 mA results in almost no increase in I_C. This is so because, once the emitter current becomes large enough to move the operating point into the saturation region, further increases in I_E do not affect I_C. We say that the transistor output has become *saturated*, since I_C can increase no further even if I_E increases. The extra emitter current increases I_B but not I_C. For instance, in part a, I_B is $I_E - I_C = 5$ mA, while in part b, I_B is $I_E - I_C = 15$ mA.   ◀

   In Example 9.8, it may not be clear how saturation can occur, since the V_{CC} source acts to reverse bias the *C-B* junction. As I_E increases, causing a corresponding increase in I_C, the voltage drop across R_C also increases. The resistor voltage is opposed in polarity to the V_{CC} voltage. Thus a point is reached where V_{CB} will change polarity as I_E is increased. At this point, saturation occurs. Once saturation is reached, V_{CB} remains at approximately zero.
   The examples above illustrate the use of the load line in finding the currents and voltages in a Com.-B transistor circuit. This method will gain importance in later coverage of transistor amplifiers.

9.7.   Analyzing Common-Base Circuits: Approximate Method

There are always situations in which analysis using the load line method is impossible or impractical. In many cases the necessary transistor curves are not immediately available and often the circuit being analyzed contains more than one transistor, which makes load-line analysis very difficult, at best. At other times, the accuracy of the load-line technique may not be required and faster, less accurate methods are more useful. The approximate method that we shall discuss here is easier to use than the load-line method since it takes advantage of certain properties common to all transistors and does not utilize the characteristic curves. The Com.-B circuit of Figure 9.16 will be analyzed using approximate techniques. Examples 9.6–9.8 will be redone using these techniques and the results compared to those of the previous section to see how "approximate" the approximate method really is.

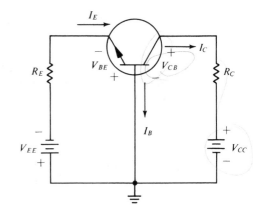

Figure 9.16.   DC COMMON-BASE CIRCUIT

Cutoff condition

In using the approximate method to analyze a circuit such as that in Figure 9.16, it is advisable to first determine in what region of its operation the transistor is biased. When the *E-B* junction is reverse biased or unbiased ($I_E = 0$), the transistor is in *cutoff* and only a small leakage current flows in the collector. In low-power silicon transistors this leakage current may be typically a fraction of 1 μA at room temperature, while for corresponding germanium units it may be typically 1 μA. This leakage current normally can be neglected. In this case, the voltage drop across R_C is essentially zero. Thus the entire collector supply voltage, V_{CC}, is dropped across the *C-B* junction. That is, $V_{CB} = V_{CC}$.

▶ EXAMPLE 9.9

Using $V_{CC} = 40$ volts and $R_C = 1.6$ kΩ in the circuit of Figure 9.16, determine I_C and V_{CB} when $I_E = 0$.

    With $I_E = 0$, the only collector current flowing will be a small leakage current and $I_C \approx 0$. The voltage drop across the 1.6-kΩ resistor is zero. Thus $V_{CB} = 40$ V. Compare these results with those of Example 9.7.   ◀

Transistor in active or saturation region

When the *E-B* junction is forward biased, emitter current will flow, causing a corresponding collector current to flow. The value of I_E is determined in the manner outlined in the last section and used in Example 9.5. When the value of I_E is known, whether the transistor is in the *active* region or *saturated* region of operation remains to be determined. In the *active* region, the collector current is approximately equal to $\alpha_{DC} \times I_E$. If the value of α_{DC} is not known, the collector current may be assumed to be equal to I_E, since α_{DC} is usually very close to unity. After determining I_C in this manner, its value is used to calculate V_{CB} using Kirchhoff's voltage law (Equation 9.7). If the calculated value of V_{CB} indicates that the *C-B* junction is *reverse biased*, then the transistor is operating in the *active* region. In this case, the previously determined value of I_C is correct and the solution is complete. If, however, V_{CB} turns out to be of the opposite polarity, so that the *C-B* junction is forward biased, then the transistor is not operating in the active region, but, instead, in the saturation region. In this case, the previously determined value of I_C is incorrect, since that value assumed *active* operation. When the transistor is *saturated*, the collector current will not equal $\alpha_{DC} \times I_E$, but essentially will be limited to that value which causes the *C-B* junction to become forward biased by a few tenths of a volt (recall Example 9.8). This value of collector current at *saturation* is given the symbol $I_{C(sat)}$ and is approximately determined by assuming a V_{CB} of zero in Equation 9.7 and calculating the collector current. This gives

$$I_{C(sat)} \approx \frac{V_{CC}}{R_C} \qquad (9.8)$$

The actual value of $I_{C(sat)}$ is slightly smaller than that given by Equation 9.8 since V_{CB} is not zero, but, instead, a few tenths of a volt at saturation. In most calculations, however, Equation 9.8 is sufficient. Equation 9.8 indicates that once the transistor becomes saturated the collector current is determined by the external circuit (V_{CC} and R_C) rather than by the transistor. Once the value of I_E is increased to the value needed to cause I_C to equal $I_{C(sat)}$, further increases in I_E cause I_C to increase only slightly (recall Example 9.8) and the transistor has become saturated.

To summarize the procedure to be followed when calculating I_E, I_C and V_{CB} when the transistor is *not* in cutoff:

(a) Calculate I_E using

$$I_E = \frac{V_{EE} - V_{BE}}{R_E}$$

where $V_{BE} = 0.7$ V for Si and 0.3 V for Ge.

(b) Assume $I_C = \alpha_{DC} \times I_E$ if α_{DC} is known; otherwise assume $I_C \approx I_E$.

(c) Using this value of I_C, calculate V_{CB} using

$$V_{CB} = V_{CC} - I_C \times R_C$$

(d) If V_{CB} is positive ($V_{CC} > I_C \times R_C$), then the transistor is operating in the *active* region and the calculated values of I_C and V_{CB} are correct.

(e) If V_{CB} is negative ($V_{CC} < I_C \times R_C$), then the transistor is operating in the *saturation* region and the values of I_C and V_{CB} calculated above are incorrect. The correct collector current is given by

$$I_C = I_{C(sat)} \approx \frac{V_{CC}}{R_C}$$

and V_{CB} is very small (≈ 0 V).

The following examples illustrate this procedure:

▶EXAMPLE 9.10

In the circuit of Figure 9.16, find I_C and V_{CB} if V_{CC} = 40 V, R_C = 1.6 kΩ and I_E = 20 mA.

Since α_{DC} is not given, assume I_C = 20 mA. Using this value, the voltage drop across the 1.6-kΩ resistor is 32 V. Thus V_{CB} = 40 − 32 = 8 V. The *C-B* junction is reverse biased and the transistor is in the *active* region. Thus the approximate operating point is I_C = **20 mA**, V_{CB} = **8 V**. Compare this to the results of Example 9.6.   ◀

▶EXAMPLE 9.11

Repeat Example 9.10 for I_E = 30 mA and I_E = 40 mA.

(a) Assume I_C = I_E = 30 mA. Calculate V_{CB} = 40 V − 30 mA × 1.6 kΩ = −8 V. This indicates forward bias on the *C-B* junction; the transistor is in the *saturation* region. Thus the value of I_C is not 30 mA, but is given by

$$I_{C(sat)} = \frac{40 \text{ V}}{1.6 \text{ k}\Omega} = 25 \text{ mA}$$

and V_{CB} ≈ **0**. Compare this to the solution of Example 9.8a.

(b) If I_E = 30 mA is enough to saturate the transistor, I_E = 40 mA will certainly cause saturation. Therefore, the value of I_C is still 25 mA. Compare to Example 9.8b.   ◀

The results of these examples are not in exact agreement with the corresponding results in the last section. However, this slight inaccuracy is not critical in most circuit calculations. After all, even the resistors and power supply voltage in a circuit have tolerances for inaccuracies, so that any calculation is really only an approximation.

▶EXAMPLE 9.12

Figure 9.17 shows a silicon PNP transistor in a Com.-B circuit. Calculate the value of voltage V_{EE} which just saturates the transistor.

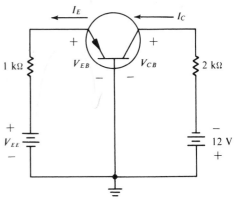

Figure 9.17.

At saturation, the *C-B* junction becomes forward biased. Since this is a PNP transistor, V_{CB} will be slightly positive at saturation. To calculate $I_{C(\text{sat})}$ for this transistor circuit, we have

$$I_{C(\text{sat})} \approx \frac{12 \text{ V}}{2 \text{ k}\Omega} = 6 \text{ mA} \tag{9.8}$$

Thus the emitter current to just cause saturation is given by

$$I_{E(\text{sat})} = \frac{I_{C(\text{sat})}}{\alpha_{\text{DC}}} \approx \frac{6 \text{ mA}}{1} = 6 \text{ mA} \tag{9.9}$$

To find the value of V_{EE} which will supply an emitter current of 6 mA, we have

$$V_{EE} = 6 \text{ mA} \times 1 \text{ k}\Omega + V_{EB}$$
$$= 6 \text{ V} + 0.7 \text{ V}$$
$$= 6.7 \text{ V} \qquad \blacktriangleleft$$

▶ EXAMPLE 9.13

If the 2-kΩ resistor is changed to a 4-kΩ resistor, what value of V_{EE} is needed to cause saturation?

$$I_{C(\text{sat})} = \frac{12 \text{ V}}{4 \text{ k}\Omega} = 3 \text{ mA}$$

Thus we have

$$I_{E(\text{sat})} = 3 \text{ mA}$$

and

$$V_{EE} = 3 \text{ mA} \times 1 \text{ k}\Omega + 0.7 \text{ V} = 3.7 \text{ V} \qquad \blacktriangleleft$$

These examples illustrate the fact that the value of $I_{C(\text{sat})}$ does not depend on the transistor, but rather on the circuit values (V_{CC} and R_C). Changing these components changes the point at which saturation occurs.

9.8.   The Common-Emitter Configuration

An NPN transistor in the *common-emitter* configuration is shown in Figure 9.18 with the bias voltages applied for operation in the *active* region. In the common-emitter configuration (abbreviated Com.-E), the bias voltages are applied between collector and emitter (V_{CE}) and base and emitter (V_{BE}). The *E-B* junction is forward biased since the base is made more positive than the emitter by V_{BE}. The collector is more positive than the emitter by V_{CE}. The voltage at the collector with respect to the base is then given by

$$V_{CB} = V_{CE} - V_{BE} \tag{9.10}$$

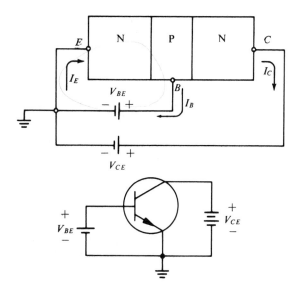

Figure 9.18.   NPN TRANSISTOR IN THE COMMON-EMITTER CONFIGURATION
AND BIASED IN THE ACTIVE REGION

To reverse bias the *C-B* junction, V_{CB} must be positive. The value of V_{CE} must then be greater than V_{BE}. When V_{CE} is less than V_{BE}, V_{CB} is negative, the *C-B* junction is forward biased and the transistor is in *saturation.*

In the Com.-E configuration, the *base* terminal is the *input* terminal and the *collector* terminal is the *output* terminal. For a particular I_B input there will be a corresponding I_C output. Although the physical operation of the transistor in this configuration is the same as for Com. B, the basic transistor equations (9.1 through 9.4) have to be rearranged so that I_C and I_E are given in terms of I_B. Equation 9.4 is repeated below, neglecting leakage I_{CBO}.

$$I_B \approx (1 - \alpha_{DC}) \times I_E \qquad (9.4)$$

which can be solved for I_E:

$$I_E \approx \frac{I_B}{(1 - \alpha_{DC})}$$

Since $I_E = I_C + I_B$, we can rewrite this as

$$I_C + I_B \approx \frac{I_B}{1 - \alpha_{DC}}$$

Solving for I_C results in

$$I_C \approx \frac{\alpha_{DC}}{1 - \alpha_{DC}} \times I_B \qquad (9.11)$$

In this equation I_C is given in terms of I_B. The equation can be simplified somewhat by defining

$$\beta_{DC} \equiv \frac{\alpha_{DC}}{1 - \alpha_{DC}} \qquad (9.12)$$

Thus Equation 9.11 becomes

$$I_C \approx \beta_{DC} \times I_B \qquad (9.13)$$

where β is the Greek letter beta.

This last formula is the most important relationship for the Com.-E configuration. It states that I_C is equal to β_{DC} multiplied by the input I_B.

β_{DC} is called the *common-emitter DC current gain* since it relates the DC output current I_C to the input current I_B. Equation 9.12 indicates that β_{DC} can be very large. For example, if $\alpha_{DC} = 0.9$, the value of β_{DC} is

$$\beta_{DC} = \frac{0.9}{1 - 0.9} = 9$$

and for $\alpha_{DC} = 0.99$, the value of β_{DC} is 99. Typically, β_{DC} can have values in the range from 10 to 500, although values as high as 1000 are not uncommon.

If we solve Equation 9.13 for β_{DC}, we obtain

$$\beta_{DC} \approx \frac{I_C}{I_B} \qquad (9.14)$$

Thus, β_{DC} is the ratio of DC collector current to DC base current. It is essentially a measure of how much more of the emitter current becomes collector current than becomes base current. β_{DC} depends on α_{DC} as shown by Equation 9.12. As α_{DC} increases, the percentage of emitter current which reaches the collector increases and, conversely, the percentage which becomes base current decreases. Thus, as α_{DC} increases, β_{DC} increases.

Now that we have I_C in terms of I_B as given by Equation 9.13, we can determine I_E in terms of I_B by using the relationship

$$\begin{aligned} I_E &= I_B + I_C \\ &= I_B + \beta_{DC} \times I_B \\ &= (\beta_{DC} + 1)I_B \end{aligned} \qquad (9.15)$$

The emitter current is essentially $(\beta_{DC} + 1)$ times the base current.

Com.-E leakage current, I_{CEO}

In the Com.-E configuration, a small collector leakage current will flow even if the base current is zero ($I_B = 0$). This is illustrated in Figure 9.19, where this leakage current is shown as I_{CEO}, *the collector-emitter current with the base open*. The value of I_{CEO} is given by

$$I_{CEO} = (\beta_{DC} + 1) \times I_{CBO} \qquad (9.16)$$

which shows that it can have a much greater value than the Com.-B leakage current, I_{CBO}. For example, if $I_{CBO} = 0.01$ μA and $\beta_{DC} = 100$, the value of I_{CEO} given by 9.16 would be 1 μA.

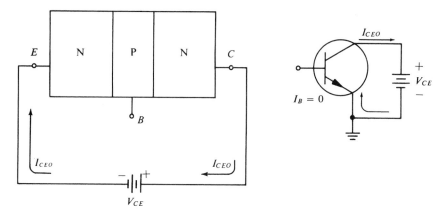

Figure 9.19.   DEFINITION OF I_{CEO}

For exactness, the term I_{CEO} should be included in the expression for I_C. That is,

$$I_C = \beta_{DC} \times I_B + I_{CEO} \qquad (9.17)$$

This expression is more correct than Equation 9.13; however, the value of I_{CEO} is usually negligible, except at elevated temperatures (especially for germanium).

A summary of the basic transistor relationships in the common-emitter configuration is given in Figure 9.20. Note that the I_{CEO} terms can often be neglected without introducing significant error.

$$I_C = \beta_{DC} \times I_B + \boxed{I_{CEO}}$$

$$I_E = (\beta_{DC} + 1)I_B + \boxed{I_{CEO}}$$

$$\beta_{DC} \equiv \frac{\alpha_{DC}}{1 - \alpha_{DC}} \approx \frac{I_C}{I_B}$$

$$\boxed{I_{CEO} \approx \beta_{DC} \times I_{CBO}}$$

Figure 9.20.   COMMON-EMITTER FORMULAS

▶ EXAMPLE 9.14

A transistor has a value of $\alpha_{DC} = 0.995$. Calculate β_{DC} for this transistor. Repeat for $\alpha_{DC} = 0.992$.

Using Equation 9.12,

$$\text{(a)} \quad \beta_{DC} = \frac{0.995}{1 - 0.995} = \frac{0.995}{0.005} = \mathbf{199}$$

$$\text{(b)} \quad \beta_{DC} = \frac{0.992}{1 - 0.992} = \frac{0.992}{0.008} = \mathbf{124}$$

Note that a small change in α_{DC} gives a large change in β_{DC}.   ◀

▶ EXAMPLE 9.15

A particular transistor has $\beta_{DC} = 100$ and $I_{CEO} = 10\ \mu A$. Determine I_C and I_E for $I_B = 0$ and for $I_B = 20\ \mu A$.

(a) For $I_B = 0$:

$$I_E = I_C = I_{CEO} = \textbf{10 } \mu\textbf{A} \qquad \text{(see Figure 9.19)}$$

Thus, with $I_B = 0$, the only collector and emitter current flowing is the leakage current I_{CEO}.

(b) Using Equation 9.13 for I_C, we have

$$I_C = \beta_{DC} I_B = 100 \times 20\ \mu A = \textbf{2000 } \mu\textbf{A}$$

and

$$I_E = (\beta_{DC} + 1)I_B = \textbf{2020 } \mu\textbf{A} \qquad ◀$$

▶ EXAMPLE 9.16 $I_E = (\beta+1)I_B$

Data on a certain transistor are shown in the following table:

I_B	I_C	I_E
0	20 μA	20 μA
100 μA	5.02 mA	5.12 mA

From this data determine (a) I_{CEO}, (b) β_{DC} and (c) I_C when $I_B = 50\ \mu A$.

(a) From the table, when $I_B = 0$, $I_C = 20\ \mu A$. Thus $I_{CEO} = 20\ \mu A$.

(b) Using Equation 9.14 for the second set of measurements,

$$\beta_{DC} = \frac{I_C}{I_B} = \frac{5.02\ \text{mA}}{100\ \mu A} \approx 50$$

(c) Using these values in Equation 9.13 for $I_B = 50\ \mu A$,

$$I_C \approx 50 \times 50\ \mu A = \textbf{2500 } \mu\textbf{A} = \textbf{2.5 mA} \qquad ◀$$

The equations presented here for the Com.-E configuration hold true for both the NPN and PNP junction transistors.

9.9. Common-Emitter Characteristic Curves

The equations given in Figure 9.20 do not completely describe the transistor operation in the Com.-E configuration. For one thing, β_{DC} is not constant, but varies with I_C and V_{CE}. β_{DC} *increases* as I_C *increases* up to a point and then decreases for further increases in I_C. It also increases as V_{CE} increases. Secondly, these equations assume that I_B is known. To find I_B, the variation of I_B with V_{BE}

must be available. For these reasons, the Com.-E characteristic curves are often utilized.

For the NPN transistor in the Com.-E configuration, typical input characteristics are shown in Figure 9.21. These are the *common-emitter input characteristics* and they relate the input current, I_B, to the voltage between emitter and base, V_{BE}, for different values of collector-to-emitter voltage, V_{CE}. The input curves are essentially P-N diode curves, as was the case on the Com.-B configuration; however, I_B is not the full forward-bias current I_E, but only the recombination current $I_E (1 - \alpha_{DC})$.

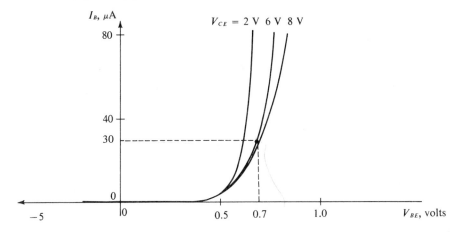

Figure 9.21.   COMMON-EMITTER INPUT CHARACTERISTICS FOR NPN TRAN-
SISTOR

In Figure 9.21, an increased value of V_{CE} causes the input curves to give a lower value of I_B for a given V_{BE}. This comes about due to the increasing of the space-charge region at the *C-B* junction as V_{CE} increases. The base width is effectively narrowed, with the result that fewer recombinations take place in the base, and I_B is lowered.

Notice that for negative values of V_{BE}, I_B is very small, since the *E-B* junction is reverse biased. This is the *cutoff* condition.

For the same NPN transistor in the Com.-E configuration, a set of output characteristics are shown in Figure 9.22. These are the *common-emitter output characteristics* and they relate the output current, I_C, to the voltage between collector and emitter, V_{CE}, for various values of input current, I_B.

A study of these output characteristics reveals several interesting points:

(a) For values of V_{CE} greater than a few tenths of a volt, I_C increases slowly as V_{CE} increases. The slope of these curves is somewhat greater than for the Com.-B output characteristics. The increase of I_C with V_{CE} for a constant I_B indicates that β_{DC} increases with V_{CE}. This region of the curves is the *active* region.

(b) For values of V_{CE} less than a few tenths of a volt, I_C decreases rapidly as V_{CE} decreases. This occurs as V_{CE} drops below the value of V_{BE}, thus causing V_{CB} to become negative ($V_{CB} = V_{CE} - V_{BE}$) and forward biasing the *C-B* junction. Since the *C-B* junction is now forward biased, the transistor is operating in the *saturation* region.

(c) The collector current is not zero when I_B is zero, but has a value of I_{CEO}. This is the current which flows when the transistor is *cut off* ($I_B = 0$).

(d) For a given base current in the active region, the collector current is β_{DC} times greater than that base current. Thus a small input current, I_B, produces a large output current, $I_C = \beta_{DC} \times I_B$.

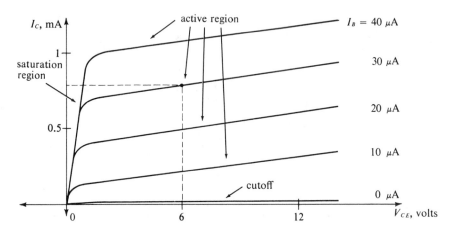

Figure 9.22.   COMMON-EMITTER OUTPUT CHARACTERISTICS FOR NPN TRANSISTOR

▶ EXAMPLE 9.17

Determine I_B and I_C for the transistor with the characteristics of Figures 9.21 and 9.22 when $V_{BE} = 0.7$ V and $V_{CE} = 6$ V. Calculate β_{DC} at this point.

(a) First, from the input characteristics, the value of I_B when $V_{BE} = 0.7$ V and $V_{CE} = 6$ V is obtained as 30 μA. Now, using the output characteristics, the value of I_C is obtained at $V_{CE} = 6$ V using the $I_B = 30$ μA curve. It is $I_C = 0.8$ mA.

(b) The value of β_{DC} at this point is

$$\beta_{DC} = \frac{0.8 \text{ mA}}{30 \text{ }\mu\text{A}} = 26.7 \qquad \blacktriangleleft$$

The Com.-E output characteristics, unlike the Com.-B output characteristics, can vary greatly from transistor to transistor. Whereas values of α_{DC} can range from 0.9 to 0.999, values of β_{DC} usually range from 10 to 1000. Even among transistors of the same type, the values of β_{DC} can differ by 200 to 300 percent.

The Com.-E output characteristics of a PNP transistor are essentially the same as those in Figures 9.21 and 9.22, except for the polarities of V_{BE} and V_{CE}.

Simply change the polarities on the V_{BE} and V_{CE} values given in these figures and the curves become PNP characteristics.

9.10.   Analyzing Common-Emitter Circuits: Load-Line Method

In analyzing a Com.-E circuit, the value of input current, I_B, must first be determined. In determining I_B, the diodelike characteristics of the *E-B* junction are again utilized. Consider the Com.-E circuit of Figure 9.23. The base current is supplied by the voltage V_{BB} through the series resistor R_B. We can write

$$V_{BB} = I_B \times R_B + V_{BE} \tag{9.18}$$

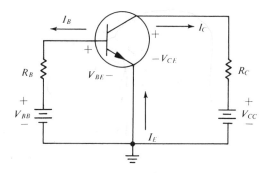

Figure 9.23.   TYPICAL DC COMMON-EMITTER CIRCUIT

The voltage drop V_{BE} across the *E-B* junction is typically 0.7 V for silicon and 0.3 V for germanium units. Using this fact makes the value of I_B readily obtainable:

$$I_B = \frac{V_{BB} - V_{BE}}{R_B} \tag{9.19}$$

▶ EXAMPLE 9.18

A silicon transistor is used in the circuit of Figure 9.24 with $V_{BB} = 10.7$ V and $R_B = 200$ kΩ. Determine I_B. Repeat for a germanium transistor.

(a) Using $V_{BE} = 0.7$ V in Equation 9.19, we have

$$I_B = \frac{10.7 \text{ V} - 0.7 \text{ V}}{200 \text{ k}\Omega} = 50 \text{ μA}$$

(b) Using $V_{BE} = 0.3$ V in Equation 9.19, we have

$$I_B = \frac{10.7 \text{ V} - 0.3 \text{ V}}{200 \text{ k}\Omega} = 52 \text{ μA} \qquad ◀$$

This procedure for calculating I_B will be the same for both the load-line and approximate methods of analysis.

The load-line method of analyzing the Com.-E circuit essentially parallels what was done in Section 9.6 for the Com.-B circuit. The procedure is to write the Kirchhoff's voltage equation for the output portion of the circuit and plot this equation (load line) on the output characteristic curves (I_C-V_{CE} axes). The intersection of this load line with the output curve corresponding to the value of I_B previously determined gives the circuit operating point. Referring to Figure 9.23, the Kirchhoff's voltage equation is

$$V_{CC} = I_C \times R_C + V_{CE} \qquad (9.20)$$

This equation relates I_C to V_{CE} and is called the load-line equation. To plot this load line, the easiest points to choose are ($I_C = 0$, $V_{CE} = V_{CC}$) and ($I_C = V_{CC}/R_C$, $V_{CE} = 0$).

▶ EXAMPLE 9.19

A transistor has the Com.-E output characteristics drawn in Figure 9.24. Find I_C and V_{CE} for $I_B = 50$ μA, with $V_{CC} = 12$ V and $R_C = 1$ kΩ. The load-line equation is

$$12 \text{ V} = I_C \times 1 \text{ kΩ} + V_{CE}$$

and is plotted as shown in the figure. The intersection of this load line and the $I_B = 50$ μA curve is thus $I_C = $ **5 mA** and $V_{CE} = $ **7 V**. The voltage across the 1-kΩ resistor is 5 mA × 1 kΩ = 5 V. These values can be checked by substituting into the load-line equation. In this case, the operating point lies in the *active* region of the transistor characteristics.   ◀

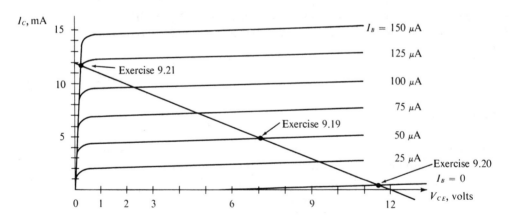

Figure 9.24.   LOAD-LINE METHOD

▶ EXAMPLE 9.20

Repeat Example 9.19 using $I_B = 0$.

The intersection of the load line and the $I_B = 0$ curve gives the new operating point as $I_C \approx 0$ and $V_{CE} = 12$ V. Since the transistor is in *cutoff*, only a small leakage current (I_{CEO}) flows and the voltage dropped across the series resistor is essentially zero. Thus $V_{CE} = V_{CC}$.   ◄

▶ EXAMPLE 9.21

Repeat Example 9.19 using $I_B = 125$ μA and $I_B = 150$ μA.
  (a) The intersection of the load line and the $I_B = 125$ μA curve gives the operating point as $I_C = 11.8$ mA and $V_{CE} = 0.2$ V. This operating point lies in the *saturation* region of the characteristics.
  (b) The operating point using $I_B = 150$ μA is the same as for $I_B = 125$ μA, since the different curves merge in the saturation region. Thus $I_C = 11.8$ mA and $V_{CE} = 0.2$ V is still the operating point. In fact, further increases in I_B will still result in the same I_C and V_{CE}. The extra base current does not produce any increase in collector current once the transistor is saturated. Notice that V_{CE} *is very small in the saturation region.*   ◄

The use of the load-line method for analyzing Com.-E circuits will gain importance in our later work on transistor amplifiers.

9.11. Analyzing Common-Emitter Circuits: Approximate Method

In using the approximate method to analyze a Com.-E circuit, it is necessary first to determine in what region of its operation the transistor is biased. When the *E-B* junction is reverse biased or unbiased ($I_B = 0$), the transistor is in *cutoff* and only a small leakage current flows in the collector. This current, I_{CEO}, is typically in the low microampere range at room temperature and normally can be neglected. In this case, the voltage drop across R_C (Figure 9.25) is essentially zero and V_{CE} is equal to the collector supply voltage, V_{CC}.

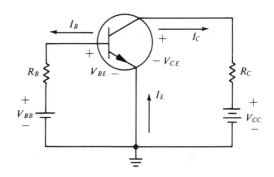

Figure 9.25.   DC COMMON-EMITTER CIRCUIT

▶ EXAMPLE 9.22

Using $V_{CC} = 12$ V and $R_C = 1$ kΩ in the circuit of Figure 9.25, determine I_C and V_{CE} when $I_B = 0$.

With $I_B = 0$, the collector current will be essentially zero. The voltage across the 1-kΩ resistor will be zero. Thus $V_{CE} = 12$ V. Compare these results with those of Example 9.20.   ◀

Transistor in active or saturation regions

When the E-B junction is forward biased, base current will flow, causing a corresponding collector current to flow. The value of I_B is determined in the manner outlined in the last section and used in Example 9.18. When the value of I_B is known, whether the transistor is in the *active* or *saturation* region remains to be determined. In the *active* region, the collector current is approximately equal to $\beta_{DC} \times I_B$ (Equation 9.13). The value of I_C thus obtained is then used to calculate V_{CE}. In the *saturation* region the approximate collector current is given by

$$I_C = I_{C(sat)} \approx \frac{V_{CC}}{R_C} \qquad (9.21)$$

since V_{CE} is very small. The procedure for finding the operating point when the E-B junction is forward biased is as follows:

(a) Calculate I_B using

$$I_B = \frac{V_{BB} - V_{BE}}{R_B}$$

where $V_{BE} = 0.7$ V for silicon and 0.3 V for germanium.

(b) Calculate $I_{C(sat)}$ using Equation 9.21.

(c) Calculate $\beta_{DC} \times I_B$.

(d) If the value of $\beta_{DC} \times I_B$ is less than $I_{C(sat)}$, then the transistor is *not* saturated. In this case, $I_C = \beta_{DC} \times I_B$ and V_{CE} can be calculated from

$$V_{CE} = V_{CC} - I_C \times R_C$$

(e) If the value of $\beta_{DC} \times I_B$ is greater than $I_{C(sat)}$, then the transistor *is* saturated. In this case, $I_C = I_{C(sat)}$ and $V_{CE} \approx 0$. The following examples illustrate this procedure.

▶ EXAMPLE 9.23

In the circuit of Figure 9.25, find I_C and V_{CE} for $I_B = 50$ μA if $V_{CC} = 12$ V and $R_C = 1$ kΩ. β_{DC} for this transistor is 100.

(a) Calculate $I_{C(sat)}$:

$$I_{C(sat)} \approx \frac{12 \text{ V}}{1 \text{ kΩ}} = 12 \text{ mA}$$

(b) Now calculate

$$\beta_{DC} \times I_B = 100 \times 50 \ \mu A = 5.0 \ mA$$

which is much less than $I_{C(sat)}$. Thus the transistor is not saturated, but is in the active region and $I_C = \beta_{DC} \times I_B = \textbf{5 mA}$. The voltage dropped across the 1-kΩ resistor is 5 mA $\times$ 1 kΩ = 5 V and $V_{CE} = 12$ V $- 5$ V $=$ **7.0 V.** Compare this to the results of Example 9.19.   ◄

▶ EXAMPLE 9.24

Repeat Example 9.23 for $I_B = 125 \ \mu A$ and $I_B = 150 \ \mu A$.

  (a) $I_{C(sat)}$ as previously calculated is 12 mA. $\beta_{DC} \times I_B = 100 \times 125 \ \mu A =$ 12.5 mA, which is greater than $I_{C(sat)}$. Thus the transistor is saturated and $I_C = I_{C(sat)} = 12$ mA; V_{CE} is approximately zero.

  (b) If $I_B = 125 \ \mu A$ is enough to saturate the transistor, increasing I_B to 150 μA will not cause I_C to increase above $I_{C(sat)} = 12$ mA. Compare these results to those of Example 9.21.   ◄

The value of base current which exactly causes the transistor to saturate is given the symbol $I_{B(sat)}$ and is obtained using

$$\beta_{DC} \times I_{B(sat)} = I_{C(sat)}$$

or

$$I_{B(sat)} = \frac{I_{C(sat)}}{\beta_{DC}} = \frac{V_{CC}}{\beta_{DC} \times R_C} \qquad (9.22)$$

Any value of base current equal to or greater than this value will produce saturation.

▶ EXAMPLE 9.25

In the circuit of Examples 9.23 and 9.24, calculate the value of base current just needed to cause saturation and determine the value of V_{BB} which will provide it. Use $R_B = 100 \ k\Omega$.

From Equation 9.22,

$$I_{B(sat)} = \frac{12 \ V}{100 \times 1 \ k\Omega} = \textbf{120 } \boldsymbol{\mu} \textbf{A}$$

The value of V_{BB} needed is given by

$$V_{BB} = I_{B(sat)} \times R_B + V_{BE}$$
$$= 120 \ \mu A \times 100 \ k\Omega + 0.7 \ V = \textbf{12.7 V} \qquad ◄$$

9.12.   The Common-Collector Configuration

The common-collector (Com.-C) configuration is much like the common-emitter configuration, except for the fact that the output is taken at the emitter rather than the collector. In either configuration the input is at the base. Figure 9.26 is a typical Com.-C circuit using an NPN transistor. Since I_B is the input

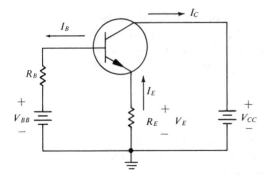

Figure 9.26.   TYPICAL COMMON-COLLECTOR CIRCUIT

current in the Com.-C configuration, it is desirable to relate I_C and I_E in terms of I_B. These equations, however, are the same as those given in Figure 9.20 for the Com.-E configuration. For convenience these equations are repeated in Figure 9.27.

$$I_E = (\beta_{DC} + 1)I_B + \boxed{I_{CEO}}$$
$$I_C = \beta_{DC}I_B + \boxed{I_{CEO}}$$
$$\beta_{DC} = I_C/I_B$$

Figure 9.27.   COMMON-COLLECTOR EQUATIONS

   In the Com.-C circuit, the DC output emitter current is approximately $(\beta_{DC} + 1)$ times the DC input base current. Thus, in this configuration, the DC *current gain* is $\beta_{DC} + 1$. This is the highest of all three configurations. Recall that for Com. B the DC current gain is α_{DC} and for Com. E it is β_{DC}.

9.13.   Analyzing Common-Collector Circuits

In the circuit of Figure 9.26, the voltage V_{BB} forward biases the E-B junction and supplies base current. For a given value of I_B, the value of I_E will be approximately $(\beta_{DC} + 1) \times I_B$. The voltage drop across resistor R_E is then given by

$$V_E = I_E \times R_E = (\beta_{DC} + 1)I_B \times R_E \tag{9.23}$$

The voltage drop across resistor R_B is simply $I_B \times R_B$. If we sum the voltages around the input loop, we have

$$V_{BB} = I_B \times R_B + [(\beta_{DC} + 1)I_B \times R_E] + V_{BE} \tag{9.24}$$

where V_{BE} is the forward voltage drop across the *E-B* junction. Rearranging Equation 9.24,

$$V_{BB} = I_B[R_B + (\beta_{DC} + 1)R_E] + V_{BE} \tag{9.25}$$

We can solve this equation for I_B, resulting in

$$I_B = \frac{V_{BB} - V_{BE}}{R_B + (\beta_{DC} + 1)R_E} \tag{9.26}$$

▶ EXAMPLE 9.26

Calculate I_B, I_E and V_E in the circuit of Figure 9.26 if $V_{BB} = 6$ V, $R_B = 10$ kΩ, $R_E = 1$ kΩ, $V_{CC} = 12$ V and the transistor has a β_{DC} of 50 and is silicon.

Since the transistor is silicon, V_{BE} will be approximately 0.7 V. Using Equation 9.26, then, we have

$$I_B = \frac{6\text{ V} - 0.7\text{ V}}{10\text{ k}\Omega + 51 \times 1\text{ k}\Omega} = \frac{5.3\text{ V}}{61\text{ k}\Omega} = \textbf{0.087 mA}$$

To calculate I_E, we have

$$I_E = 51 \times I_B = \textbf{4.44 mA}$$

The voltage V_E will then be

$$V_E = 4.44\text{ mA} \times 1\text{ k}\Omega = \textbf{4.44 V} \qquad ◀$$

If we examine Equation 9.26 for the base current in the Com.-C circuit, we see that, as far as the base current is concerned, the resistor R_E *appears* to have a resistance of $(\beta_{DC} + 1) \times R_E$. This is due to the current gain of the transistor. The V_{BB} voltage supply is essentially supplying current to a total resistance of $R_B + (\beta_{DC} + 1)R_E$. If β_{DC} is high, this total resistance will be high and the value of I_B which is drawn from V_{BB} will be small. The current in the emitter, on the other hand, will be large due to the transistor current gain. In our later study of transistor signal amplifiers, this characteristic of Com.-C circuits will be used to great advantage.

Cutoff operation

If the *E-B* junction is reverse biased or if the base lead is left open in the Com.-C circuit, the transistor will be in cutoff. In this situation (Figure 9.28), the only

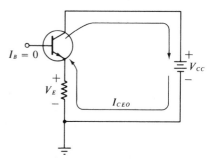

Figure 9.28.   CUTOFF CONDITION FOR THE COMMON-COLLECTOR CIRCUIT

current flowing is I_{CEO}, so that $I_E = I_{CEO}$ and $V_E = R_E \times I_{CEO}$, both equation values being close to zero.

Saturation

In the Com.-C configuration, saturation occurs when the base supply, V_{BB}, exceeds the collector supply. When this happens, both the *E-B* junction and the *C-B* junction become forward biased. As a result, further increases in V_{BB} produce a large increase in base current. This large increase in I_B is generally not desirable, since most transistor-base regions cannot safely handle large currents. For this reason, operation for the Com.-C circuit is usually limited to values of V_{BB} which are less than V_{CC}.

9.14.   Some Important Transistor Maximum Ratings

Transistor manufacturers supply data sheets which usually contain operating characteristics, parameter values and maximum ratings that apply to a particular transistor. The information will usually include a description of the device, mechanical data, maximum ratings, electrical characteristics, typical characteristic curves and parameter information.

Before inserting a transistor into a circuit design, it is necessary to avoid using it where its maximum ratings may be exceeded. The manufacturer establishes these ratings, which are based on the semiconductor material, the manufacturing process and the physical construction. If these ratings are exceeded, deterioration or destruction of a transistor will occur.

Table 9.2 lists typical maximum ratings for a junction transistor. The ratings are given at 25°C.

TABLE 9.2

ABSOLUTE MAXIMUM RATINGS (25°C)

BV_{EBO}:	emitter-base voltage: 6 V
BV_{CBO}:	collector-base voltage: 25 V
BV_{CEO}:	collector-emitter voltage: 20 V
$I_{C(max)}$:	collector current: 300 mA
$P_{D(max)}$:	total dissipation: 150 mW
$T_{J(max)}$:	junction temperature: 150°C

The first three ratings are maximum voltage ratings between the various transistor terminals. These ratings are a result of reverse breakdown occurring at the two junctions. BV_{EBO} is usually the lowest of these ratings due to the heavy doping of the emitter. BV_{CBO} is normally higher than BV_{CEO}, indicating that a larger collector supply voltage, V_{CC}, can be used in the Com.-B connection than in the Com.-E connection. $I_{C(max)}$ is the maximum allowable collector current and depends mainly on the physical size of the device. Transistors with collector current ratings up to the high ampere range are presently available. $P_{D(max)}$ is the limitation on the *total* power dissipation across both transistor junctions. In most cases, the power dissipated across the *E-B* junction is low, so that $P_{D(max)}$ essentially is the limit on *C-B* junction dissipation. This limit is determined by the transistor's maximum junction temperature, $T_{J(max)}$, and its ability to radiate heat away from its junction. This latter ability is characterized by its thermal resistance, θ_{JA}. The discussion in Chapter 5 on this topic is equally applicable here. Power transistors with $P_{D(max)}$ ratings in the hundreds-of-watts range are presently available.

▶ EXAMPLE 9.27

The circuits in Figure 9.29 use the transistor with the ratings given in Table 9.2. In each case, one of the ratings is exceeded. Determine which one for each circuit.

    (a) In circuit A, the *E-B* junction is reverse biased by 10 volts. This is above the maximum BV_{EBO} of 6 volts for this transistor.

    (b) In circuit B, a collector current of 50 mA is flowing. This is well below $I_{C(max)}$. However, the voltage from collector to emitter, V_{CE}, is 5 volts. Thus the power dissipated by the transistor is $V_{CE} \times I_C = 5 \text{ V} \times 50 \text{ mA} = 250 \text{ mW}$, which is greater than $P_{D(max)}$.

    (c) In circuit C, the supply voltage between collector and base is 30 volts. This is greater than BV_{CBO}, which is 25 volts.

    (d) In circuit D, the supply voltage between emitter and collector is 25 volts. This is greater than BV_{CEO}, which is 20 volts.   ◀

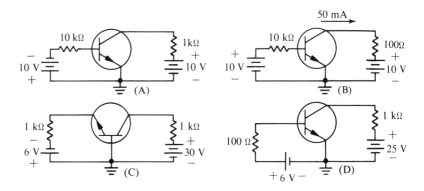

Figure 9.29.

9.15. Temperature Effects

The transistor parameters which are most affected by temperature are Com.-E DC current gain, β_{DC}, leakage currents I_{CBO} and I_{CEO}, and base-emitter forward voltage, V_{BE}. All vary with temperature and must be accounted for in most circuit designs and especially in amplifiers.

Figure 9.30 shows typical variation of β_{DC} with temperature, using 25°C as the reference temperature. The value of β_{DC} increases rapidly with temperature above 25°C and decreases less rapidly with decreases in temperature below 25°C. For example, the graph indicates that at 0°C the value of β_{DC} is down by approximately 15 percent from its value at 25°C, and at 50°C, β_{DC} is up by 50 percent. Most germanium and silicon transistors show approximately this variation of β_{DC} with temperature and the information is usually supplied by the manufacturer. The Com.-B DC current gain, α_{DC}, shows much less of a variation with temperature, rarely more than a few percentage points over the useful temperature range.

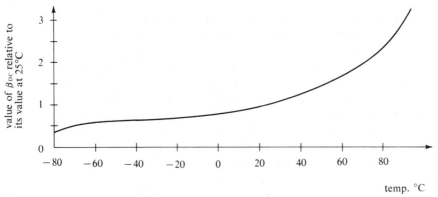

Figure 9.30.   VARIATION OF β_{DC} WITH TEMPERATURE

The transistor leakage currents, I_{CBO} and I_{CEO}, both increase with temperature. The value of I_{CBO} approximately doubles for every 10°C increase in temperature. Since the value of I_{CEO} depends on I_{CBO} according to

$$I_{CEO} = (\beta_{DC} + 1) \times I_{CBO} \qquad (9.16)$$

it will more than double every 10°C. This temperature effect is not as critical in silicon units as it is in germanium since the leakages are much smaller to begin with in silicon (I_{CBO} is typically 0.01 μA).

The variation of base-emitter forward voltage, V_{BE}, is typically 2 mV per °C, with V_{BE} decreasing as temperature increases. For example, a silicon transistor with a V_{BE} of 0.7 V at 25°C would typically have a V_{BE} of 0.6 V at 75°C.

The above effects all serve to increase the output current in transistor circuits. This can be seen in the following example:

▶ EXAMPLE 9.28

Calculate I_B and I_C in the circuit of Figure 9.31 at 25°C and at 55°C if the transistor is silicon and has $\beta_{DC} = 100$ and $I_{CBO} = 0.1$ μA at 25°C.

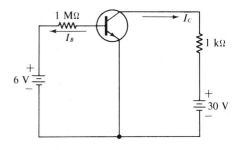

Figure 9.31.   EXAMPLE 9.28

(a) At 25°C:

$$I_B = \frac{6\text{ V} - 0.7\text{ V}}{1\text{ M}\Omega} = 5.3\ \mu\text{A}$$

$$I_C = \beta_{DC} \times I_B + I_{CEO} \qquad\qquad (9.17)$$
$$= 100 \times 5.3\ \mu\text{A} + 101 \times 0.1\ \mu\text{A}$$
$$= 530\ \mu\text{A} + 10.1\ \mu\text{A} = 540.1\ \mu\text{A}$$

(b) At 55°C:

$$V_{BE} = 0.7\text{ V} - (30) \times (2\text{ mV})$$
$$= 0.64\text{ V}$$
$$I_{CBO} = 0.1\ \mu\text{A doubled 3 times}$$
$$= 0.8\ \mu\text{A}$$
$$\beta_{DC} = 1.45 \times 100 = 145 \quad \text{(from Figure 9.30)}$$

Thus

$$I_B = \frac{6\text{ V} - 0.64\text{ V}}{1\text{ M}\Omega} = 5.36\ \mu\text{A}$$

$$I_C = 145 \times 5.36\ \mu\text{A} + 146 \times 0.8\ \mu\text{A}$$
$$= 777\ \mu\text{A} + 116.8\ \mu\text{A} = 893.8\ \mu\text{A} \qquad \blacktriangleleft$$

This increase in collector current with temperature is very important in the design of transistor amplifiers, especially in the Com.-E circuits. We shall see that, with proper circuit design procedures, the effect of temperature on circuit operation can be minimized.

9.16.   The Transistor as a Switch

An important application of the transistor is in the ever-increasing area of switching circuits. This broad category of circuits includes applications in such fields as digital computers, control systems, counting and timing systems, data-processing systems, digital instrumentation, pulse communications, radar, telemetry and television. When it is used as a switch, the transistor is operated in one of two states: conducting or nonconducting. The nonconducting state is *cutoff*, while the conducting state may be either in the *active* region or in the *saturation* region. The following discussion will concentrate on these last two states of operation and, more importantly, on the transition from one state to the other. In the process, many switching parameters of the transistor will be introduced as means of evaluating a switching transistor's performance.

Mechanical switch

In Figure 9.32, a pair of mechanical contacts form a switch. When the switch is open (as shown), the current through the resistor is equal to zero and $V_R = 0$. When the switch is closed, the current becomes E/R and $V_R = E$. These results assume that the open switch resistance R_{OFF}, between points X and Y is infinite and the closed switch resistance, R_{ON}, between X and Y is zero. This would be true only for an ideal switch. That is, an ideal switch would have an R_{OFF}/R_{ON} ratio of infinity. Mechanical switches come close to providing this ideal ratio, but suffer from shortcomings such as slow switching speed, low reliability and contact bounce. Transistor switches cannot provide such a high R_{OFF}/R_{ON} ratio, but they can, to a great extent, overcome the disadvantages of the mechanical switch.

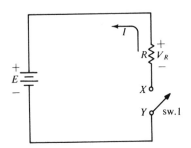

Figure 9.32.   MECHANICAL SWITCH

The voltage (V_R) in the circuit has two discrete levels or states, zero and E volts. The current in the circuit also has two states, zero and E/R. In digital computer circuitry, these states are used to identify bits of information and the *transition time* (see Figure 9.33) required to switch from one state to the other determines the speed at which computer operations can be performed. Thus, in our discussion of transistor switches, there are two areas to be investigated: the *steady-state* operation, which concerns operations in the "on" and "off" states, and the *transient* operation, which is present during the transition from one state to the other.

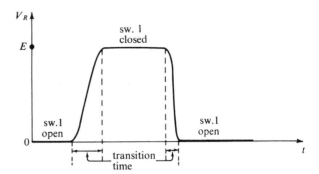

Figure 9.33.   TRANSITION TIME BETWEEN STATES

Transistor switch

In the circuit of Figure 9.32, a mechanical switch was used to control the current through the circuit. Figure 9.34 shows a corresponding arrangement in which the mechanical switch has been replaced by a transistor. The transistor is connected in the Com.-E configuration which is used in practically all switching applications, since its high current gain allows large collector currents to be switched by means of a relatively small base current. As we learned earlier, the transistor can operate in either the cutoff, active or saturation mode, depending on the status of the input base current. In switching circuits, the cutoff condition is referred to as the "off" condition and the saturated condition as the "on" condition. The *active* condition retains its name. With zero base current, or with the *E-B* junction reverse biased, the transistor is turned "off" and only a small leakage current will flow from emitter to collector. With a forward-biased *E-B* junction, a base current will flow, resulting in a substantial collector current. The transistor is then *active* and will remain that way unless I_B is large enough to cause the collector current to saturate and enter the "on" state. The amount of base current, then, controls the degree to which the transistor conducts collector current. Thus the value of I_B controls the resistance between the collector and emitter (the terminals of the switch).

If $I_B = 0$, the transistor is "off" and its resistance between collector and emitter will be very high; typically, R_{OFF} will be 100–1000 megohms for a sili-

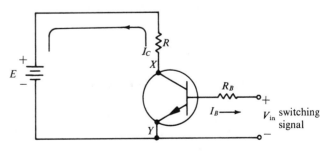

Figure 9.34.   THE TRANSISTOR USED AS A SWITCH

con switching transistor. If I_B is sufficient to cause saturation, the transistor will be at its maximum current and, therefore, minimum resistance; typically, R_{ON} will be 0.1–10 Ω. If I_B produces operation in the "active" region, then the transistor resistance could have any value between R_{OFF} and R_{ON}.

Figure 9.35 summarizes this behavior. Note that in the "off" state the transistor resistance can be considered essentially infinite (open circuit), while in the "on" state it can be considered almost a short circuit.

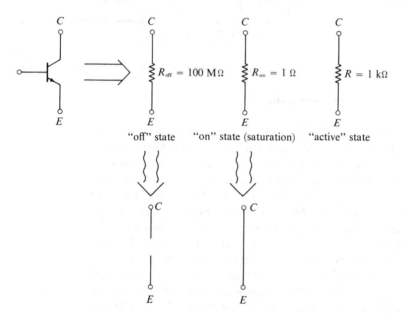

Figure 9.35.   VARIOUS TRANSISTOR STATES

It was stated previously that the conducting state of a transistor switch is sometimes chosen in the "active" region (the nonconducting state is always cut off or "off"). In such cases, the transistor is said to be operating as a *nonsaturated switch* as opposed to a *saturated switch*, when the conducting state is in the "on" region. Circuits utilizing the transistor as a nonsaturated switch typically have faster switching speeds. However, they dissipate more power, are usually more difficult to design and are less immune to noise than circuits using the transistor as a saturated switch. Some of these differences will be discussed in the following sections.

9.17.   Transient Operation

The transition time between states of a transistor switch is the principal factor limiting the maximum frequency at which switching can occur. As such, it is important to understand the response of a transistor to an input pulse. Let us consider the response of the transistor switch to a pulse of input voltage, V_{in}, as shown in Figure 9.36. At $t = 0$, V_{in} is negative, reverse biasing the *E-B* junction, and the transistor is "off," corresponding to the condition in which $I_B \approx 0$ and $I_C \approx 0$. The value of V_{BE} is simply -10 V of reverse bias. At $t = t_1$, V_{in} is suddenly increased to $+10$ V. If we assume R_B is 47 kΩ, this value of V_{in} will eventually provide a base current of 200 μA to cause the transistor to turn "on." However, the transistor will not respond immediately. As shown in the figure, the base-emitter voltage increases *gradually* toward its forward-bias value after V_{in} has switched. As a result, the collector current will not begin increasing toward its saturation value of 10 mA until V_{BE} becomes forward biased. Even then, I_C will increase only gradually toward 10 mA. The time interval from $t = t_1$, when V_{in} switches, to the time when the collector current rises to 10 percent of its final value is called *delay time*, t_d, as indicated in the figure. The time it takes the collector current to rise from 10 percent to 90 percent of its final value is called *rise time*, t_r. The sum of delay time and rise time is the amount of time it takes the transistor to turn "on" and is called *turn-on time*, t_{on}. That is,

$$t_{on} = t_d + t_r \qquad\qquad (9.27)$$

The turn-on time depends on many factors, including transistor parameters and external circuit components. The principal factors determining t_{on} are:

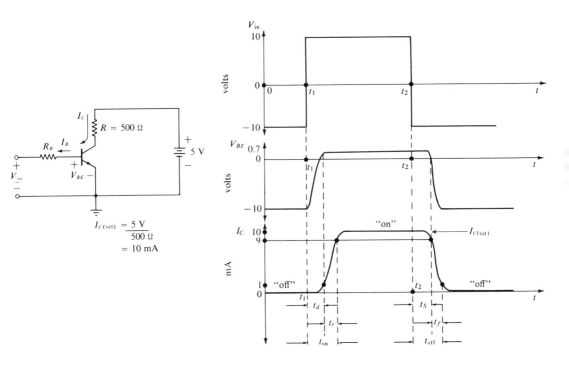

Figure 9.36.   TRANSISTOR IN RESPONSE TO A FAST-SWITCHING INPUT PULSE

(1) the transistor junction capacitances which prevent the transistor voltages from changing instantaneously (these junction capacitances must charge and discharge gradually through the circuit resistances); and (2) the time required for the emitter current to diffuse across the base region into the collector region once forward bias is established.

Values of t_{on} ranging from 75–300 nanoseconds (10^{-9} seconds) are typical and can vary even for the same transistor under different circuit conditions.

Once the collector current reaches its steady-state value of 10 mA, it will remain there as long as V_{in} stays at 10 V. At $t = t_2$, when V_{in} suddenly returns to -10 V, the transistor does not respond immediately. As can be seen in Figure 9.36, a certain amount of time elapses before I_C even begins to decrease. The *storage time*, t_s, is defined as the time it takes for the collector current to drop to 90 percent of its "on" value once the input has switched. The *fall time*, t_f, is defined as the time it takes the collector current to drop from 90 percent to 10 percent. The total *turn-off time*, t_{off}, is the sum of storage time and fall time. That is,

$$t_{off} = t_s + t_f \tag{9.28}$$

The fall time is determined mainly by the same factors which were mentioned in conjunction with turn-on time. The storage time, however, deserves special consideration. It comes about because of the fact that when the transistor is in the "on" state, before switching has occurred, it is saturated. In saturation, the C-B junction is forward biased and the collector will inject carriers into the base region in the same way that the emitter does. These charges injected into the base recombine there and are called *stored charges*. When the transistor is being turned "off," these stored charges must return to the collector region. This process takes time and the movement of these charges maintains the flow of collector current. After all of the stored charge has been removed from the base, the collector current will begin to decrease. The storage time depends on the transistor doping profile but is also dependent on how saturated the transistor is in the "on" state. That is, if I_B is greater than the minimum value needed [$I_{B(sat)}$] to cause I_C to saturate, the transistor will be essentially *overdriven*, or *oversaturated*, and the stored charge will be greater, resulting in a longer storage time. As I_B is made larger, the value of t_s and, therefore, t_{off} will increase.

Typical values for t_{off} are in the range of 100–300 nanoseconds and, again, depend not only on the transistor but also on other circuit parameters. The total circuit switching time, T_t, is defined as:

$$T_t = t_{on} + t_{off} \tag{9.29}$$

and is a figure of merit. Its value essentially limits how many times per second the circuit can be pulsed. The smaller the value of T_t, the higher the frequency of switching which the circuit can follow. For example, if the value of T_t is 1 μs, then we cannot expect the circuit to be able to switch more often than once every 1 μs, or one million times per second. Mathematically, this can be written as

$$f_{\max} = \frac{1}{T_t} \qquad (9.30)$$

where $f_{\max}$ is the maximum circuit switching frequency.

The previous description of switching action concerned a saturated switch. If, instead, a nonsaturated switching circuit were considered, such as one which switched from cutoff to somewhere in the active region, the description would be slightly altered. The parameters t_d, t_r and t_f would be essentially unchanged but t_s would be very small since no saturation occurs. As such, the turn-off time and thus T_t would be shorter for unsaturated switches. Thus unsaturated switching circuits can operate at higher switching frequencies, which is the main reason for their use.

9.18.   Switching Circuit Applications

A transistor switch can be turned "off" and "on" by controlling the base current. Normally, it takes a relatively small value of I_B to produce saturation and the resulting large value of collector current. This collector current can be made to flow through a load which requires a relatively large current. Thus a small swing in the I_B input can cause the load current to switch from essentially zero to $I_{C(\text{sat})}$.

Figure 9.37 shows a transistor switch used to control the current through a lamp, L1, which is used as the collector load. The input signal source is normally at zero volts; it steps up to 6 V when the lamp is to be turned on.

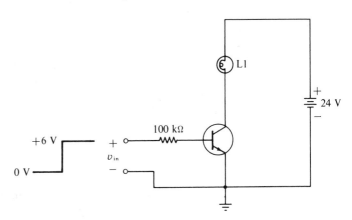

Figure 9.37.

▶ EXAMPLE 9.29

The lamp in Figure 9.37 is rated at 20 mA @ 24 V. Determine: (a) the minimum value of β_{DC} for the transistor to ensure saturation when $V_{in} = +6$ V; (b) the power delivered to the lamp load and the power drawn from the input signal when the transistor switch is "on."

(a) When the transistor is "on," the entire collector supply of 24 V will be across the lamp, thereby producing 20 mA of current. Thus $I_{C(sat)} = 20$ mA. The base current produced when $V_{IN} = +6$ V is

$$I_B = \frac{6 \text{ V} - 0.7 \text{ V}}{100 \text{ k}\Omega} = 53 \ \mu A$$

assuming a silicon transistor. The transistor Com.-E current gain, β_{DC}, has to be large enough to produce $I_C = I_{C(sat)} = 20$ mA for $I_B = 53 \ \mu A$. Therefore,

$$\beta_{DC(min)} = \frac{20 \text{ mA}}{53 \ \mu A}$$
$$= 377$$

Any value of β_{DC} larger than this value will ensure that the transistor is fully "on" when $V_{IN} = +6$ V.

(b) The lamp power is the product of its voltage and current.

$$P_{lamp} = 24 \text{ V} \times 20 \text{ mA}$$
$$= 480 \text{ mW}$$

The power drawn from the input signal is the product of V_{IN} and I_B, the current drawn from V_{IN}.

$$P_{IN} = 6 \text{ V} \times 53 \ \mu A$$
$$= 318 \ \mu W = 0.318 \text{ mW}$$

As these calculations show, the power drawn from the input is much smaller than the power switched to the load. The current gain (β_{DC}) of the transistor switch allows a small input power to control a large amount of load power. It should be noted that the load power is actually drawn from the 24-V source; the input signal simply controls whether or not this power will be delivered to the load by switching the transistor "on" or "off." ◀

As the preceding example illustrates, the current gain (β_{DC}) of the transistor allows a very low current input to switch a much larger load current. It may be recalled from our discussion of photodiodes that these devices produce relatively low values of current and, therefore, cannot be used to directly control a relay. Figure 9.38 shows a circuit where the current gain of a transistor is used to amplify the photodiode current to the level necessary to control a relay.

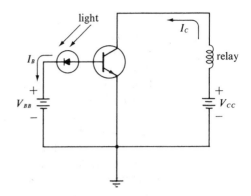

Figure 9.38.   TRANSISTOR SWITCH USED TO AMPLIFY PHOTODIODE CURRENT

The photodiode is connected in series with the base of the transistor (in place of R_B). Note that the direction of the photodiode is such that it is reverse biased by the V_{BB} source, since that is its normal state of operation. The application of light to the photodiode causes its current to increase. This current serves as the base current for the transistor, which then amplifies it by β_{DC} to produce a much larger collector current for the relay load.

9.19.   The Phototransistor

A *phototransistor* operates much like a photodiode, except that it provides built-in current amplification. Figure 9.39A shows the symbols for phototransistors. Light is focused on the *E-B* junction and causes a base current to flow (internally). The collector current is greater than the light-produced base current by a factor of β_{DC}. Note that only collector and emitter leads are used.

It should be obvious that the phototransistor can produce a greater current than a photodiode for the same illumination. Typically, the phototransistor has

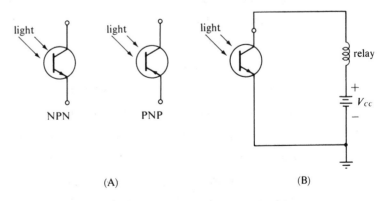

(A)   (B)

Figure 9.39.   (A) PHOTOTRANSISTOR SYMBOLS; (B) PHOTOTRANSISTOR CIRCUIT current gains between 100 and 1000. Figure 9.39B shows a typical phototransistor circuit. This circuit performs the same function as the circuit in Figure 9.38 but requires fewer components and fewer connections. However, a phototransistor has a somewhat slower response than a photodiode (1 μs versus 0.1 μs). Thus, for high-speed applications, a photodiode combined with a high-speed switching transistor is more useful.

In certain phototransistors the base lead is used to allow external connections to partially turn "on" the transistor. In other words, a small base current is applied to the phototransistor in the conventional manner (V_{BB} and R_B). This serves to keep the transistor conducting lightly. Now any light applied to the *E-B* junction will increase I_B and move the transistor toward saturation. This method is used in applications where very low light levels are encountered.

GLOSSARY

Emitter: most heavily doped region of the junction transistor.

Collector: region of the junction transistor which is doped with the same impurity type as the emitter.

Base: most lightly doped region of the transistor. The type of its doping impurity is opposite to that of the emitter and collector.

Emitter-base junction: P-N junction formed by emitter and base regions.

Collector-base junction: P-N junction formed by collector and base regions.

Active region: transistor operation with *E-R* forward biased and *C-B* reverse biased.

Cutoff region: transistor operation with *E-B* reverse biased (or open) and *C-B* reverse biased.

Saturation: transistor operation with both junctions forward biased.

I_{CBO}:   collector leakage current which flows when $I_E = 0$ in the Com.-B configuration.

α_{DC}:   DC Com.-B current gain.

I_{CEO}:   collector leakage current which flows when $I_B = 0$ in the Com.-E configuration.

β_{DC}:   DC Com.-E current gain.

$I_{C(sat)}$:   value of collector current at saturation.

$I_{B(sat)}$:   value of base current needed to produce $I_{C(sat)}$.

$BV_{EBO}, BV_{CBO}, BV_{CEO}$:   maximum allowable transistor voltages.

Transition time: time required for a switch to switch from one state to another.

Steady-state operation: when the switch is resting in one of its states.

Transient operation: when the switch is in the process of changing states.

"On" state: transistor saturated.

"Off" state: transistor cut off.

Saturated switch: transistor switching between "off" and "on" states.

Nonsaturated switch: transistor switching between "off" and "active" states.

Delay time, t_d: time required for collector current to rise to 10 percent of its final value.

Rise time, t_r: time required for collector current to rise from 10 percent to 90 percent of its final value.

Turn-on time, t_{on}:   sum of delay time and rise time.

Storage time, t_s: time required for collector current to drop to 90 percent of its "on" value.

Fall time, t_f: time required for collector current to drop from 90 percent to 10 percent of its "on" value.

Turn-off time, t_{off}:   sum of storage time and fall time.

Total switching time, T_t: sum of t_{on} and t_{off}.

PROBLEMS

9.1   Sketch the structure of an NPN junction transistor and label the emitter, base and collector regions. Also label the *E-B* and *C-B* junctions.

9.2   Repeat Problem 9.1 for a PNP junction transistor.

9.3   Explain the function of the emitter in the operation of a junction transistor.

9.4   What is done to the base region of a transistor to improve its operation?

9.5   Why is the *C-B* junction of a transistor reverse biased for active-region operation?

9.6   What causes collector current to flow when there is no emitter current? What is this collector current called?

9.7   Why is base current in a transistor usually much smaller than I_C or I_E in active operation?

9.8   Why is collector current in a transistor usually about the same as emitter current in active operation?

9.9   What happens to I_C as the value of reverse bias on the *C-B* junction increases? Explain.

9.10   Indicate in which region (active, saturation or cutoff) the transistors in Figure 9.40 are biased.

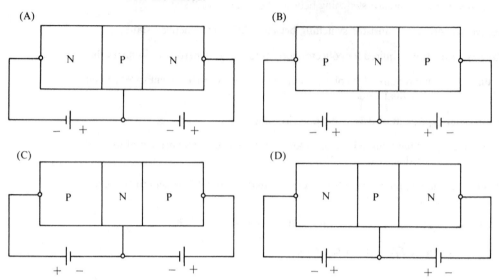

Figure 9.40.

9.11   Fill in the blanks in the following data table taken for a certain transistor:

I_B (mA)	I_C (mA)	I_E (mA)
1.0	19.0	——
—	18.0	18.95
—	+0.01	0
1.5	——	30

9.12   From the table for Problem 9.11, determine I_{CBO} for this transistor. Calculate α_{DC} at $I_C = 19$ mA.

9.13   A certain transistor has $\alpha_{DC} = 0.99$ and $I_{CBO} = 10$ μA. Calculate I_C and I_B when $I_E = 10$ mA. Repeat for $I_E = 20$ mA. Assume operation in the active region.

9.14   Indicate which of the following statements pertain to NPN transistors and which pertain to PNP transistors:
(a) The emitter injects *holes* into the base region.
(b) When biased in the active region, current flows *into* the emitter terminal.
(c) The collector is biased *positively* relative to the base for active operation.
(d) The principal current carriers are electrons.
(e) The *E-B* junction is forward biased for active operation.

9.15   What happens to the ability of the collector region to collect injected carriers in the saturated condition?

9.16 How much emitter current flows in the cutoff condition?

9.17 Draw a PNP transistor in the Com.-B configuration biased for operation in the *active* region.

9.18 What is considered the input terminal and what is considered the output terminal in the Com.-B configuration?

9.19 Sketch typical Com.-B *input* characteristic curves for a PNP transistor. Label all variables.

9.20 Sketch typical Com.-B *output* characteristic curves for a PNP transistor. Label all variables and indicate active, cutoff and saturation regions.

9.21 Does the value of α_{DC} increase or decrease with V_{CB}?

9.22 Using the characteristic curves of Figures 9.12 and 9.13, determine the approximate values of I_C and I_E when $V_{BE} = 0.7$ V and $V_{CB} = 10$ V.

9.23 Determine I_E in the circuits in Figure 9.41.

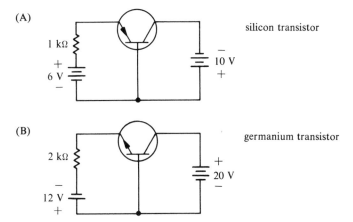

(A)

1 kΩ

+
6 V
−

10 V

silicon transistor

(B)

2 kΩ

12 V

20 V

germanium transistor

Figure 9.41

9.24 The germanium transistor in the circuit in Figure 9.42 has the characteristic curves of Figure 9.15.
(a) Determine I_E, I_C and V_{CB} using the load-line method when $R_C = 1$ kΩ.
(b) Determine what value of R_C will just cause saturation.

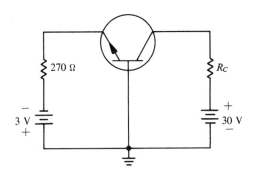

270 Ω

R_C

3 V

30 V

Figure 9.42.

9.25 The circuits in Figure 9.43 use a silicon transistor. Using the approximate method, determine I_E, I_C and V_{CB} if $\alpha_{DC} = 0.95$.

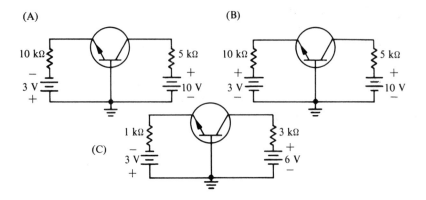

(A)   (B)

(C)

Figure 9.43.

9.26 In the circuit in Figure 9.44 which uses a germanium transistor, determine what value of V_{EE} will cause saturation.

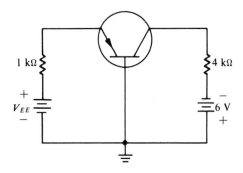

Figure 9.44.

9.27 A certain transistor circuit has $I_{C(sat)} = 10$ mA. If I_E is increased to 15 mA, what will be the values of I_C and I_B?

9.28 Draw a PNP transistor in the Com.-E configuration biased for operation in the *active* region.

9.29 What are the input and output terminals in the Com.-E configuration?

9.30 Sketch typical Com.-E *input* characteristics for a PNP transistor. Label all variables.

9.31 Sketch typical Com.-E *output* characteristics for a PNP transistor. Label all variables and indicate active, cutoff and saturation regions.

9.32 Does the value of β_{DC} increase or decrease with V_{CE}?

9.33 Determine I_B in the circuit in Figure 9.45, which utilizes a silicon transistor.

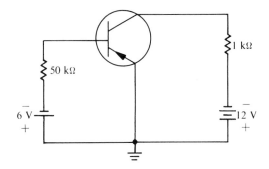

Figure 9.45.

9.34   The silicon transistor in Figure 9.46 has the characteristic curves of Figure 9.24.
(a) Determine I_B, I_C and V_{CE} using the load-line method when $R_C = 600\ \Omega$.
(b) Determine what value of R_C will cause saturation.

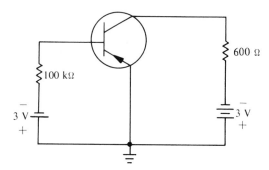

Figure 9.46.

9.35   The circuits in Figure 9.47 use a germanium transistor. Determine I_B, I_C and
V_{CE} in each using the approximate method. Use $\beta_{DC} = 50$.

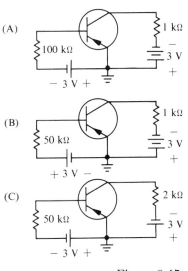

Figure 9.47.

9.36 In the circuit in Figure 9.48, which uses a silicon transistor, determine what value of V_{BB} will cause saturation.

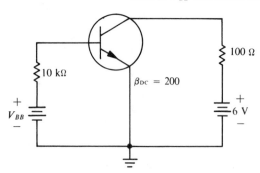

Figure 9.48.

9.37 In Figure 9.48, if $V_{BB} = 5$ V, what value of R_C will just cause saturation?

9.38 Calculate I_B, I_E and V_E in the circuit in Figure 9.49.

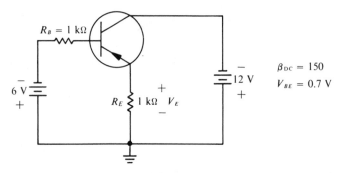

Figure 9.49.

9.39 In the circuit of Figure 9.49, calculate I_B, I_E and V_E when the 1-kΩ base resistor is reduced to zero. The results should bear out the fact that the output current, I_E, and voltage, V_E, do not depend on the base resistor when it is small compared to $\beta_{DC} \times R_E$. That is, the value of I_B and the voltage across R_B are negligible when $R_B \ll \beta_{DC} R_E$.

9.40 Using the fact brought out in Problem 9.39, find I_E and V_E in Figure 9.50.

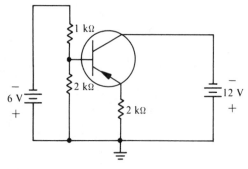

Figure 9.50.

9.41 A certain transistor has the following ratings at 25°C:

$$BV_{EBO}: \quad 6 \text{ V}$$
$$BV_{CBO}: \quad 40 \text{ V}$$
$$BV_{CEO}: \quad 30 \text{ V}$$
$$I_{C(max)}: \quad 1 \text{ A}$$
$$P_{D(max)}: \quad 1 \text{ W}$$
$$T_{J(max)}: \quad 150°\text{C}$$

If the transistor is used in the circuit in Figure 9.51:

(a) What is the largest value of V_{CC} which should be used?

(b) If $V_{CC} = 20$ V is used, what is the smallest R_C which could be used if the transistor is to be saturated?

(c) If $I_E = 0.5$ A, what minimum value of R_C is needed to insure that $P_{D(max)}$ is not exceeded? (Use $V_{CC} = 20$ V.)

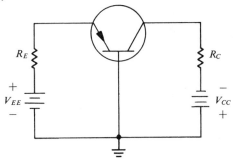

Figure 9.51

9.42 A transistor has $P_{D(max)} = 1$ W at 25°C and $\theta_{JA} = 0.1°\text{C/mW}$. Calculate $P_{D(max)}$ at 55°C.

9.43 The transistor in the circuit in Figure 9.52 has the following parameters at 25°C:

$$\beta_{DC} = 200$$
$$I_{CEO} = 1 \text{ } \mu\text{A}$$
$$V_{BE} = 0.25 \text{ V}$$

Calculate I_C and V_{CE} at 25°C and 75°C.

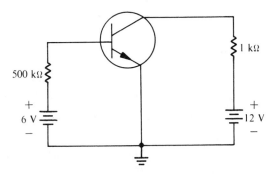

Figure 9.52.

9.44   In the photocell-relay circuit studied in Chapter 8, the same current flowed
through both the photocell and relay. This circuit would be useless for driving
high-current relays since photocells are generally low-current devices. If we use
a transistor, we can still use a low-current photocell to control a high-current
relay. Figure 9.53 is such a circuit. It uses an RCA–4413 photocell (Appendix
II).

(a) If the transistor has $\beta_{DC} = 100$ and is made of germanium, determine the
relay current when the cell is in the dark and when it is illuminated with
10 fc.

(b) Modify the circuit so that the relay pulls in when the cell is in the dark
and drops out when it is illuminated.

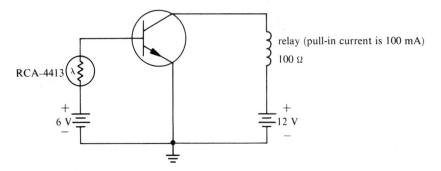

Figure 9.53.

9.45   What are the comparative advantages and disadvantages of mechanical and
transistor switches?

9.46   In Figure 9.54, what are the two states of the output voltage if the transistor
is operated as a saturated switch?

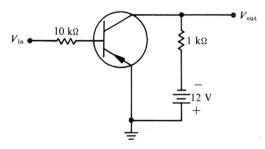

Figure 9.54.

9.47   If the transistor in the circuit of Figure 9.54 has an $I_{CBO} = 0.1$ μA and
$V_{CE(sat)} = 0.1$ V, determine its R_{OFF}/R_{ON} ratio.

9.48   Sketch the response of I_C as V_{in} switches instantaneously from $+10$ V to
-10 V in the circuit of Figure 9.54. Label t_d, t_r and t_{on}.

9.49   What causes the delay in the collector current as the transistor switches from
"off" to "on"?

9.50 Sketch the response of I_C as V_{in} switches instantaneously from -10 V to $+10$ V in the circuit of Figure 9.54. Label t_s, t_f and t_{off}.

9.51 What causes the turn-off delay in the collector current?

9.52 Why would the turn-off delay increase if V_{in} were switching from -15 V to $+10$ V instead of from -10 V to $+10$ V?

9.53 Why is t_{off} less in an unsaturated circuit?

9.54 In the circuit of Figure 9.55, what is the minimum value of β_{DC} that the silicon transistor needs to ensure that it will saturate when V_{in} switches from 0 V to -6 V?

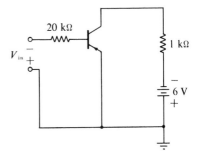

Figure 9.55.

9.55 Calculate the power switched to the 1-kΩ load in Figure 9.55. Also, calculate the power input from V_{IN}.

9.56 Refer to the circuit of Figure 9.38. The transistor has $\beta_{DC} = 150$. The relay has 20 kΩ resistance and a pull-in current of 0.3 mA. Using $V_{CC} = 6$ V, determine the amount of illumination required to pull in the relay. Refer to the photodiode characteristics in Chapter 8.

9.57 Refer again to the circuit in Figure 9.38. Which of the following changes would cause the relay to pull in at a lower level of illumination?
(a) Increase in V_{CC}.
(b) Increase in V_{BB}.
(c) Increase in β_{DC}.
(d) Increase in temperature of the circuit.

REFERENCES

Foster, J. F., *Semiconductors, Diodes and Transistors*, Vol. 3. Beaverton, Oregon: Programmed Instruction Group, Tetronix, Inc., 1964.

Malvino, A. P., *Transistor Circuit Approximations*, 2nd ed. New York: McGraw-Hill Book Company, 1973.

Oppenheimer, S. L., *Semiconductor Logic and Switching Circuits*, 2nd ed. Columbus, Ohio: Charles E. Merrill Publishing Co., 1973.

Tocci, R. J., *Fundamentals of Pulse and Digital Circuits*. Columbus, Ohio: Charles E. Merrill Publishing Co., 1971.

10

Amplifier Principles

10.1. Introduction

We understand amplifiers when we can predict their performance by careful analysis of the circuit. As such, we must be prepared to calculate input and output impedances and amplifier gains. This chapter introduces many of the fundamental relationships that characterize amplifier circuits. The material in this chapter is general, in that no particular amplifying device is mentioned, and applies to vacuum-tube amplifiers as well as the junction transistor and field-effect transistor amplifiers which will be covered in later chapters.

10.2. Reproduction and Amplification

Figure 10.1 represents a general amplifier circuit. The amplifier input voltage or current (or both) can be considered the *stimulus*. The amplifier output current or output voltage (or both) can be considered the amplifier's *response* to the input stimulus. When input voltage or current variations, called the *input signal*, are applied to the amplifier, the amplifier circuit acts on them to produce output current and voltage variations, called the *output signal*. If the output

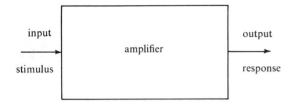

Figure 10.1.   GENERAL AMPLIFIER

signal possesses the same variations as the input signal, the amplifier has dupli-
cated the input variations in its output; this is called *reproduction*. If the output
signal variations are made larger than the input variations which produced
them, the process is called *amplification*.

Figure 10.2A illustrates the process of reproduction, while 10.2B illustrates
the process of amplification. In part A, the output and input signals have exactly the same shape. In part B, the output signal is much larger than the input
signal, but is not of the same shape. Reproduction is a requirement in applications where the shape of the input has to be preserved. In many applications,
such as in the audio amplifier stage of radio receivers, faithful reproduction and
amplification are both desired. Figure 10.2C illustrates this. There are a considerable number of applications in which only amplification is required. This
is usually the case in photodetection circuits, for example, where a small signal
from a photocell is amplified to a level necessary to drive control circuitry.

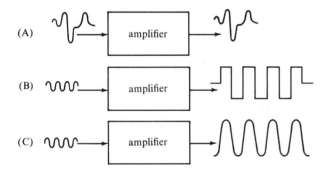

Figure 10.2.   (A) REPRODUCTION;   (B) AMPLIFICATION;   (C) COMBINED
REPRODUCTION AND AMPLIFICATION

In conclusion, reproduction essentially means duplication, and amplification means enlargement. It is essential that the difference between these two
processes be clear in the mind of the reader.

10.3.   Voltage Amplification

A circuit which is specifically designed to develop an output-voltage signal
greater than the input-voltage signal, is called a *voltage amplifier*. A voltage

amplifier is said to have a *voltage gain*, which is defined as the ratio of the change in output voltage to the change in input voltage which produced it. In equation form,

$$A_v = \frac{\Delta e_{out}}{\Delta e_{in}} \qquad (10.1)$$

where A_v = voltage gain.
Δe_{out} = change in output voltage.
Δe_{in} = change in input voltage.

In the remaining chapters of the text, lowercase letters will be used to denote signal (varying) voltages and currents such as e_{in} and e_{out}, while uppercase letters will denote DC (unchanging) voltages and currents such as E_{IN}.

▶ EXAMPLE 10.1

When the input voltage to a certain amplifier changes from 2 V to 3 V, its output voltage increases from 10 V to 20 V. What is the voltage gain of this amplifier?
   Using expression 10.1, we have

$$A_v = \frac{\Delta e_{out}}{\Delta e_{in}} = \frac{20 \text{ V} - 10 \text{ V}}{3 \text{ V} - 2 \text{ V}}$$
$$= 10 \qquad ◀$$

▶ EXAMPLE 10.2

The output voltage of a certain amplifier changes from 20 V to 10 V as its input increases from 2 V to 3 V. What is the voltage gain of this amplifier?
   In this case,

$$A_v = \frac{\Delta e_{out}}{\Delta e_{in}} = \frac{10 \text{ V} - 20 \text{ V}}{3 \text{ V} - 2 \text{ V}} = -10$$

Thus A_v is a negative quantity indicating that an *increase* in e_{in} causes a *decrease* in e_{out}.   ◀

   In the foregoing examples, the output signal was amplified by a factor of 10. Values of A_v from approximately one on up to hundreds of thousands can be encountered in electronic amplifiers.

10.4.   Current Amplification

A circuit which is specifically designed to develop an output-current signal greater than the input-current signal is called a *current amplifier*. A current amplifier is said to have a *current gain*, which is defined as the ratio of the change in output current to the change in input current which produces it. In equation form,

$$A_i = \frac{\Delta i_{\text{out}}}{\Delta i_{\text{in}}} \qquad \text{(10.2)}$$

where A_i = current gain.
Δi_{out} = change in output current.
Δi_{in} = change in input current.

Current amplifiers are often called *power amplifiers*, since the large output currents result in a substantial output power. Devices to be used in power amplifiers are specially manufactured so that they can withstand large currents.

▶ EXAMPLE 10.3

A certain amplifier has its output current increased from 10 mA to 20 mA in response to an input current change from 100 μA to 200 μA. What is the current gain of this amplifier?
Using Equation 10.2, we have

$$A_i = \frac{\Delta i_{\text{out}}}{\Delta i_{\text{in}}} = \frac{20 \text{ mA} - 10 \text{ mA}}{200 \text{ μA} - 100 \text{ μA}}$$

$$= 100 \qquad \blacktriangleleft$$

10.5. Power Amplification

In amplifiers the ratio of output signal power to input signal power is called *power gain* and is given by

$$A_p = \frac{p_{\text{osig}}}{p_{\text{isig}}} \qquad \text{(10.3)}$$

where A_p = power gain.
p_{osig} = signal power output.
p_{isig} = signal power input.

Power gain can also be expressed as the product of voltage gain and current gain. That is,

$$A_p = A_v \times A_i \qquad \text{(10.4)}$$

From Equation 10.4 it is apparent that it is not necessary to have both voltage gain and current gain greater than unity in order to achieve large power gains. In fact, some of the most widely used power amplifier circuits have large current gains along with voltage gains which are less than unity.

It is important to note that power gain as defined in Equations 10.3 and 10.4 is for signal power only. Many amplifiers also have DC output and input power dissipation which is not included in the expression for power gain.

▶ EXAMPLE 10.4

A certain amplifier has a signal output power dissipation of 10 W and a signal input power of 10 mW. Find the power gain.

$$A_p = \frac{p_{\text{osig}}}{p_{\text{isig}}} = \frac{10 \text{ W}}{10 \text{ mW}} = 1000 \qquad \blacktriangleleft$$

▶ EXAMPLE 10.5

The same amplifier has a voltage gain of 10. What is the current gain of the amplifier? Since

$$A_p = A_v \times A_i$$

then

$$A_i = \frac{A_p}{A_v} = \frac{1000}{10} = 100 \qquad \blacktriangleleft$$

Devices which are to be used in high-power amplifiers are usually of special construction so as to permit the large values of power dissipation without damage.

10.6. The Impedance Concept

In the circuit analysis of amplifiers, the term *impedance* is used to denote the effective AC resistance. The impedance of a linear resistor is simply equal to its resistance. However, the impedance of a nonlinear device is dependent on the operating point (recall zener impedance in Chapter 6). In general, two impedance values are important in amplifier work: *input impedance* and *output impedance*.

The input impedance, Z_{in}, of an amplifier is the ratio of voltage change to the resulting input current change. In equation form,

$$Z_{\text{in}} = \frac{\Delta e_{\text{in}}}{\Delta i_{\text{in}}} \qquad (10.5)$$

The input impedance is a measure of how much signal current the amplifier draws for a given input signal voltage.

▶ EXAMPLE 10.6

A certain amplifier has its input current increased from 1 mA to 2 mA when its input voltage increases from 0.5 V to 0.55 V. Determine Z_{in}.

Using Equation 10.5,

$$Z_{in} = \frac{\Delta e_{in}}{\Delta i_{in}} = \frac{0.55 \text{ V} - 0.50 \text{ V}}{2 \text{ mA} - 1 \text{ mA}}$$

$$= \frac{0.050 \text{ V}}{0.001 \text{ A}}$$

$$= 50 \text{ } \Omega \qquad \blacktriangleleft$$

The output impedance, Z_{out}, of an amplifier is the ratio of the change in output voltage to the change in output current. In equation form,

$$Z_{out} = \frac{\Delta e_{out}}{\Delta i_{out}} \qquad (10.6)$$

The output impedance of an amplifier is a measure of its ability to provide the same output signal (voltage or current) to different sized loads. An amplifier with a very high output impedance (say, 100 kΩ) will provide a fairly constant output-signal current, while one with a very low output impedance (say, 50 Ω) will provide a fairly constant output-signal voltage. These characteristics will be examined shortly.

10.7. Amplifier Relationships

Figure 10.3A represents a voltage amplifier. The input-signal voltage is applied between terminals 1 and 2 and the output terminals are 3 and 4. Most often, terminals 2 and 4 will be one and the same, but to preserve generality they will

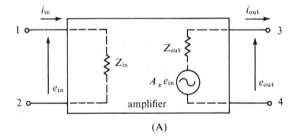

(A)

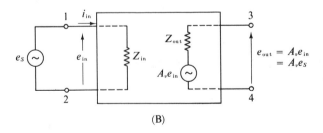

(B)

Figure 10.3. (A) GENERAL REPRESENTATION OF VOLTAGE AMPLIFIER;
(B) AMPLIFIER DRIVEN BY SIGNAL SOURCE e_S

be treated separately. Included in the representation are the amplifier input impedance, Z_{in}, output impedance, Z_{out}, and voltage gain, A_v. Since Z_{in}, Z_{out} and A_v are all AC parameters, *this representation is useful for AC signal voltages only.*

The input to the amplifier is represented by its input impedance, Z_{in}, which is seen looking into terminals 1–2. The output of the amplifier is shown as a *practical* signal voltage generator represented by a voltage source, $A_v \times e_{in}$, in series with the amplifier output impedance, Z_{out}. Part B of the figure shows the amplifier being driven by a signal source e_S. As a result of the input-signal voltage, an input-signal current i_{in} will flow through the input impedance. Using Ohm's Law, it is clear that

$$i_{in} = \frac{e_{in}}{Z_{in}} \qquad (10.7)$$

This is the current which the amplifier draws from the input-signal source. The output voltage is easily seen to be equal to A_v times the input e_S; that is, with no load across the output, there is no output current and, therefore, no loss in voltage across Z_{out}. The situation in Figure 10.3B is somewhat idealized, since the input-source resistance has been neglected and no load has been connected across the amplifier output.

Practical amplifier situation

Figure 10.4 shows the more practical situation, in which the source resistance of the signal source (represented by R_S) is included along with the load, R_L, which is connected to the output terminals. The inclusion of R_S and R_L changes the picture somewhat. First, it should be noticed that the voltage e_{in}, which is actually present at the input 1–2, will be *less than* the input signal, e_S, due to the loss of voltage across R_S. Stated differently, there is a voltage-divider action between R_S and Z_{in} so that only a fraction of e_S will appear at the input terminals of the amplifier. Using the voltage-divider rule,

$$e_{in} = \left(\frac{Z_{in}}{Z_{in} + R_S} \right) \cdot e_S \qquad (10.8)$$

Secondly, it can be seen that an output-signal current i_{out} will flow, thereby causing a voltage to be developed across Z_{out}. As a result, the actual output

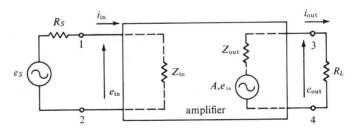

Figure 10.4.   PRACTICAL AMPLIFIER SITUATION

voltage e_{out} at terminals 3–4 will be *less than* $A_v e_{in}$. In other words, R_L and Z_{out} form a voltage divider so that

$$e_{out} = \left(\frac{R_L}{R_L + Z_{out}}\right) \cdot A_v e_{in} \qquad (10.9)$$

Because of these voltage-divider effects at the input and output of the amplifier, the output, e_{out}, will generally be less than $A_v \times e_S$ (the ideal situation of Figure 10.3). The *overall circuit voltage gain* is defined by

$$G_v = \frac{e_{out}}{e_S} \qquad (10.10)$$

and represents the actual voltage gain between the load and the input signal in the practical situation.

▶ EXAMPLE 10.7

An amplifier circuit has $Z_{in} = 10$ kΩ, $Z_{out} = 1$ kΩ and $A_v = 10$. The amplifier is driven by a 1-V peak-peak input signal with a source resistance, R_S, of 1 kΩ. Calculate e_{out} and G_v when $R_L = 1$ kΩ.

Using Equation 10.8, we have

$$e_{in} = \left(\frac{10 \text{ k}\Omega}{10 \text{ k}\Omega + 1 \text{ k}\Omega}\right)(1 \text{ V p-p})$$
$$= 0.91 \text{ V p-p}$$

Using Equation 10.9, we have

$$e_{out} = \left(\frac{1 \text{ k}\Omega}{1 \text{ k}\Omega + 1 \text{ k}\Omega}\right)(A_v e_{in}) = 5 e_{in}$$
$$= 4.55 \text{ V (p-p)}$$

Thus

$$G_v = \frac{e_{out}}{e_S} = \frac{4.55 \text{ V}}{1.000 \text{ V}} = 4.55 \qquad ◀$$

As the preceding example illustrates, the overall *circuit* voltage gain, G_v, is less than the *amplifier* voltage gain, A_v. This is usually the case, since G_v includes the effects of the external circuit components R_S and R_L, while A_v is essentially the ideal amplifier gain ($R_S = 0$ and $R_L = \infty$, no load). Under certain conditions, G_v can have a value which approaches A_v. Referring to Figure 10.4, it can be reasoned that, when $R_S \ll Z_{in}$, e_{in} will be approximately e_S; similarly, when $R_L \gg Z_{out}$, e_{out} will be approximately $A_v e_{in} = A_v e_S$. Thus, when these two conditions are met we have $G_v \simeq A_v$. That is,

$$G_v \simeq A_v \quad \text{when} \quad \begin{cases} R_S \ll Z_{in} \\ R_L \gg Z_{out} \end{cases}$$

▶ EXAMPLE 10.8

Repeat Example 10.7 using $R_S = 100 \ \Omega$ and $R_L = 100 \ \text{k}\Omega$.

$$e_{\text{in}} = \left[\frac{10 \ \text{k}\Omega}{10 \ \text{k}\Omega + 100}\right](e_S) = 0.99 \ \text{V} \quad \text{(p-p)}$$

$$e_{\text{out}} = \left[\frac{100 \ \text{k}\Omega}{100 \ \text{k}\Omega + 1 \ \text{k}\Omega}\right](10e_{\text{in}}) = (0.99)(10 \times 0.99 \ \text{V}) \quad \text{(p-p)}$$

$$= 9.8 \ \text{V (p-p)}$$

therefore,

$$G_v = \frac{e_{\text{out}}}{e_S} = \frac{9.8 \ \text{V}}{1 \ \text{V}} = 9.8 \simeq A_v \qquad \blacktriangleleft$$

Amplifier current gain

The overall current gain, G_i, of the amplifier in Figure 10.4 can be calculated simply by determining i_{in} and i_{out} and taking their ratio:

$$G_i = \frac{i_{\text{out}}}{i_{\text{in}}} \qquad \qquad \text{(10.11)}$$

▶ EXAMPLE 10.9

Determine G_i for the amplifier of Example 10.7.

$$i_{\text{in}} = \left(\frac{e_S}{R_S + Z_{\text{in}}}\right) = \frac{1 \ \text{V} \quad \text{(p-p)}}{11 \ \text{k}\Omega} \qquad \text{(10.12)}$$

$$= 0.09 \ \text{mA} \quad \text{(p-p)}$$

$$i_{\text{out}} = \frac{e_{\text{out}}}{R_L} = \frac{4.55 \ \text{V} \quad \text{(p-p)}}{1 \ \text{k}\Omega}$$

$$= 4.55 \ \text{mA} \quad \text{(p-p)} \qquad \text{(10.13)}$$

Thus

$$G_i = \frac{4.55 \ \text{mA}}{0.09 \ \text{mA}} = 50.5 \qquad \blacktriangleleft$$

Amplifier overall power gain

The overall power gain, G_p, of the amplifier in Figure 10.4 can be calculated as

$$G_p = \frac{p_{\text{load}}}{p_{\text{isig}}} = G_v \times G_i \qquad \text{(10.14)}$$

This power gain relates the actual signal power delivered to the load to the power supplied by the input signal.

10.8.   Impedance Matching in Amplifiers

The *maximum power transfer principle* states that the power transferred from a source to a load is at a maximum when the load impedance is equal to the source impedance. Applying this principle to the general amplifier of Figure 10.4, it can be seen that, for maximum power transfer from the signal source, e_S, to the amplifier input, the input impedance, Z_{in}, must match the source resistance, R_S. Similarly, for maximum power transfer from the amplifier output to the load, R_L, the value of R_L must match Z_{out}.

Thus, for maximum power transfer from the signal source to the amplifier load, Z_{in} must equal R_S, and R_L must equal Z_{out}. In practice, these conditions often are not met. For example, transistor audio amplifiers have output impedances typically in the range of 1–10 kΩ and must frequently drive a low impedance (typically 8 Ω) load. To provide impedance matching at the input and output of an amplifier, *impedance matching transformers* or *impedance matching circuits* are normally inserted as illustrated in Figure 10.5. The impedance matching circuit on the input serves to transform the amplifier's input impedance to a value equal to R_S so that the signal source essentially is driving a load impedance equal to its source impedance, allowing maximum power transfer from the source. A similar function is performed by the output matching circuit. It serves to transform R_L to a value equal to Z_{out} so that the maximum power is drawn from the amplifier output. An important requirement of the matching circuit is that it should transfer to the load as much of the power delivered to its input as possible, with minimum internal power loss.

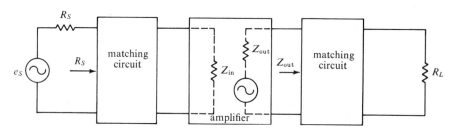

Figure 10.5.   IMPEDANCE MATCHING

Often, the output of an amplifier is fed into the input of another amplifier. In this case, the input impedance of the second amplifier serves as the load for the first amplifier. Here impedance matching is used to match Z_{in} of the second amplifier to the Z_{out} of the first amplifier, if maximum power is desired.

GLOSSARY

Input signal: variations in input voltage or current (AC portion of input).

Output signal: variations in output voltage or current (AC portion of output).

Reproduction: duplication of input signal by an amplifier.

Amplification :   enlargement of input signal by an amplifier.

Voltage (current, power) amplifier: circuit specifically designed to develop an output-signal voltage (current, power) greater than the input-signal voltage (current, power).

Impedance: effective AC resistance.

Input (output) impedance: impedance seen looking into the input (output) terminals of a circuit.

Impedance matching: matching of source and load impedances for maximum power transfer.

PROBLEMS

10.1   Will the circuit in Figure 10.6 perform reproduction and/or amplification of the input signal?

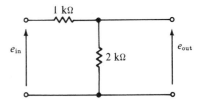

Figure 10.6.

10.2   What is the voltage gain, A_v, in each of the amplifiers in Figure 10.7?

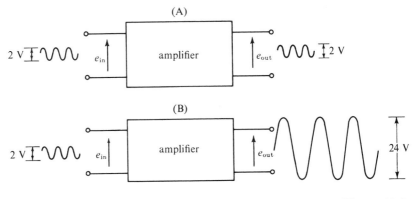

Figure 10.7.

10.3   Does reproduction take place in both amplifiers of Figure 10.7?

10.4   The amplifier in Figure 10.7A has an input current signal of 1 mA p-p, causing an output signal current of 10 mA p-p. What is the amplifier's current gain, A_i, and power gain, A_p?

10.5   The input and output curves of a certain amplifier are shown in Figure 10.8. From these curves, calculate Z_{in} and Z_{out} around the points indicated. (Use $\Delta i_{in} = 2$ mA and $\Delta i_{out} = 50$ mA.)

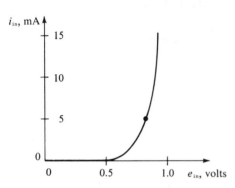 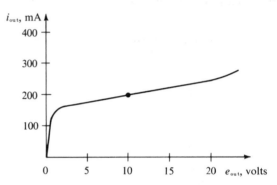

Figure 10.8.

10.6   Consider the amplifier in Figure 10.4. This amplifier has $Z_{in} = 100$ kΩ $A_v = 1000$ and $Z_{out} = 100$ Ω.
(a) Determine e_{out} and G_v for $e_S = 1$ mV p-p and $R_S = 1$ kΩ when a 10-kΩ load is present at the output.
(b) Repeat (a) for $R_L = 1$ kΩ, $R_L = 100$ Ω and $R_L = 10$ Ω. Plot G_v versus R_L.
(c) Determine G_i for each of the values of R_L above. Plot G_i versus R_L.
(d) Determine G_p for each value of R_L. Plot G_p versus R_L. Note that G_p is maximum when $R_L = Z_{out}$.

10.7   Using Equations 10.10–10.13, derive a formula that relates G_i in terms of R_S, Z_{in}, R_L and G_v.

10.8   Indicate *true* or *false* for each of the following statements concerning voltage amplifiers:
(a) Overall voltage gain, G_v, will be maximum when R_s is very large.
(b) Increasing Z_{in} will increase both G_v and G_i.
(c) In order to produce maximum voltage gain, it is desirable to make Z_{out} low.
(d) The output power from an amplifier will *always* increase as the source resistance is decreased toward zero.
(e) An amplifier will supply maximum current when $R_L = 0$.

11

Transistor Amplifiers: The Common-Base Amplifier

11.1.   Introduction

Thus far all of our work on junction transistors has dealt with DC currents and voltages. In this and following chapters, we shall investigate the operation of transistors when driven by AC voltages and currents. In particular, emphasis shall be placed on the transistor's ability to amplify an input signal. The application of the transistor as an *amplifier* is of such importance that devoting whole chapters to this subject is warranted even in a text concerned primarily with devices.

In the treatment of transistor amplifiers, the major differences among Com.-B, Com.-E and Com.-C amplifier configurations will be brought out. The emphasis will be on fundamental concepts rather than on exact circuit relationships.

Introduction of biasing methods, small-signal equivalent circuits and high-frequency effects should give the student sufficient background to pursue more thorough coverage of transistor amplifier circuits in later courses.

11.2.   Transistor Small-Signal Equivalent Circuits

In our study of transistor amplifiers we will be interested in how the transistor responds to small AC signals. To understand and analyze transistor amplifiers, it is helpful to employ an *equivalent circuit* in place of the transistor. The equivalent circuit represents the behavior of the transistor in response to *small** AC signals. Figure 11.1A shows the general equivalent circuit which will be used principally in our analysis of the Com.-B and Com.-E amplifiers. Terminals X, Y and Z represent the three transistor connections (E, B and C, though not necessarily in that order). Terminal X is the input terminal; Y is the output terminal and Z is the common, or reference, terminal.

The resistance connected between the input terminal, X, and the common terminal, Z, is the transistor's input resistance, h_i. The symbol h_i actually represents the transistor's AC *input* resistance; that is, its resistance to AC input signals. The circled arrow is a signal current source representing the signal current flowing through the output terminal, Y. The value of this current source is shown as h_f times the input signal current, i_{in}. The symbol h_f represents the transistor's *forward* AC current gain between input and output.

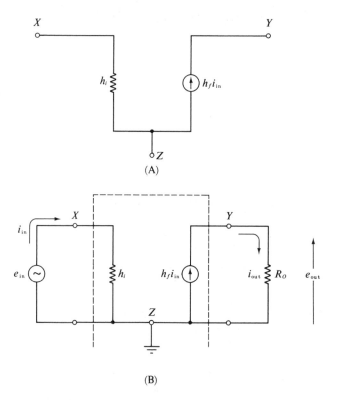

Figure 11.1.   (A) GENERAL TRANSISTOR EQUIVALENT CIRCUIT; (B) GENERAL TRANSISTOR EQUIVALENT CIRCUIT WITH APPLIED SIGNAL AND LOAD

---

*The equivalent circuits will be assumed to be valid for AC output signals whose amplitude (p-p) is no larger than 10 percent of the collector supply voltage.

The usefulness of the transistor equivalent circuit can be illustrated with the aid of Figure 11.1B, which shows an AC signal input, e_{in}, connected between input terminal, X, and common terminal, Z. An external resistance, R_O, has been connected between output terminal, Y, and common terminal, Z. The transistor equivalent circuit is enclosed in dotted lines. The input e_{in} produces an input current signal i_{in} through the input resistance, h_i. The output current source produces a current of $h_f \times i_{in}$ through the load, which in turn produces an output signal voltage e_{out} across the load. This action is summarized in the following relationships:

$$i_{in} = \frac{e_{in}}{h_i} \tag{11.1}$$

$$i_{out} = h_f i_{in} = h_f \left(\frac{e_{in}}{h_i} \right) \tag{11.2}$$

$$e_{out} = i_{out} \times R_O = \frac{h_f R_O}{h_i} \times e_{in} \tag{11.3}$$

The circuit's voltage gain, A_v, can be determined from Equation 11.3 as

$$A_v = \frac{e_{out}}{e_{in}} = \frac{h_f R_O}{h_i} \tag{11.4}$$

This expression shows that a voltage gain greater than one is possible if $h_f R_O$ is larger than h_i.

The circuit's current gain is easily obtained from examining the circuit or from Equation 11.2 as

$$A_i = \frac{i_{out}}{i_{in}} = h_f \tag{11.5}$$

This indicates that the current gain of the circuit is the same as the transistor's current gain, h_f. The value of h_f as well as h_i will depend on which transistor configuration is being used (Com.-B, -C or -E).

Now that the basic idea of the transistor equivalent circuit has been developed, we will turn our attention to the various transistor amplifier configurations. The equivalent circuit will be a useful part of the analyses to follow. The important points to remember are:

(a) The small-signal equivalent circuit is valid for small AC signals only.
(b) The values of the equivalent circuit parameters h_f and h_i will depend on the transistor configuration.

11.3.   Amplifier Analysis Procedure

The many amplifier circuits which will be discussed have several characteristics in common; we will, therefore, be able to use the same general procedure in analyzing their operation. This general procedure essentially consists of three basic steps: (1) DC analysis of the amplifier circuit; (2) AC signal analysis of the amplifier circuit; (3) superposition of the results of (1) and (2).

Figure 11.2 is a typical transistor amplifier circuit. Again the transistor terminals are labelled X, Y, Z for generality. The circuit contains DC bias sources, V_{xx} and V_{yy}, which are used to establish the transistor's DC currents and voltages. The input signal source produces the AC voltages and currents for the transistor. Coupling capacitor C_c is used to block any DC current from the V_{xx} bias supply from flowing back into the signal source. At the same time C_c acts as a very low impedance to the AC signal es and allows the AC current to flow into the transistor input.

Output coupling capacitor C_o is used to block any DC current from the V_{yy} bias supply from flowing through the load. At the same time it acts as a low impedance to AC so that AC signal current can flow through the load, producing a signal voltage output.

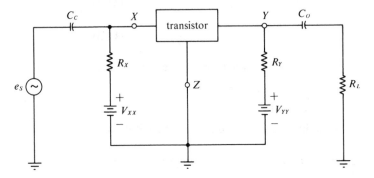

Figure 11.2.   TRANSISTOR AMPLIFIER CIRCUIT

DC (quiescent) analysis

The first step in analyzing the amplifier operation is to determine the DC currents and voltages in the circuit which are present as a result of the DC bias supplies. These currents and voltages constitute the *DC operating point* of the amplifier, or, as it is commonly called, the *quiescent operating point* (Q point for short). In calculating the DC operating point values, the amplifier circuit can be simplified somewhat by utilizing the fact that C_c and C_o will be open circuits for DC. Figure 11.3A shows the simplified amplifier circuit to be used for DC calculations. This simplified circuit contains only resistors, DC sources and the transistor. As such, it can be analyzed using the techniques developed in Chapter 9.

One might wonder why there is any need for a DC operating point at all, since the amplifier is used to amplify AC signals. The bias supplies are needed to bias the transistor in the *active* region (where its characteristics are relatively linear) so that it will not produce significant distortion of the AC signal. Once the Q point is established, the AC input signal will cause the transistor voltages and currents to vary above and below their DC values while still remaining in the *active* region.

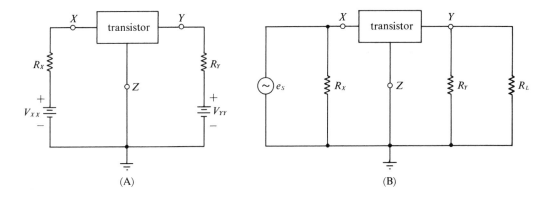

Figure 11.3.   (A) AMPLIFIER CIRCUIT SIMPLIFIED FOR DC ANALYSIS;
(B) AMPLIFIER SIMPLIFIED FOR AC ANALYSIS

AC analysis

The second step in the amplifier analysis involves the determination of the signal voltages and currents. In calculating these AC values, the amplifier circuit can be simplified somewhat by assuming that capacitors C_c and C_o both have reactances $(X_c = 1/2\pi fC)$ which are very low at the frequency of the input signal and can therefore be considered short circuits to the AC signals. In addition, the DC supplies, V_{xx} and V_{yy}, are also considered short circuits to AC since they cannot have an AC voltage across their terminals. These assumptions result in the simplified circuit of Figure 11.3B which is used for calculating the amplifier signal voltages and currents. In this AC circuit, the transistor is also usually replaced by its AC equivalent circuit (introduced in the preceding section). The parameters h_i and h_f used in this circuit usually depend on the DC operating point (Q point) at which the transistor is biased.

Superposition of DC and AC results

The final step in the amplifier analysis involves superimposing (combining) the results of the DC and AC analyses of the previous two steps. In other words, the complete picture of the amplifier operation consists of the AC variations around the Q point. To illustrate, consider Figure 11.4, which shows how the DC and AC components of the transistor's collector-to-emitter voltage are combined. The DC component is labelled V_{CE}, while the signal component is labelled v_{CE}, consistent with our use of uppercase symbols for DC and lowercase for AC. In this example, a 1-volt, peak-peak AC component is added to a DC component of 18 volts to produce a total collector-emitter voltage waveform that is a DC level of 18 V with a 1-V, p-p signal riding on it.

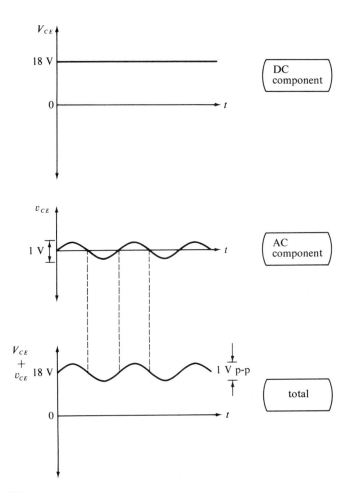

Figure 11.4.   EXAMPLE OF SUPERPOSITION OF DC AND AC VALUES

Summary

To summarize, the steps involved in analyzing transistor amplifier circuits are the following:

(a) Determine the DC values of currents and voltage (Q point) while treating all capacitors as *open* circuits.

(b) Determine the AC values of currents and voltages while treating all capacitors and DC sources as *short* circuits.

(c) Superimpose the results of steps (a) and (b).

11.4.   Common-Base Amplification

The *E-B* junction is the controlling circuit of the transistor in the same way that the control grid is the controlling element of the vacuum tube. The voltage-versus-current curve for the *E-B* junction is shown in Figure 11.5. The current and voltage in the *E-B* junction have a *nonlinear* relationship in that the emitter (or base) current will not change linearly as the voltage from the base to emitter changes.

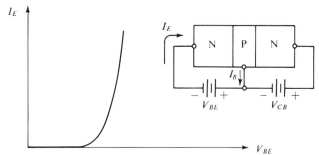

Figure 11.5.   NONLINEAR RELATIONSHIP BETWEEN I_E (OR I_B) AND V_{BE}

Amplification with a transistor in the Com.-B configuration is accomplished by varying the *E-B* junction forward bias, V_{BE}, which in turn causes a variation in the emitter current. This will be accompanied by a change in the collector current flowing across the reverse-biased *C-B* junction. Since the *E-B* junction is a forward-biased junction, it has a relatively low resistance so that small changes in V_{BE} will cause large changes in current. The same current variation occurring across the reverse-biased C-B junction, which has a high resistance, can result in voltage amplification.

In other words, if the *emitter* current is changed by some amount through a small change in V_{BE}, we can expect the *collector* current to change by approximately the same amount. Recall that the current gain from emitter to collector is always close to *one*, since only a small portion of emitter current flows through the base lead. It is possible to place a large load resistor in the collector circuit; across this load resistor, the change in collector current will develop a large change in voltage. This results in voltage amplification since the load signal voltage will usually be much greater than the small change in V_{BE} which produced it.

The amount by which the *collector* current will change in response to a change in *emitter* current is indicated by the transistor parameter α (alpha), which is the *AC common-base current gain*, and is given by

$$\alpha = \frac{\Delta I_C}{\Delta I_E}\bigg|V_{CB} = \text{constant} \qquad \textbf{(11.6)}$$

This formula states that α is obtained by taking the ratio of the collector current change to the emitter current change while holding the collector-base voltage constant since the value of V_{CB} does have a small effect on collector current. α is the AC counterpart of α_{DC} and pertains only to changing (AC) currents. Like α_{DC}, the value of α is also very close to unity. Figure 11.6 illustrates how the value of α is obtained from the Com.-B output curves. In general,

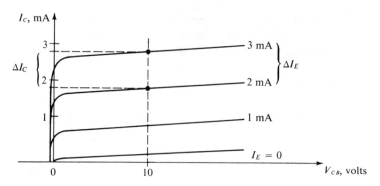

Figure 11.6.   CALCULATION OF α

the value of α will vary slightly depending on where in the active region it is measured. It is usually in the range of 0.95 to 0.999.

▶ EXAMPLE 11.1

From the data table below determine α. Also determine α_{DC} at $I_E = 10$ mA.

I_E	I_C	V_{CB}
10 mA	9.9 mA	10 V
20 mA	19.75 mA	10 V

Using Equation 11.1,

$$\alpha = \frac{19.75 - 9.9}{20 - 10} = \mathbf{0.985}$$

The value of α_{DC} is obtained by taking the ratio I_C/I_E:

$$\alpha_{DC} = \frac{9.9 \text{ mA}}{10 \text{ mA}} = \mathbf{0.99}$$

Note that $\alpha_{DC} \neq \alpha$. This is generally true, although both values are so close to unity that the difference is unimportant.   ◀

▶ EXAMPLE 11.2

A transistor has $\alpha = 0.97$. If the emitter current increases by 2 mA, by how much will the collector current increase?

    Since $\alpha = \Delta I_C/\Delta I_E$, then $\Delta I_C = \alpha(\Delta I_E)$. In this case, $\Delta I_E = 2$ mA. Thus

$$\Delta I_C = 0.97 \times 2 \text{ mA} = \mathbf{1.94 \text{ mA}}$$

This indicates that I_C will change by almost the same amount as the emitter current since α is almost one.   ◄

▶ **EXAMPLE** 11.3

For a certain transistor, an increase in emitter-base forward bias, V_{BE}, of 10 mV produces a 1-mA increase in emitter current. If a 10-kΩ resistor is connected in the collector circuit, what will be the resultant increase in the resistor voltage?

Since we are not given the value of α, we can assume that $\alpha \approx 1$ so that

$$\Delta I_C \simeq \Delta I_E = 1 \text{ mA}$$

With $R_C = 10$ kΩ, then

$$\Delta V_{RC} = \Delta I_C \times 10 \text{ k}\Omega \approx 1 \text{ mA} \times 10 \text{ k}\Omega$$
$$\approx \textbf{10 V}$$

The change in voltage across the collector resistor is 10 V, while the applied change in V_{BE} was only 10 mV. This represents a voltage gain of 10 V/10 mV = 1000. This gives some indication of the type of voltage amplification possible in a Com.-B amplifier.   ◄

11.5.   Common-Base Amplifiers: Basic Circuit

The basic Com.-B amplifier circuit is shown in Figure 11.7. Expect for the input signal and coupling capacitor, the circuit is identical to the Com.-B circuit studied in Chapter 9. The input signal is applied to the transistor emitter through the coupling capacitor which serves to block any DC current from flowing from the emitter supply, V_{EE}, back through the signal source. As far as the DC currents and voltages in this circuit are concerned, the methods for determining their values are exactly those outlined in Chapter 9. The DC values of I_E, I_C and V_{CB} are called the *bias values*, or *quiescent values*, and they represent a certain point on the transistor characteristic curves. This point is called the *bias point*, or *quiescent point* (Q point), and in most cases is in the *active*

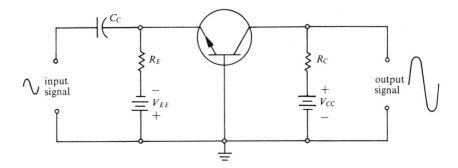

Figure 11.7.   COMMON-BASE AMPLIFIER CIRCUIT

region. The transistor operates at this point when no signal is applied. The Q point, which depends only on the DC voltage sources and resistors R_E and R_C and is in no way dependent on the input signal, places the transistor in the *active* region of its operation.

The input signal, as we shall see, will cause the transistor circuit values to change around the Q point, although these values will remain in the active region. As a result of the input signal, the transistor currents and voltages will have signal, or AC, portions as well as their DC values. In all the work to follow, uppercase symbols such as I_E will be used to represent DC (Q point) values and lower case symbols such as i_E will represent AC (signal) values. The complete representation of any circuit value will be the superposition of its DC value and its AC value.

DC analysis

Let us assume the following values for the circuit of Figure 11.7: $V_{EE} = 6$ V; $R_E = 1$ kΩ; $V_{CC} = 12$ V; $R_C = 1$ kΩ. For purposes of calculating the circuit Q point, we can neglect the input signal source and the coupling capacitor. This situation is redrawn in Figure 11.8 where a silicon transistor is assumed. From the values chosen we can calculate:

$$I_E = \frac{6\text{ V} - 0.7\text{ V}}{1\text{ k}\Omega} = 5.3\text{ mA}$$

$$I_C = \alpha_{DC}I_E \simeq I_E = 5.3\text{ mA}$$

$$V_{CB} = 12\text{ V} - 5.3\text{ mA} \times 1\text{ k}\Omega$$

$$= 6.7\text{ V}$$

These values of I_E, I_C and V_{CB} constitute the bias point, or Q point. Note that the transistor is biased in the *active* region since $I_C < I_{C(\text{sat})}$.

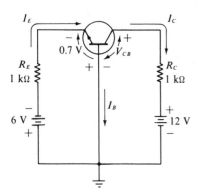

Figure 11.8.   DC ANALYSIS OF COMMON-BASE AMPLIFIER CIRCUIT

AC analysis

In order to analyze the AC operation of the Com.-B amplifier, the circuit is redrawn in Figure 11.9A with the coupling capacitor and DC sources replaced by short circuits. In this AC circuit we can see that the signal source, e_S, supplies a signal current, i_S. A portion of this signal current will flow through the emitter bias resistor, R_E, as i_X; however, in properly designed amplifiers this portion will be so small that it can be assumed that most of i_S will flow into the emitter as i_E. In other words, $i_E \simeq i_S$.

The emitter signal current, i_E, is superimposed on the emitter DC bias current, I_E (previously calculated as 5.3 mA). This changing emitter current produces a changing collector current in the form of the AC collector current signal, i_C. We saw in Section 11.4 that the change in collector current is equal to α multiplied by the change in emitter current. In terms of signal currents, then,

$$i_C = \alpha i_E \qquad (11.7)$$

We know that α is very close to unity, so it is apparent that in the common-base amplifier the output current, i_C, is almost equal to the input signal current, i_E.

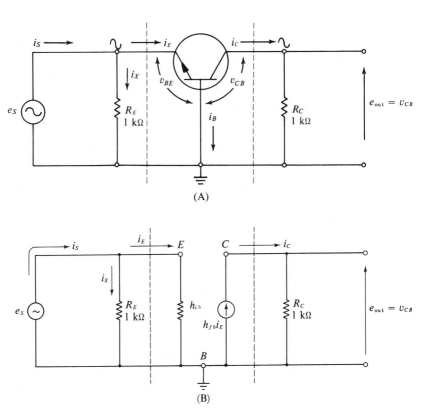

Figure 11.9.   (A) COMMON-BASE AMPLIFIER WITH DC SUPPLIES AND COUPLING CAPACITOR REPLACED BY SHORT CIRCUITS FOR AC ANALYSIS; (B) COMMON-BASE AMPLIFIER WITH TRANSISTOR AC EQUIVALENT CIRCUIT

11.6.   Common-Base Gains

Let us now replace the transistor in Figure 11.9A by its AC equivalent circuit (introduced in Section 11.2), resulting in the circuit of Figure 11.9B. Note that the AC input resistance of the transistor in the Com.-B configuration is labelled as h_{ib}, the subscript i denoting *input* resistance and the subscript b denoting common-*base* configuration. Similarly, the transistor's AC current gain has been denoted as h_{fb}, where f stands for *forward* current gain and b, again, for common *base*. From what was said earlier concerning the relationship between the input current i_E and the output i_C, it should be obvious that h_{fb} is the same as α. That is,

$$h_{fb} = \alpha = \frac{i_C}{i_E} \qquad (11.8)$$

In all of our work to follow we will use h_{fb} and α interchangeably since they both represent the same quantity.

Amplifier current gain, A_i

Examination of Figure 11.9B shows that the current gain of the amplifier, A_i, is the same as h_{fb}. This is so because the input source current, i_S, is approximately the same as i_E if we assume $i_X \simeq 0$. This is a valid assumption whenever R_E is much larger than h_{ib}, which is usually the case. This means that the Com.-B amplifier current gain, A_i, is the same as the transistor's Com.-B current gain, h_{fb} (α). That is,

$$A_i = \frac{i_{\text{out}}}{i_{\text{in}}} \approx \frac{i_C}{i_S} = \frac{i_C}{i_E} \approx h_{fb} \qquad (11.9)$$

whenever $R_E \gg h_{ib}$. In cases where R_E and h_{ib} are similar in magnitude, i_X cannot be neglected and A_i will actually be less than h_{fb}. In any case, A_i for a Com.-B amplifier is always less than one.

Amplifier voltage gain, A_v

The amplifier output signal voltage is produced by the collector signal current flowing through R_C. Thus

$$e_{\text{out}} = i_C \times R_C \qquad (11.10)$$

so that once i_C is known, the output voltage can be calculated and the amplifier voltage gain (e_{out}/e_S) can be determined. Note that e_{out} is the same as the AC collector-base voltage, v_{CB}, since R_C is connected between collector and base.

To illustrate the calculation of A_v let us assume values for the transistor parameters as $h_{ib} = 5\,\Omega$ and $h_{fb} = 0.98$. Now, note that the input signal source, e_S, is across the E-B junction and, therefore, across h_{ib}. Using $e_S = 1$ mV peak-peak, then, we can calculate

$$i_E = \frac{e_S}{h_{ib}} = \frac{1 \text{ mV p-p}}{5\,\Omega} = 0.2 \text{ mA p-p}$$

Therefore,

$$i_C = h_{fb}i_E = 0.98 \times 0.2 = \textbf{0.196 mA p-p}$$

so that

$$e_{\text{out}} = i_C R_C = 0.196 \text{ mA} \times 1 \text{ k}\Omega$$
$$= 0.196 \text{ V p-p}$$
$$= \textbf{196 mV p-p}$$

Thus an input signal of 1 mV p-p produces an output signal of 196 mV p-p, so that

$$A_v = \frac{e_{\text{out}}}{e_S} = \textbf{196}$$

This procedure can be used to calculate A_v for any Com.-B amplifier.

A general formula for A_v can be derived as follows:

$$i_E = \frac{e_S}{h_{ib}}$$

$$i_C = h_{fb}i_E = h_{fb}\left(\frac{e_S}{h_{ib}}\right)$$

$$e_{\text{out}} = i_C R_C$$
$$= \left(\frac{h_{fb}e_S}{h_{ib}}\right) \times R_C$$

$$A_v = \frac{e_{\text{out}}}{e_S} = \frac{h_{fb}R_C}{h_{ib}} \qquad\qquad \textbf{(11.11)}$$

Since h_{fb} is normally close to unity, an approximate value for A_v is given by

$$A_v \simeq \frac{R_C}{h_{ib}} \qquad\qquad \textbf{(11.12)}$$

This last relationship shows that A_v is equal to the ratio of the external collector resistor, R_C, and the transistor Com.-B input resistance, h_{ib}. Relatively large values of R_C can result in voltage gains as high as 2500 since h_{ib} is usually relatively low, as we shall see later.

Amplifier power gain, A_p

Since $A_p = A_v \times A_i$, it should be obvious that a Com.-B amplifier can have considerable power gain even if its current gain is less than one. For example, in the previous illustration, A_i was 0.98 and A_v was 196, so that

$$A_p = 196 \times 0.98 = 192$$

which indicates an output signal power 192 times greater than the signal power produced by the input source e_S. Typically, Com.-B amplifiers can have power gains between 100 and 2500.

11.7.   Determination of h_{ib}

The Com.-B input resistance, h_{ib}, is essentially the AC resistance of the *E-B* junction. As such, it can be calculated as the ratio of a change in base-emitter voltage to the resulting change in emitter current.* Thus

$$h_{ib} = \frac{\Delta V_{BE}}{\Delta I_E} \qquad\qquad (11.13)$$

h_{ib} can be obtained from the transistor's Com.-B input curve as illustrated in Figure 11.10. The value of h_{ib} is a measure of how much the emitter current will change for a change in V_{BE}. Since the curve in Figure 11.10 is nonlinear, it should be apparent that h_{ib} will be different on different parts of the curve. As the curve gets steeper for higher values of I_E, h_{ib} gets smaller, since a larger ΔI_E occurs for a given ΔV_{BE}.

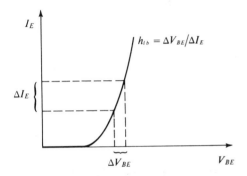

Figure 11.10.   CALCULATION OF h_{ib}

▶ EXAMPLE 11.4

A certain transistor has an input curve with the following points:

	A	*B*	*C*	*D*
I_E	0.1 mA	2 mA	5 mA	10 mA
V_{BE}	0.5 V	0.6 V	0.65 V	0.66 V

*To be exact, h_{ib} should be calculated or measured while holding V_{CB} constant. However, since V_{CB} affects h_{ib} only slightly, its effect will be neglected here.

Calculate h_{ib} between points A and B. Repeat between points C and D:

$$h_{ib} (A\text{--}B) = \frac{0.6 \text{ V} - 0.5 \text{ V}}{2 \text{ mA} - 0.1 \text{ mA}} = \frac{0.1 \text{ V}}{1.9 \text{ mA}}$$

$$= 526 \; \Omega$$

$$h_{ib} (C\text{--}D) = \frac{0.66 \text{ V} - 0.65 \text{ V}}{10 \text{ mA} - 5 \text{ mA}} = \frac{0.01 \text{ V}}{5 \text{ mA}}$$

$$= 2 \; \Omega$$

This illustrates that h_{ib} is lower for higher values of I_E.   ◄

In many situations it is often inconvenient or impractical to calculate h_{ib} from the input curve. Because of this situation, a useful approximation for the value of h_{ib} has come into popular usage:

$$h_{ib} \simeq \frac{0.025 \text{ V}}{I_E} \tag{11.14}$$

where I_E is the value of the DC emitter current at which the transistor is biased. In other words, the value of transistor AC input resistance is *inversely* proportional to the value of Q-point emitter current. For example, if a Com.-B transistor amplifier has a DC emitter current of 0.1 mA, formula 11.14 indicates that

$$h_{ib} \simeq \frac{0.025 \text{ V}}{0.0001 \text{ A}} = 250 \; \Omega$$

Similarly, for $I_E = 1$ mA, a value of 25 Ω is indicated for h_{ib}, and so on. Typically, h_{ib} values range from 10 Ω to 300 Ω for most Com.-B amplifiers.

As we saw in the last section, the voltage gain of a Com.-B amplifier depends on h_{ib}. On the other hand, h_{ib} depends on the DC emitter current bias, I_E. This indicates that A_v can be varied by changing the DC bias on the transistor. This will be illustrated in the next section.

11.8.   Complete Analysis Example

We shall now perform a thorough analysis of the Com.-B amplifier shown in Figure 11.11A. The circuit uses a germanium transistor and we will assume that the transistor's α_{DC} and α (h_{fb}) are both approximately *one*.

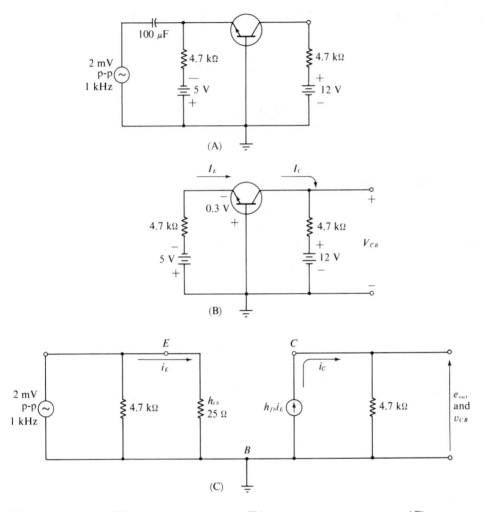

Figure 11.11.   (A) AMPLIFIER CIRCUIT; (B) DC EQUIVALENT CIRCUIT; (C) AC
EQUIVALENT CIRCUIT

DC analysis

The circuit can be simplified for purposes of DC analysis as redrawn in Figure
11.11B. Since the transistor is germanium, we have

$$I_E = \frac{5 \text{ V} - 0.3 \text{ V}}{4.7 \text{ k}\Omega} = 1 \text{ mA}$$

$$I_C = \alpha_{DC} I_E \simeq I_E = 1 \text{ mA}$$

$$V_{CB} = 12 \text{ V} - 1 \text{ mA} \times 4.7 \text{ k}\Omega$$

$$= 7.3 \text{ V}$$

These values represent the amplifier Q point.

AC analysis

The AC equivalent circuit for the amplifier is drawn in Figure 11.11C. The 100-μF coupling capacitor is treated as a short circuit. Actually, its impedance can be calculated at a frequency of 1 kHz as

$$X_C = \frac{1}{2\pi f C} = \frac{1}{2\pi \times 10^3 \times 100 \times 10^{-6}} \simeq 1.6 \; \Omega$$

which is small enough to be neglected. The transistor input resistance, h_{ib}, can be calculated from Equation 11.14 using $I_E = 1$ mA, the Q-point value:

$$h_{ib} \simeq \frac{0.025}{0.001} = 25 \; \Omega$$

The signal source is directly across h_{ib} and produces

$$i_E = \frac{2 \text{ mV p-p}}{25 \; \Omega} = 80 \; \mu\text{A p-p}$$

Since $h_{fb} \approx 1$, then $i_C \approx i_E = 80 \; \mu$A p-p. This i_C flows through $R_C = 4.7$ kΩ, producing

$$v_{CB} = e_{\text{out}} = 80 \; \mu\text{A} \times 4.7 \text{ k}\Omega = 376 \text{ mV p-p}$$

The amplifier voltage gain is therefore

$$A_v = \frac{e_{\text{out}}}{e_S} = \frac{376}{2} = 188$$

Figure 11.12 shows the relationship between the input signal and the complete output waveform consisting of V_{CB} (7.3 V) and v_{CB} (376 mV p-p). The AC

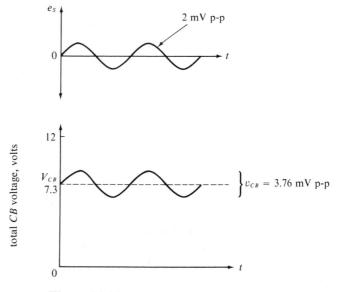

Figure 11.12.   INPUT AND OUTPUT WAVEFORMS

portion rides on the Q-point DC level. Note that the v_{CB} signal is *in phase* with the input signal, e_S. This is generally true of the Com.-B amplifier circuit except at very low and extremely high frequencies.

Increasing A_v

The voltage gain in the Com.-B amplifier depends essentially on the ratio R_C/h_{ib}. We could have calculated the values of A_v for the circuit of Figure 11.11 using Equation 11.12:

$$\frac{R_C}{h_{ib}} = \frac{4.7 \text{ k}\Omega}{25 \text{ }\Omega} = 188$$
$$\therefore A_v \approx 188$$

From this relationship it is apparent that A_v can be increased by either increasing R_C or by decreasing h_{ib}. Let us examine these two possibilities.

An increase in R_C will have practically no effect on the Q-point values of emitter current or collector current. However, the DC value of *C-B* voltage will *decrease* because of the larger DC voltage dropped across R_C. For example, with $R_C = 6.8$ kΩ, the value of V_{CB} drops to 12 V $-$ 6.8 V $= 5.2$ V. In addition, the increase in R_C will have no effect on the signal portions of i_E and i_C. However, the larger R_C results in a larger AC output-voltage signal for the same i_C. For example, with $R_C = 6.8$ kΩ, the value of e_{out} will be 544 mV p-p, indicating a voltage gain of $544/2 = 277$. Figure 11.13A shows the complete *C-B* voltage obtained by increasing R_C from 4.7 kΩ to 6.8 kΩ. Compare it to Figure 11.12.

To decrease h_{ib} it is necessary to bias the transistor at a higher value of I_E since, as we have seen, h_{ib} decreases as I_E increases (Equation 11.14). The Q-point value of I_E can be increased by either increasing the V_{EE} source voltage or by decreasing the emitter bias resistor, R_E. It is usually easier to change R_E. If R_E is decreased from 4.7 kΩ to 3.3 kΩ, I_E will become 1.42 mA. This, of course, will cause I_C to increase to approximately the same value. The new value of I_E will result in

$$h_{ib} \simeq \frac{0.025 \text{ V}}{1.42 \text{ mA}} = 18 \text{ }\Omega$$

With $R_C = 4.7$ kΩ, this causes A_v to increase to

$$A_v = \frac{4.7 \text{ k}\Omega}{18 \text{ }\Omega} = 260$$

The student should verify that the complete *C-B* output waveform resulting from decreasing R_E to 3.3 kΩ is as shown in Figure 11.13B. Note that the Q-point value, V_{CB}, has shifted downward due to the increase in I_C. Compare this to Figure 11.12.

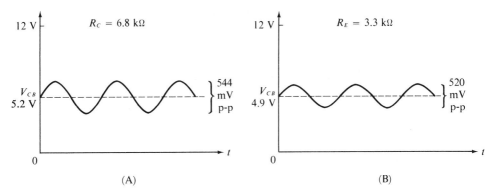

Figure 11.13.

11.9.   Common-Base Amplifier Input Impedance

When discussing input impedance of the Com.-B amplifier we must distinguish it from the transistor's input resistance, h_{ib}. If we refer to Figure 11.11C, it can be seen that h_{ib} is in parallel with the emitter bias resistor, R_E. Thus the total input impedance seen by the signal source is the parallel combination of R_E and h_{ib}. Denoting the *input impedance of the Com.-B amplifier* by the symbol Z_{ib}, we have

$$Z_{ib} = R_E \parallel h_{ib} = \frac{R_E \times h_{ib}}{R_E + h_{ib}} \qquad (11.15)$$

Normally the value of R_E is much larger than h_{ib} so that $Z_{ib} \simeq h_{ib}$. For example, for the amplifier of Figure 11.12,

$$Z_{ib} = \frac{4.7 \text{ k}\Omega \times 25 \text{ }\Omega}{4.7 \text{ k}\Omega + 25 \text{ }\Omega} = 24.8 \text{ }\Omega \simeq h_{ib}$$

It is apparent that the input impedance of a Com.-B amplifier can be very low. This characteristic makes it difficult to drive a Com.-B amplifier from moderate or high impedance signal sources without some sort of impedance matching network (Chapter 10). On the other hand, the low input impedance makes it ideal for amplifying signals from low impedance sources such as magnetic phonograph pickups.

11.10.   Common-Base Amplifier Output Impedance

The output impedance of an amplifier is the total impedance seen looking back into the output terminals of the amplifier. Referring to Figure 11.12A, when we look from the output back into the *C-B* terminals, the only impedance that we can see is the collector resistor, R_C. Thus we have

$$Z_{ob} \simeq R_C \qquad (11.16)$$

where Z_{ob} is the *output impedance of the Com.-B amplifier.*

The relationship above is only an approximate one since there is another factor, heretofore neglected, which will have an effect on the value of Z_{ob}. This factor is the *transistor's Com.-B output resistance*, r_{ob}. Since the output is taken across the *C-B* junction in the Com.-B configuration, r_{ob} represents the AC resistance of the *C-B* junction. Since it usually can be neglected, the effect of r_{ob} was not included in the transistor's AC equivalent circuit used in our previous analyses.

Figure 11.14 shows a more exact representation of the AC equivalent circuit for the Com.-B amplifier of Figure 11.12A. It includes r_{ob}, which is shown connected between collector and base. It is in parallel with R_C; as such, the *amplifier* output impedance Z_{ob} is more exactly given by

$$Z_{ob} = r_{ob} \parallel R_C = \frac{r_{ob} \times R_C}{r_{ob} + R_C} \qquad (11.17)$$

The value of r_{ob} is typically very large, usually in the range of 250 kΩ to 1 MΩ, and depends somewhat on the transistor Q point. In many cases, r_{ob} is much larger than R_C and so can be neglected.

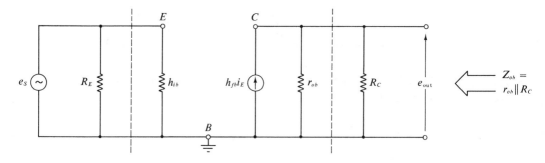

Figure 11.14.   COMMON-BASE AMPLIFIER AC EQUIVALENT CIRCUIT, IN-CLUDING THE EFFECT OF r_{ob}

Calculation of r_{ob}

The transistor's Com.-B *output resistance* is a measure of the effect of the collector-base voltage, V_{CB}, on the collector current. Its value is given by the formula

$$r_{ob} = \frac{\Delta V_{CB}}{\Delta I_C}\bigg|_{I_E} = \text{constant} \qquad (11.18)$$

Figure 11.15 illustrates how r_{ob} may be obtained from the Com.-B output characteristics. It is obvious from the figure that the value of r_{ob} can be very high,

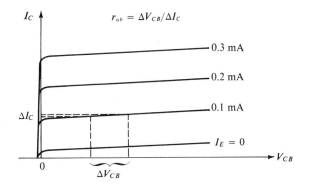

$$r_{ob} = \Delta V_{CB}/\Delta I_C$$

Figure 11.15.   CALCULATION OF r_{ob}

since large changes in V_{CB} produce only very small changes in collector current in the active region. High values of r_{ob} indicate output curves which are flatter (have smaller slopes).

Output conductance, h_{ob}

Most transistor specification sheets do not specify r_{ob} directly; instead, they usually provide the value of h_{ob}, which is the transistor's *output conductance in the Com.-B configuration*. Recall that conductance is the reciprocal of resistance and is measured in *mhos* (1/ohms). Thus

$$r_{ob} = \frac{1}{h_{ob}} \tag{11.19}$$

with h_{ob} in mhos and r_{ob} in ohms.

▶ EXAMPLE 11.5

A certain Com.-B amplifier has $R_C = 50$ kΩ and uses a transistor with $h_{ob} = 1$ µmho. Calculate the amplifier's output impedance.

$$r_{ob} = \frac{1}{h_{ob}} = \frac{1}{10^{-6} \text{ mho}} = 10^6 \text{ ohms}$$
$$= 1 \text{ M}\Omega$$

Thus

$$Z_{ob} = r_{ob} \parallel R_C = \frac{1 \text{ M}\Omega \times 50 \text{ k}\Omega}{1 \text{ M}\Omega + 50 \text{ k}\Omega}$$
$$= \textbf{47.6 k}\boldsymbol{\Omega}$$

If r_{ob} were neglected, $Z_{ob} \simeq R_C = 50$ kΩ. When r_{ob} is ten or more times larger than R_C, it can usually be neglected without introducing more than a 10 percent error.   ◀

In summary, the output impedance Z_{ob} depends mainly on R_C. As such, it can be relatively large since a large R_C is often used for a high voltage gain. Typically, Z_{ob} ranges from 10 kΩ–100 kΩ. This high output impedance of the Com.-B amplifier makes it unsuitable for driving low impedance loads unless some impedance matching circuit is used. However, it can work directly into high impedance loads such as the input to vacuum tube amplifiers and FET amplifiers.

11.11.   Effects of R_S and R_L

Now that we have analyzed the basic Com.-B amplifier, we can investigate the effect of including the signal source resistance R_S and the effect of a load on the amplifier output.

Let us consider the Com.-B amplifier circuit in Figure 11.16A and let us assume that the transistor has the following parameters at the Q point:

$$h_{fb} = \alpha = 0.99$$
$$h_{ib} = 50 \ \Omega$$
$$h_{ob} = 1 \ \mu\text{mho} \quad (r_{ob} = 1 \ \text{M}\Omega)$$

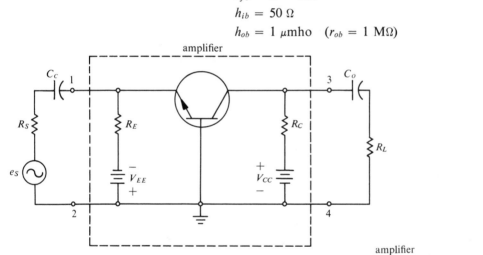

(A)

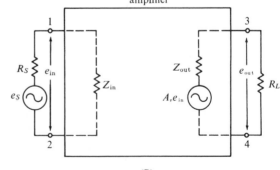

(B)

Figure 11.16.   (A) COMMON-BASE AMPLIFIER CIRCUIT WITH LOAD; (B) AC EQUIVALENT CIRCUIT

The circuit values are:

$$R_E = 10 \text{ k}\Omega$$
$$V_{EE} = 6 \text{ V}$$
$$R_C = 10 \text{ k}\Omega$$
$$V_{CC} = 12 \text{ V}$$

The portion of the circuit enclosed by the dotted line is the amplifier proper, which was analyzed in the previous sections. Feeding the input terminals (1–2) of the amplifier is the signal source e_S which is coupled to the amplifier input through C_c. Also included is the source resistance, R_S, which until now has been neglected. The amplifier output terminals (3–4) are connected to a load resistance R_L through an output coupling capacitor, C_o. The coupling capacitor prevents DC current from flowing to the load, allowing only AC voltage to appear across R_L.

Once Z_{in}, Z_{out} and A_v are known, the amplifier proper can be replaced by the general voltage amplifier representation introduced in Chapter 10. This representation is shown in Figure 11.16B. Since our interest is in AC signals, the coupling capacitors have both been replaced by short circuits. The amplifier parameters can be calculated from the values given above as:

$$Z_{\text{in}} = Z_{ib} = R_E \parallel h_{ib} \tag{11.15}$$
$$\simeq 50 \ \Omega$$

$$Z_{\text{out}} = Z_{ob} = R_C \parallel r_{ob} \tag{11.17}$$
$$= 10 \text{ k}\Omega \parallel 1 \text{ M}\Omega$$
$$\simeq 10 \text{ k}\Omega$$

$$A_v = \frac{\alpha R_C}{h_{ib}} = \frac{0.99 \times 10 \text{ k}\Omega}{50} \tag{11.11}$$
$$= 198$$

With these values inserted into the circuit of Figure 11.16B, the *overall* circuit operation can be analyzed for any values of R_S and R_L.

▶ EXAMPLE 11.6

Determine e_{out} and G_v for the amplifier of Figure 11.16 for $R_S = 150 \ \Omega$, $R_L = 5 \text{ k}\Omega$ and $e_S = 1 \text{ mV p-p}$.

$$e_{\text{in}} = \left(\frac{Z_{\text{in}}}{R_S + Z_{\text{in}}} \right) \times e_S$$
$$= \frac{50 \ \Omega}{150 \ \Omega + 50 \ \Omega} \times 1 \text{ mV}$$
$$= 0.25 \text{ mV p-p}$$

$$e_{out} = \left(\frac{R_L}{R_L + Z_{out}}\right) \times A_v e_{in}$$

$$= \left(\frac{5 \text{ k}\Omega}{5 \text{ k}\Omega + 10 \text{ k}\Omega}\right) \times 198 \times 0.25 \text{ mV}$$

$$= \textbf{16.5 mV p-p}$$

$$G_v = \frac{e_{out}}{e_S} = \frac{16.5}{1} = \textbf{16.5}$$

Note that the *overall* voltage gain, G_v, is much less than the amplifier's gain, A_v, due to the voltage divider action at the input and at the output.   ◀

It should be apparent that the overall voltage gain, G_v, in Figure 11.16 will be *maximum* when $R_S \ll Z_{ib}$ and when $R_L \gg Z_{ob}$. Since $Z_{ib} \simeq h_{ib}$ and $Z_{ob} \simeq R_C$, then we can say that

$$G_v \simeq A_v \quad \text{when} \quad \begin{cases} R_S \ll h_{ib} \\ R_L \gg R_C \end{cases}$$

In practice, these requirements are difficult to meet: a low value of R_S means that the signal source must be close to ideal; and such a large value of R_L is generally impractical. For this reason, Com.-B amplifiers are only useful in applications which take advantage of their desirable characteristics, such as a good high-frequency response and stable operation under varying temperature conditions.

11.12.   Graphical Analysis

Our study of the Com.-B amplifier has thus far utilized the small-signal AC equivalent circuit for the transistor. If we are interested in "large-signal" operation of the amplifier ($e_{out} > 10$ percent of V_{CC}), a more accurate technique must be used which takes into account the variations in the transistor characteristics that accompany large voltage and current variations. This technique is a graphical analysis utilizing the transistor's Com.-B output curves and load lines. This graphical method also provides a means of examining the waveforms more closely; it is especially useful in studying the effects of the Q point and input signal amplitude on the shape of the output waveform.

Let us consider the amplifier circuit in Figure 11.17. The circuit values are as shown:

$$e_S = 25 \text{ mV p-p}$$
$$R_E = 4.7 \text{ k}\Omega$$
$$V_{EE} = 5 \text{ V}$$
$$R_C = 10 \text{ k}\Omega$$
$$V_{CC} = 20 \text{ V}$$
$$R_L = 10 \text{ k}\Omega$$

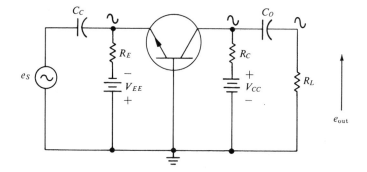

Figure 11.17.   COMMON-BASE AMPLIFIER

The transistor is germanium and has a value of $h_{ib} = 25\ \Omega$ at $I_E = 1$ mA. Its Com.-B output curves are shown in Figure 11.18.

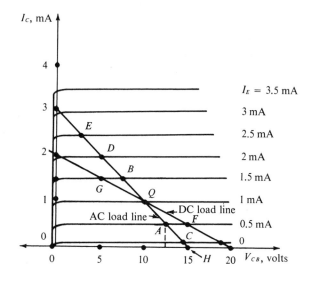

Figure 11.18.   OUTPUT CURVES FOR TRANSISTOR IN FIGURE 11.17

The procedure will be first to determine the Q point (DC operating point). Since only DC values are to be considered in finding the Q point, the signal source e_S and the load resistor R_L do not affect its determination because of the DC isolation provided by the coupling capacitors. The load line to be used to find the Q point will be called the *DC load line*. It is determined by V_{CC} and R_C as outlined in Chapter 9. The load line for $V_{CC} = 20$ V and $R_C = 10$ kΩ is superimposed on the curves of Figure 11.18. The Q point will reside at the intersection of the DC load line and the transistor curve corresponding to the DC value of emitter current. The latter quantity is easily calculated as

$$I_E = \frac{V_{EE} - V_{BE}}{R_E}$$

$$= \frac{5 \text{ V} - 0.3 \text{ V}}{4.7 \text{ k}\Omega} = 1 \text{ mA}$$

The resulting Q point is at $I_C = 0.99$ mA and $V_{CB} = 10.1$ V.

The AC operation can now be determined. In doing so we must now include the effect of R_L. In the collector circuit there is essentially a total resistance of R_C in parallel with R_L for AC current, assuming C_o is a short circuit to AC. Thus, for the AC analysis, our load line will be determined by

$$R' = R_C \parallel R_L = \frac{R_C \times R_L}{R_C + R_L}$$

This load line, called the *AC load line*, will pass through the Q point and will have a slope of $-1/R'$. For the circuit under consideration, R' is 5 kΩ and the resulting AC load line has been constructed in Figure 11.18.

To plot the AC load line, another point is needed in addition to the Q point. A convenient point to choose is the intersection of the V_{CB} axis ($I_C = 0$). To find V_{CB} at this point, simply take the Q-point value of I_C and multiply it by R'. Then add this value to the Q-point value of V_{CB}. For our example,

$$I_C \text{ (at Q point)} \simeq 1 \text{ mA}$$
$$V_{CB} \text{ (at Q point)} \simeq 10.1 \text{ V}$$
$$R' = 10 \text{ k}\Omega \parallel 10 \text{ k}\Omega = 5 \text{ k}\Omega$$

Thus

$$1 \text{ mA} \times 5 \text{ k}\Omega = 5 \text{ V}$$

so that

$$V_{CB} \text{ (axis intersection)} = 10.1 \text{ V} + 5 \text{ V}$$
$$= 15.1 \text{ V}$$

This is the value of V_{CB} where the AC load line intersects the V_{CB} axis. Connecting this point (point H in figure) and the Q point with a straight line results in the desired AC load line.

The AC load line represents the effects of both R_C and R_L. As such, it is always *steeper* than the DC load line since R' is always less than R_C.

Once the AC load line has been constructed, it remains to determine the variation in input emitter current produced by the signal source. This can be calculated using Ohm's law. Using $e_S = 25$ mV p-p we have

$$i_E = \frac{e_S}{h_{ib}} = \frac{25 \text{ mV p-p}}{25 \text{ }\Omega} = 1 \text{ mA p-p}$$

This tells us that the emitter current variation is 0.5 mA on either side of the DC emitter current (1 mA in this case). That is, the emitter current will vary

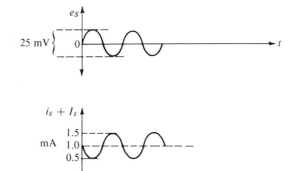

Figure 11.19.   VARIATION IN EMITTER CURRENT DUE TO INPUT SIGNAL

from 0.5 mA to 1.5 mA as the signal voltage is applied. This is illustrated in Figure 11.19. * As a result of this variation in emitter current the operating point will move up and down the AC load line. At the lower limit of emitter current (0.5 mA) the operating point moves down to point A in Figure 11.18, while at the upper limit of emitter current (1.5 mA) the operating point moves up to point B. The current and voltage values at these points are recorded in Table 11.1. From the table it is seen that as the emitter current makes a complete excursion from 0.5 mA to 1.5 mA the collector-base voltage makes a corresponding excursion from 12.6 V to 7.6 V. Figure 11.20 shows the various waveforms. The waveform of collector-base voltage contains both a DC level (10.1 V) and an AC component (5 V p-p). Only the AC component appears across R_L becoming e_{out}. Thus e_{out} is 5 V p-p and the overall amplifier gain is

$$G_v = \frac{e_{out}}{e_S} = \frac{5 \text{ V p-p}}{25 \text{ mV p-p}} = 200$$

TABLE 11.1

Point	I_E	I_C	V_{CB}	e_S
B	1.5 mA	1.48 mA	7.6 V	−12.5 mV
Q	1 mA	0.99 mA	10.1 V	0
A	0.5 mA	≈0.5 mA	12.6 V	+12.5 mV

From Figure 11.20 an important characteristic of Com.-B amplifiers is observed: the output signal is *in phase* with the input signal. This is always the

---

*Note that the i_E waveform is 180° out-of-phase with e_S. This is so because as e_S goes positive it causes the voltage at the emitter relative to base to become *less* negative; this reduces the *E-B* forward bias and causes I_E to decrease. The opposite occurs when e_S goes negative. See Figure 11.17.

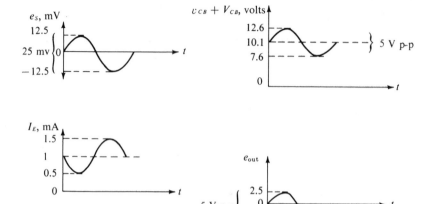

Figure 11.20.   AMPLIFIER WAVEFORMS

case at medium frequencies of operation for Com.-B amplifiers with a resistive load.

Signal distortion (clipping)

In all the discussions up to this point the input signal has been restricted to values small enough to keep the transistor operating in the active region. It is not unusual to have the input signal voltage become large enough to cause the transistor to become cut off or saturated during a portion of the input swing. For example, referring to Figure 11.18, if the input signal were increased so that the emitter current variation became 2 mA p-p around the Q point, the circuit operating point would move between the extremes of points C and D. This gives an output voltage swing of approximately 10 volts. The point C is right at cutoff, at which point $V_{CB} = 15$ V* and $I_C = 0$. Further increase in the input signal may cause the upper excursion of emitter current to increase, say to 2.5 mA, but it will not affect the lower excursion since I_E cannot go below zero. This would cause the V_{CB} waveform to stay at 15 V during the interval in which $I_E = 0$. This may be more easily understood with the help of Figure 11.21, in which the e_S, I_E and V_{CB} waveforms are shown for this case. The V_{CB} waveform is seen to be "clipped" at the 15-volt level during the interval when the transistor is driven into cutoff by the input signal.

A similar result occurs when the input signal causes the transistor to become saturated. In this case, the V_{CB} waveform will be clipped at $V_{CB} = 0$ V during the interval in which the transistor is saturated. Referring to Figure 11.17, this

_____

*For interested students, $V_{CB} \neq V_{CC} = 20$ V at cutoff because of the charged capacitor C_o which acts like a DC source of 10 V (V_{CB} at Q point).

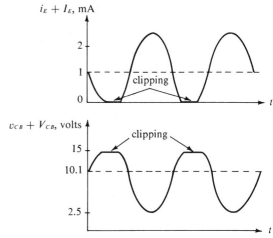

Figure 11.21.   WAVEFORMS ILLUSTRATING CUTOFF CLIPPING

would occur when the input signal is large enough to cause the emitter current to reach approximately 3.0 mA and greater, causing saturation. Both types of clipping are normally undesirable since the output waveform becomes highly distorted. To minimize the possibility that clipping will occur, the Q point should be chosen about halfway between saturation and cutoff along the AC load line at point *B*. It is obvious that in the circuit under consideration the maximum output voltage swing is 15 V p-p, between $V_{CB} = 15$ V at cutoff and $V_{CB} \approx 0$ at saturation. An increase in R_L or in V_{CC} will increase the possible output voltage swing.

▶ EXAMPLE 11.7

For the circuit of Figure 11.17 and the transistor curves of Figure 11.18, determine the voltage gain using the circuit values given previously, except let $R_L = \infty$.

In this case, the AC load line is the same as the DC load line determined by V_{CC} and R_C in Figure 11.18. Using this load line and the $I_E = 1.5$ mA curve to locate point *G* and the $I_E = 0.5$ mA curve to locate point *F*, the swing in V_{CB} is from 5 volts to 15 volts. In other words, the output signal is 10 V p-p, giving a voltage gain of

$$G_v = \frac{10 \text{ V}}{25 \text{ mV}} = \textbf{400} \qquad ◀$$

▶ EXAMPLE 11.8

For the conditions of Example 11.7, determine the maximum possible output voltage swing.

Referring to the DC load line in Figure 11.18, it is apparent that V_{CB} can range from 20 V to approximately 0 V, giving a possible **20-V** swing.   ◀

11.13.   Common-Base Amplifiers: Single-Supply Biasing

Up to now, all of the DC biasing for a Com.-B circuit has utilized two separate supplies: V_{EE} for the *E-B* forward bias and V_{CC} for the *C-B* reverse bias. This is very wasteful, since the same biasing can be accomplished using only a single power supply. The single power supply is usually the V_{CC} supply which is the largest DC voltage in the circuit. This V_{CC} supply is used, as before, as the collector bias voltage. In addition, the V_{CC} supply is divided down to a lower value using a voltage divider. This lower DC level is called V_{BB} and is applied to the *base* rather than to the emitter. This is so because V_{BB} will be positive relative to ground since V_{CC} is positive relative to ground. See Figure 11.22A.

The voltage divider *R1-R2* serves to reduce V_{CC} to a smaller DC supply level V_{BB} which is applied between base and ground. For clarity, this situation can be visualized as being the same as that in Figure 11.22B. The V_{BB} supply provides the *E-B* forward bias. That is,

$$I_E = \frac{V_{BB} - V_{BE}}{R_E}$$

The important difference between this method and the one which we used previously is that the *C-B* reverse bias is now equal to V_{CC} *minus* V_{BB}, since the base is positive relative to ground rather than at ground. Typically, the V_{BB} used is one-quarter to one-tenth the value of V_{CC}.

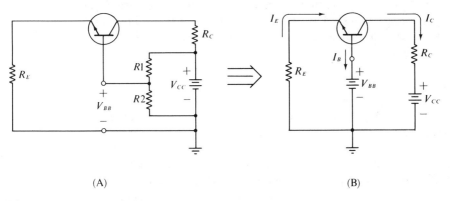

(A)   (B)

Figure 11.22.   COMMON-BASE BIASING USING SINGLE DC SUPPLY

If a better regulated voltage V_{BB} is required, *R2* can be replaced by a zener diode of appropriate value. This will keep V_{BB} relatively constant even for large changes in base current which might occur under signal conditions.

11.14.   Common-Base Amplifiers: Summary

A brief summary of the Com.-B transistor amplifier is presented in Table 11.2 listing its major characteristics with an eye toward comparing it to the Com.-E and Com.-C amplifiers in subsequent sections.

TABLE 11.2

COMMON-BASE AMPLIFIER

Current gain A_i: approximately 1
Voltage again A_v: very high; 100–2500 typical
Power gain A_p: high; 100–2500 typical
Input impedance Z_{ib}: very low; 10 Ω–200 Ω typical
Output impedance Z_{ob}: very high; 10 kΩ typical
Phase shift (input to output);   0°

GLOSSARY

h_f: transistor's forward AC current gain between input and output.

h_i: transistor's AC input resistance.

Q point: quiescent operating point; DC bias values of amplifier voltages and currents.

α (alpha):   AC current gain between emitter and collector.

h_{fb}:   forward AC current gain in common-base configuration; same as α.

h_{ib}:   AC input resistance in common-base configuration.

Z_{ib}:   total input impedance of Com.-B amplifier.

Z_{ob}:   total output impedance of Com.-B amplifier.

r_{ob}:   transistor's AC output resistance in Com.-B configuration.

h_{ob}:   transistor's AC output conductance in Com.-B configuration.

PROBLEMS

11.1   Refer to Figure 11.1B. Determine e_{out} if $h_i = 100 \ \Omega$, $h_f = 0.96$, $R_{out} = 8$ kΩ and $e_{in} = 2$ mV p-p.

11.2   Explain the functions of C_c and C_o in the amplifier of Figure 11.2.

11.3   Define *Q point*.

11.4   What transistor parameter measures AC current gain from emitter to collector?

11.5   The following set of data is for a given transistor:

V_{BE}	I_E	I_C	V_{CB}
0.70 V	1 mA	0.98 mA	10 V
0.69 V	0.6 mA	0.59 mA	10 V

Determine α for this transistor.

11.6   What is the difference between α and α_{DC}?

11.7   In the Com.-B amplifier, what are the purposes of V_{EE} and V_{CC}?

11.8   Define h_{fb}. What is its value for the transistor of Problem 11.5?

11.9   Define h_{ib}. Determine its value for the transistor of Problem 11.5.

11.10   Refer to the amplifier of Figure 11.7. The circuit uses a silicon transistor and has the following values: $V_{EE} = 7$ V, $R_E = 2.5$ kΩ, $V_{CC} = 40$ V, $R_C = 10$ kΩ and $e_s = 1$ mV p-p.
(a) Determine the amplifier's Q-point values (I_E, I_C, V_{CB}).
(b) Determine the output signal amplitude.
(c) Determine A_i, A_v and A_p.
(d) Sketch the total waveforms (AC + DC) for the emitter current, collector current and C-B voltage.

11.11   Indicate what would happen to the total C-B waveform of the amplifier of the previous problem for each of the following changes:
(a) Increase in e_s.
(b) Decrease in V_{EE}.
(c) Decrease in R_C.
(d) Inclusion of R_S (source resistance).
(e) Increase in V_{CC}.
(f) Replacement of transistor with one of the same type.
(g) Connect output to a load R_L through a coupling capacitor C_o.

11.12   Determine the input and output impedances of the amplifier of Problem 11.10. Use the general voltage amplifier representation for this amplifier and calculate G_v if a signal source with $e_s = 1$ mV p-p and $R_S = 15$ Ω is used, and the amplifier output is capacitively coupled to a 15-kΩ load. Assume $h_{ob} = 1$ μmho.

11.13   Answer *true* or *false:*
(a) The input impedance of a Com.-B amplifier depends *only* on the transistor used.
(b) Z_{ob} is typically the same value as R_C.
(c) A_v is typically 10–200 for Com.-B amplifiers.
(d) The Q point of the Com.-B amplifier has very little effect on the voltage gain.
(e) A_v can usually be increased by reducing R_E.
(f) e_{out} is in phase with e_S.
(g) The output of one Com.-B amplifier could be used as the input to a second Com.-B amplifier to produce even larger voltage gains.

(h) *Overall* voltage gain for a *practical* Com.-B amplifier can always be increased by reducing h_{ib}.

11.14 The amplifier drawn in Figure 11.23 uses a germanium transistor with output characteristics of Figure 11.18.
(a) Plot DC load line and determine Q point.
(b) Plot AC load line and determine complete waveform of *C-B* voltage using a graphical technique.
(c) Determine e_{out} and G_v.

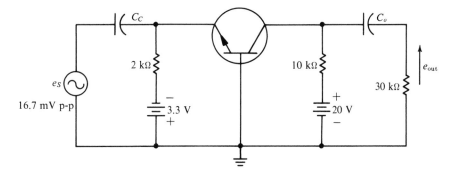

Figure 11.23.

11.15 Repeat Problem 11.14 using $e_S = 33.3$ mV p-p.

11.16 The circuit in Figure 11.22 has $V_{CC} = 25$ V, $V_{BB} = 5$ V, $R_E = 10$ kΩ and $R_C = 18$ kΩ. The transistor is silicon. Determine I_E, I_C and V_{CB}.

11.17 Consider the amplifier in Figure 11.24. The circuit Q point is calculated as: $I_C \approx I_E = 1$ mA; $V_{CB} = 18$ V. The voltage gain e_{out}/e_S is calculated as 160. However, when the amplifier is constructed and tested in the lab, the following measurements were recorded: $e_S = 1$ mV p-p; $V_{CB} = 18$ V; $v_{cb} = 480$ mV p-p; $e_{out} = 0$. Which of the following could be possible causes of the discrepancy between the calculated and measured values? Explain your choice.
(a) Transistor is faulty; shorted or open.
(b) The load is shorted out.
(c) The output capacitor C_o is open.
(d) The 12-kΩ resistor is faulty (too large).
(e) The input capacitor is shorted out.

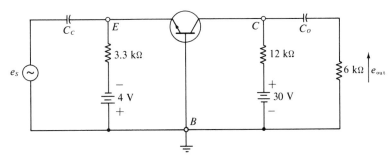

Figure 11.24.

12

The Common-Emitter Amplifier

12.1. Introduction

In the Com.-E amplifier, the input signal is applied to the *base* and the output signal is taken at the *collector*. In this chapter, the Com.-E amplifier will be analyzed in the same manner as was the Com.-B amplifier in the preceding chapter. We shall see that there are several major differences between these two types of amplifiers. For instance, the Com.-E amplifier has a larger input impedance (typically 1 kΩ) and possesses a large current gain as well as voltage gain. These characteristics serve to make the Com.-E amplifier generally more useful than the Com.-B amplifier.

12.2. AC Current Gain: β

In the Com.-E amplifier, the input signal is applied to the base and will cause the *base* current to change above and below its DC bias value. As such, we can expect the *collector* current to also change, but by a much greater amount. Recall that the current gain from base to collector is always much greater than one.

The amount by which the collector current will change in response to a change in base current is indicated by the transistor parameter β, which is the *AC common emitter current gain* and is given by

$$\beta = \frac{\Delta I_C}{\Delta I_B}\bigg|V_{CE} = \text{constant} \qquad (12.1)$$

This formula states that β is obtained by taking the ratio of the collector current change to the base current change while holding the collector-emitter voltage constant. β is the AC counterpart of β_{DC} and pertains only to changing (AC) currents. The value of β can be determined from the Com.-E output curves as shown in Figure 12.1 and will vary depending on where in the active region it is measured. Typically, it is in the range of 30–300, although it can reach 1000. β is normally slightly below the value of β_{DC} at the same point on the characteristics.

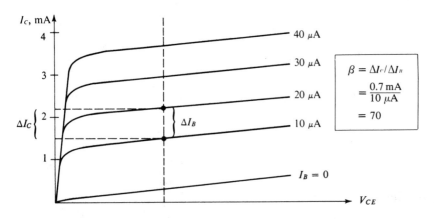

Figure 12.1.   CALCULATION OF β

▶ EXAMPLE 12.1

Calculate β from the data below:

I_B	I_C	V_{CE}
100 μA	6 mA	10 V
125 μA	7 mA	10 V

Using Equation 12.1,

$$\beta = \frac{\Delta I_C}{\Delta I_B} = \frac{7 \text{ mA} - 6 \text{ mA}}{125 \ \mu\text{A} - 100 \ \mu\text{A}} = 40 \qquad ◀$$

It is important to remember that β is the AC current gain and is the ratio of *changes* in current. It can also be defined as

$$\beta = \frac{i_C}{i_B} \qquad (12.2)$$

where i_C and i_B are signal values of collector and base currents.

▶ EXAMPLE 12.2

A signal applied to the input of a common-emitter amplifier produces a signal base current i_B equal to 20 μA p-p. Determine i_C if $\beta = 75$.

$$i_C = \beta \times i_B$$
$$= 75 \times 20 \ \mu A$$
$$= 1500 \ \mu A \ \text{p-p}$$

This relatively large AC collector current can produce a relatively large AC voltage across a resistor placed in the collector circuit, resulting in a large voltage gain. ◀

12.3.   Common-Emitter Amplifier: Basic Circuit

The basic Com.-E amplifier circuit is shown in Figure 12.2. Except for the input signal and coupling capacitor, the circuit is identical to the circuit studied in Chapter 9. As far as the DC currents and voltages are concerned, the methods for determining their values are exactly the same as those outlined in Chapter 9. The DC values of I_B, I_C and V_{CE} represent the amplifier Q point. The Q point depends only on V_{CE}, R_C and R_B and in no way depends on the input signal. The input signal e_S will cause the circuit values to change around the Q point so that each voltage and current in the circuit will have an AC component added to its DC value.

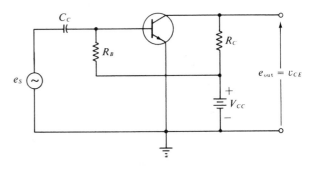

Figure 12.2.   BASIC COMMON-EMITER AMPLIFIER

DC analysis

Let us assume the following values for the circuit of Figure 12.2: $V_{CC} = 12$ V; $R_B = 700$ kΩ; $R_C = 12$ kΩ. The transistor is silicon and has $\beta_{DC} = 40$. For Q-point calculation, the input source and the coupling capacitor can be neglected, as shown in Figure 12.3. The Q-point values can be calculated as follows:

$$I_B = \frac{12 \text{ V} - 0.7 \text{ V}}{700 \text{ k}\Omega} \approx \textbf{16 } \boldsymbol{\mu}\textbf{A}$$

$$I_C = \beta_{DC}I_B = \textbf{640 } \boldsymbol{\mu}\textbf{A}$$

$$V_{CE} = 12 \text{ V} - 640 \text{ } \mu\text{A} \times 12 \text{ k}\Omega$$

$$= \textbf{4.3 V}$$

These values indicate that the transistor is biased in the *active* region since $I_C < I_{C(\text{sat})}$.

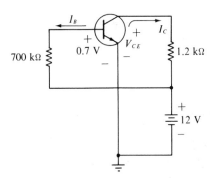

Figure 12.3.   CIRCUIT FOR DC ANALYSIS

AC analysis

To analyze the AC operation, the circuit of Figure 12.2 is redrawn as in Figure 12.4A, with the coupling capacitor and V_{CC} supply replaced by short circuits. In this AC circuit, we can see that the signal source supplies a signal current, i_S. A portion of i_S will flow through R_B as i_X; however, in properly designed amplifiers, this portion will be very small so that it can be assumed that most of i_S will become base current. That is, $i_B \approx i_S$.

To further aid in the AC analysis, the transistor can be replaced by its AC equivalent circuit as shown in Figure 12.4B. Note that the AC input resistance of the transistor in the Com.-E configuration is labeled h_{ie}, the subscript *i* denoting *input* resistance and the subscript *e* denoting common-*emitter* configuration. Similarly, the transistor's AC current gain has been denoted as h_{fe}, where *f* stands for *forward* current gain and *e* again for common-*emitter* configuration. From what was said earlier concerning the relationship between in-

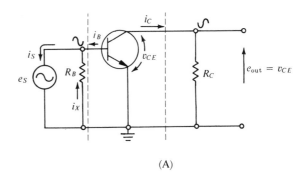

(A)

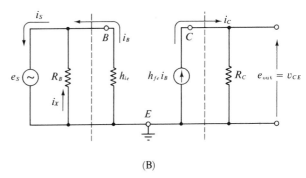

(B)

Figure 12.4.   (A) COMMON-EMITTER AMPLIFIER SIMPLIFIED FOR AC ANALYSIS; (B) COMMON-EMITTER AMPLIFIER WITH TRANSISTOR REPLACED BY ITS SMALL-SIGNAL EQUIVALENT CIRCUIT

put current i_B and output current i_C, it should be obvious that h_{fe} is the same as β. That is,

$$h_{fe} \equiv \beta = \frac{i_C}{i_B} \qquad (12.3)$$

In all of our work to follow, the symbols h_{fe} and β will be used interchangeably since they represent the same thing.

Amplifier current gain, A_i

Examination of Figure 12.4B shows that the amplifier's current gain, A_i, is essentially equal to h_{fe} if we assume that i_X is negligible so that $i_B \approx i_S$. This is a valid assumption whenever R_B is much larger than h_{ie}. In other words,

$$A_i \simeq h_{fe} \quad \text{for} \quad R_B \gg h_{ie}$$

In cases where R_B is similar in magnitude to h_{ie}, i_X cannot be neglected and A_i will be less than h_{fe} since some of i_S is wasted in R_B and does not reach the transistor. The *transistor* itself still has a current gain of h_{fe}, but the amplifier A_i is the ratio of i_C to i_S.

The value of h_{fe} typically ranges from 20–300 and can sometimes be as high as 1000. The Com.-E amplifier, unlike the Com.-B amplifier, has a substantial current gain. This fact will be important, as we shall see.

Amplifier voltage gain, A_v

The amplifier output voltage is produced by the collector signal current flowing through R_C so that

$$e_{\text{out}} = v_{CE} = i_C \times R_C \qquad (12.4)$$

Once i_C is known, e_{out} and A_v can be found. To illustrate, let us assume $h_{fe} = 40$ and $h_{ie} = 2$ kΩ for the circuit in Figure 12.4B. Using $e_S = 1$ mV p-p,

$$i_B = \frac{e_S}{h_{ie}} = \frac{1 \text{ mV}}{2 \text{ k}\Omega} = 0.5 \text{ µA p-p}$$

Therefore,

$$i_C = h_{fe}i_B = 40 \times 0.5$$
$$= 20 \text{ µA p-p}$$

and

$$e_{\text{out}} = 20 \text{ µA} \times 12 \text{ k}\Omega = 240 \text{ mV p-p}$$

Thus, for $e_S = 1$ mV, an output of 240 mV is produced:

$$A_v = \frac{e_{\text{out}}}{e_S} = \frac{240}{1} = 240$$

A general formula for A_v can be derived as follows:

$$i_B = \frac{e_S}{h_{ie}}$$
$$i_C = h_{fe}i_B = \frac{h_{fe}e_S}{h_{ie}}$$
$$e_{\text{out}} = i_C R_C = \frac{h_{fe}e_S R_C}{h_{ie}}$$
$$A_v = \frac{e_{\text{out}}}{e_S} = h_{fe}\frac{R_C}{h_{ie}} \qquad (12.5)$$

This last relationship shows that A_v is equal to h_{fe} times the ratio of the collector resistor, R_C, to the Com.-E input resistance, h_{ie}. Typically, values of A_v can range as high as 2500 for the Com.-E amplifier, *the same as for the Com.-B amplifier*.

Amplifier power gain, A_p

With $A_p = A_v \times A_i$, it is clear that a Com.-E amplifier can have considerable power gain since *both* A_v and A_i can be large. For example, in our illustration above, $A_i = h_{fe} = 40$ and $A_v = 240$, so that $A_p = 9600$. Power gains as high as 200,000 are not uncommon when h_{fe} is fairly high. The Com.-E amplifier enjoys a much larger power gain than the Com.-B amplifier, which has no current gain.

12.4. Determination of h_{ie}

The Com.-E input resistance h_{ie} is the AC resistance of the E-B junction with the base current as input. As such, it can be calculated as the ratio of a change in V_{BE} to the resulting change in I_B. Thus,

$$h_{ie} \equiv \frac{\Delta V_{BE}}{\Delta I_B}\bigg|V_{CE} = \text{constant} \qquad (12.6)$$

The value of h_{ie} can be obtained from the Com.-E input curve as illustrated in Figure 12.5 in much the same fashion as h_{ib} is obtained from the Com.-B input curve.

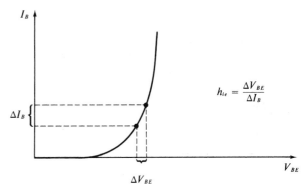

Figure 12.5.   CALCULATION OF h_{ie}

In many instances, it is inconvenient or impractical to calculate h_{ie} from the input curve. Instead, the relationship between h_{ie} and h_{ib} can be used. It is true that

$$h_{ie} = (h_{fe} + 1)h_{ib} \approx h_{fe}h_{ib} \qquad (12.7)$$

Therefore, h_{ie} can be determined once h_{ib} is known.

▶ EXAMPLE 12.3

A certain Com.-E amplifier is biased at $I_B = 50\ \mu A$, $I_C = 12$ mA and $V_{CE} = 8$ V. The transistor has $h_{fe} = 240$. Determine h_{ie}.
    Since $I_C = 12$ mA, then $I_E \simeq 12$ mA; therefore,

$$h_{ib} = \frac{0.025}{0.012} = 2.1\ \Omega$$

Thus

$$h_{ie} \simeq h_{fe} \times h_{ib} = 240 \times 2.1 \ \Omega$$
$$= 504 \ \Omega \qquad \blacktriangleleft$$

As the preceding example illustrates, h_{ie} depends both on the Q point (which determines h_{ib}) and on h_{fe}. Values of h_{ie} typically range from a few hundred ohms up to 10 kΩ. This, of course, is much larger than h_{ib}, the Com.-B input resistance, by the factor h_{fe}.

12.5.   Complete Analysis Example

We shall now perform a thorough analysis of the Com.-E amplifier in Figure 12.6A. The circuit uses a silicon transistor with β_{DC} and β (same as h_{fe}) both approximately equal to 200.

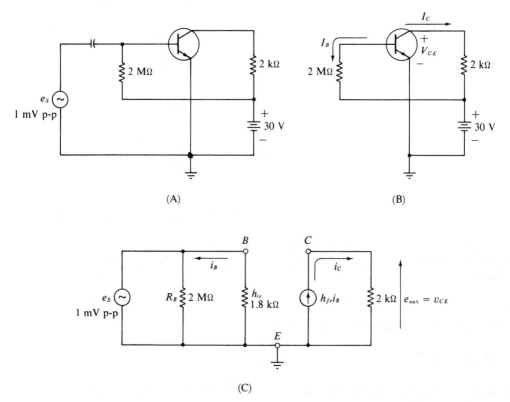

Figure 12.6.   (A) common-emitter amplifier; (B) dc analysis; (C) ac analysis

DC analysis

For DC analysis the circuit can be simplified as in Figure 12.6B. The Q-point values are calculated as follows:

$$I_B = \frac{30 \text{ V} - 0.7 \text{ V}}{2 \text{ M}\Omega}$$

$$= \textbf{14.65 } \boldsymbol{\mu}\textbf{A}$$

$$I_C = \beta_{DC} \times I_B$$

$$= 200 \times 14.65$$

$$= 2930 \text{ } \mu\text{A} = \textbf{2.93 mA}$$

$$I_E \simeq I_C = \textbf{2.93 mA}$$

$$V_{CE} = 30 \text{ V} - 2.93 \text{ mA} \times 2 \text{ k}\Omega$$

$$\simeq \textbf{24.1 V}$$

AC analysis

The AC equivalent circuit for the amplifier is drawn in Figure 12.6C. The value of h_{ie} is determined first by calculating h_{ib} and then by using Equation 12.7:

$$h_{ib} = \frac{0.025}{I_E} = \frac{0.02500}{0.00293}$$

$$\simeq 9 \text{ } \Omega$$

$$h_{ie} = h_{fe} \times h_{ib} = 1800 \text{ } \Omega$$

Since e_S is across h_{ie}, then

$$i_B = \frac{e_S}{h_{ie}} = \frac{1 \text{ mV}}{1.8 \text{ k}\Omega}$$

$$= 0.555 \text{ } \mu\text{A p-p}$$

Thus

$$i_C = h_{fe}i_B = 200 \times 0.555 \text{ } \mu\text{A}$$

$$= 111 \text{ } \mu\text{A p-p}$$

and

$$v_{CE} = e_{\text{out}} = i_C R_C = 111 \text{ } \mu\text{A} \times 2 \text{ k}\Omega$$

$$= 222 \text{ mV p-p}$$

The amplifier voltage gain is

$$A_v = \frac{e_{\text{out}}}{e_S} = \frac{222}{1} = 222$$

Figure 12.7 shows the complete waveform for the output collector-emitter voltage consisting of $V_{CE} = 24.1$ V and $v_{CE} = 222$ mV p-p. Note that the v_{CE} *signal is 180° out-of-phase with e_S*. This is a characteristic of Com.-E amplifiers not shared with Com.-B or Com.-C amplifiers.

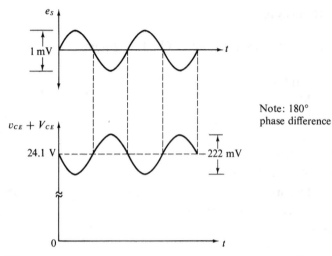

Note: 180°
phase difference

Figure 12.7.   INPUT AND OUTPUT WAVEFORMS

The power gain for this amplifier is simply $A_v \times A_i$. A_i is equal to h_{fe} since most of i_S will flow as base current. Thus

$$A_p = 222 \times 200 = 44,400$$

This very large power gain is typical of Com.-E amplifiers.

Increasing A_v

If we examine Equation 12.5, it might appear that A_v can be increased by increasing h_{fe}:

$$A_v = h_{fe}\frac{R_C}{h_{ie}} \tag{12.5}$$

However, an increase in h_{fe} will produce a corresponding increase in h_{ie} since $h_{ie} \simeq h_{fe} \times h_{ib}$; A_v will remain unaffected by the increase in h_{fe}. In other words, A_v is independent of h_{fe}. This independence can also be shown as follows:

$$A_v = \frac{h_{fe}R_C}{h_{ie}} \simeq \frac{h_{fe}R_C}{h_{fe} \times h_{ib}}$$

$$\approx \frac{R_C}{h_{ib}} \tag{12.8}$$

This last result should be recognized as the same expression as for the A_v of the Com.-B amplifier (Equation 11.12). This should not be surprising, since in both Com.-E and Com.-B the input e_S is applied across the E-B junction and the output signal is developed across the collector resistor. This means that A_v is the *same* for both types of amplifiers. As such, everything that was said in Chapter 11 concerning A_v is equally valid for the Com.-E amplifier. A_v can be increased by increasing R_C or by decreasing h_{ib}. Again, a smaller h_{ib} is obtained by biasing the transistor at a higher DC current.

12.6. Common-Emitter Amplifiers: Input and Output Impedance

The transistor's Com.-E AC input resistance, h_{ie}, is in parallel with R_B as seen in Figure 12.6C. Thus the total input impedance seen by the signal source is the parallel combination of R_B and h_{ie}. Denoting the *input impedance of the Com.-E amplifier* by Z_{ie}, we have

$$Z_{ie} = \frac{R_B \times h_{ie}}{R_B + h_{ie}} \tag{12.9}$$

The amplifier in Figure 12.6 has

$$Z_{ie} = 2 \text{ M}\Omega \parallel 1800 \text{ }\Omega$$
$$\simeq 1800 \text{ }\Omega$$

In this case, R_B is much larger than h_{ie} and has negligible effect on the input impedance. Whenever it is more than twenty times larger than h_{ie}, R_B can be neglected, and $Z_{ie} \simeq h_{ie}$. The value of Z_{ie} will typically range from 500 Ω to 10 kΩ. The Com.-E input impedance is somewhat higher than the Com.-B input impedance, making the Com.-E amplifier more readily driven by moderate impedance signal sources. In fact, an important feature of the Com.-E amplifier is its ability to drive other Com.-E amplifier stages without special impedance matching methods as a result of its comparable input and output impedances.

Output impedance

Referring again to Figure 12.6C, it can be seen that, looking back from the output to the *C-E* terminals, the only impedance is the collector resistor, R_C. Thus we have

$$Z_{oe} \simeq R_C \tag{12.10}$$

where Z_{oe} is the *output impedance of the Com.-E amplifier*.

The preceding expression is only approximate since it does not include the transistor's own AC resistance across its output terminals. The *transistor's Com.-E output resistance*, r_{oe}, is the AC resistance between collector and emitter. The effect of r_{oe} was not included in the transistor's AC equivalent circuit in our previous analysis. Figure 12.8 shows a more exact representation of the AC equivalent circuit for the Com.-E amplifier of Figure 12.6A. It includes r_{oe}, which is connected between C and E and is in parallel with R_C. Thus, Z_{oe} is more exactly given by

$$Z_{oe} = R_C \parallel r_{oe} \tag{12.11}$$

The value of r_{oe} is typically in the range from 20 kΩ to 100 kΩ, which makes it somewhat lower than r_{ob} for the Com.-B. The parallel combination of R_C and r_{oe} usually results in a Com.-E output impedance of moderate value (1 kΩ to 10 kΩ).

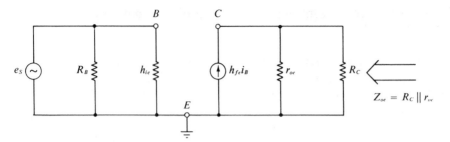

Figure 12.8.   INCLUSION OF r_{oe} IN EQUIVALENT CIRCUIT

Calculation of r_{oe}

The transistor's Com.-E output resistance r_{oe} is a measure of the effect of the collector-emitter voltage on the collector current. Its value can be obtained using

$$r_{oe} = \frac{\Delta V_{CE}}{\Delta I_C}\bigg|_{I_B} = \text{constant} \qquad (12.12)$$

Figure 12.9 shows how r_{oe} may be obtained from the Com.-E output curves. The Com.-E output curves are not as flat as the Com.-B curves, which explains why r_{oe} is less than r_{ob}.

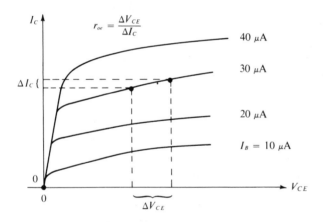

Figure 12.9.   CALCULATION OF r_{oe}

Output conductance h_{oe}

Most transistor specification sheets usually provide a typical value for h_{oe}, which is the *Com.-B AC output conductance*. h_{oe} is the reciprocal of r_{oe} and is usually given in units of micromhos. Thus

$$r_{oe} = \frac{1}{h_{oe}} \qquad (12.13)$$

▶ EXAMPLE 12.4

A certain Com.-E amplifier uses $R_C = 3$ kΩ and uses a transistor with $h_{oe} = 20$ μmhos. What is the output impedance of this amplifier?

$$r_{oe} = \frac{1}{h_{oe}} = \frac{1}{20 \times 10^{-6} \text{ mhos}}$$

$$= \frac{10^6}{20} \Omega = 50 \text{ k}\Omega$$

$$Z_{oe} = r_{oe} \parallel R_C = 50 \text{ k}\Omega \parallel 3 \text{ k}\Omega$$

$$= 2.83 \text{ k}\Omega \qquad\qquad ◀$$

12.7.   Effects of R_S and R_L

Let us consider the amplifier in Figure 12.10A and assume that the transistor has the following parameters at the Q point:

$$h_{fe} = \beta = 60$$
$$h_{ie} = 2.5\text{k}\Omega$$
$$h_{oe} = 25 \text{ }\mu\text{mhos} \qquad (r_{oe} = 40 \text{ k}\Omega)$$

Again, the portion of the circuit enclosed by the dotted lines is the amplifier proper. The values of Z_{in}, Z_{out} and A_v for the amplifier proper can be calculated for subsequent use in the general voltage amplifier representation. Using the transistor values given above, together with the circuit values, we have:

$$Z_{in} = Z_{ie} = 240 \text{ k}\Omega \parallel 2.5 \text{ k}\Omega \qquad\qquad \textbf{(12.9)}$$

$$\simeq 2.5 \text{ k}\Omega$$

$$Z_{out} = Z_{oe} = 40 \text{ k}\Omega \parallel 2 \text{ k}\Omega \qquad\qquad \textbf{(12.11)}$$

$$= 1.9 \text{ k}\Omega$$

$$A_v = \frac{h_{fe}R_C}{h_{ie}} = \frac{60 \times 2 \text{ k}\Omega}{2.5 \text{ k}\Omega} = 48 \qquad\qquad \textbf{(12.5)}$$

For this amplifier, the general representation becomes that shown in Figure 12.10B. The two circuits in the figure are equivalent for AC signals only as indicated by the absence of coupling capacitors and DC supplies. The negative sign in front of the voltage gain in Figure 12.10B is to indicate the 180° phase shift from input to output in Com.-E amplifiers. (This will be discussed in the next section.)

The circuit in Figure 12.10B can now be used to determine the amplifier operation for any values of R_S and R_L.

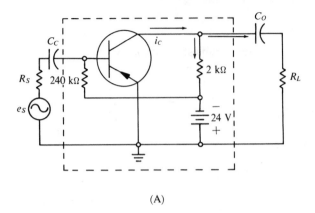

(A)

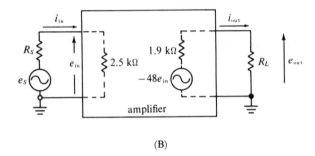

(B)

Figure 12.10.   (A) COMMON-EMITTER AMPLIFIER, INCLUDING R_S AND
                  R_L; (B) GENERAL REPRESENTATION

▶ EXAMPLE 12.5

Determine e_{out} and G_v for $R_S = 150 \ \Omega$ and $R_L = 5 \ k\Omega$. Use $e_S = 10$ mV p-p.

$$e_{in} = \frac{2.5 \ k\Omega}{150 \ \Omega + 2.5 \ k\Omega} \times e_S$$

$$\simeq 9.4 \ \text{mV p-p}$$

$$e_{out} = \left(\frac{5 \ k\Omega}{1.9 \ k\Omega + 5 \ k\Omega} \right) \times A_v e_{in}$$

$$= 0.724 \times (-48) \times 9.4 \ \text{mV}$$

$$= -326.7 \ \text{mV p-p}$$

$$G_v = \frac{e_{out}}{e_S} = -32.67 \qquad \text{(Negative sign indicates } 180° \text{ phase shift.)}$$

Overall gain G_v is less than A_v due to R_S and R_L.   ◀

▶ EXAMPLE 12.6

Calculate G_i and G_p for the amplifier of the last example.

$$i_S = \frac{e_S}{R_S + Z_{in}} = \frac{10 \text{ mV}}{2.65 \text{ k}\Omega} = 3.8 \text{ }\mu\text{A p-p}$$

$$i_{out} = \frac{e_{out}}{R_L} = \frac{326.7 \text{ mV}}{5 \text{ k}\Omega} = 65.3 \text{ }\mu\text{A p-p}$$

$$G_i = \frac{i_{out}}{i_S} = \frac{65.3}{3.8} = \textbf{17.2}$$

$$G_p = G_v \times G_i = 32.67 \times 17.2 = \textbf{562}$$

Note that the overall current gain G_i is less than β ($= 60$). This is because the output current to the load R_L is only a portion of the entire collector signal current. The collector signal current actually divides between R_C and R_L (see Figure 12.10A). ◀

12.8. Common-Emitter Amplifiers: Graphical Analysis of Amplifier Waveforms

Graphical analysis of the Com.-E amplifier is performed in much the same way as for the Com.-B. Let us consider the Com.-E circuit in Figure 12.11.

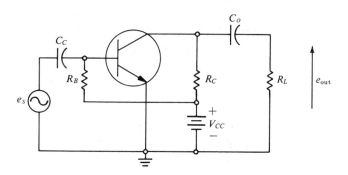

Figure 12.11.   COMMON-EMITTER AMPLIFIER

The circuit values are as follows:

$$e_S = 30 \text{ mV p-p}; \ R_S \text{ negligible}$$
$$R_B = 1 \text{ megohm}$$
$$V_{CC} = 10 \text{ V}$$
$$R_C = 10 \text{ k}\Omega$$
$$R_L = 10 \text{ k}\Omega$$

The transistor is germanium and has an input resistance $h_{ie} = 3 \text{ k}\Omega$ at $I_B = 10 \text{ }\mu\text{A}$. Its Com.-E output curves are presented in Figure 12.12. Once again, the Q point is found by constructing the DC load line, which in this case is determined by $V_{CC} = 10$ V and $R_C = 10$ kΩ. The resulting DC load line is shown in the figure. The DC value of base current is easily calculated as

$$I_B = \frac{V_{CC} - V_{BE}}{R_B}$$

$$= \frac{10 \text{ V} - 0.3 \text{ V}}{1 \text{ megohm}} = 9.7 \ \mu\text{A} \approx 10 \ \mu\text{A}$$

The resulting Q point is at $I_C = $ **0.5 mA** and $V_{CE} = $ **5 V**.

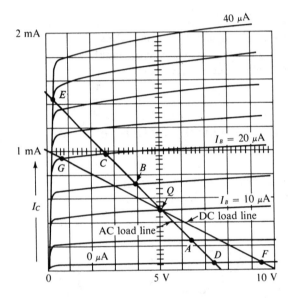

Figure 12.12.   OUTPUT CURVES FOR TRANSISTOR OF FIGURE 12.11

To determine the output signal voltage, the AC load line must be constructed. It passes through the Q point and has a slope of $-1/R'$ (recall that R' equals $R_C \parallel R_L$). In this case, $R' = 5$ kΩ, producing the AC load line shown in the figure. (The reader should review Section 11.11 on plotting the AC load line.)

The circuit operating point will move up and down the AC load line as the signal voltage is applied. The limits depend on the AC base current, which is given by

$$i_B = \frac{e_S}{h_{ie}} = \frac{30 \text{ mV p-p}}{3 \text{ k}\Omega} = 10 \ \mu\text{A p-p}$$

which indicates that the base current will vary 5 μA on either side of its Q-point value of 10 μA. That is, the base current will swing from 5 μA to 15 μA in response to the input signal. This effect is illustrated in Figure 12.13. Note that the base current waveform is *in phase* with the e_S waveform since, as e_S increases, the base-to-emitter voltage becomes more positive, which for an NPN

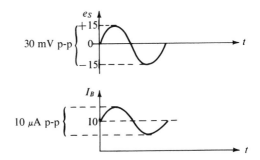

Figure 12.13.   VARIATION IN BASE CURRENT DUE TO INPUT SIGNAL

transistor means an increase in base current. This variation in base current moves the operating point between A and B along the AC load line. The collector current and voltage values at these points are recorded in Table 12.1.

TABLE 12.1

Point	I_B	I_C	V_{CE}	e_s
B	15 μA	0.72 mA	3.9 V	+15 mV
Q	10 μA	0.5 mA	5 V	0
A	5 μA	0.25 mA	6.25 V	−15 mV

Notice that as the base current makes a complete excursion from 5 μA to 15 μA the collector-emitter voltage V_{CE} makes a corresponding excursion from 6.25 V to 3.8 V. Figure 12.14 shows the various waveforms. The waveform of collec-

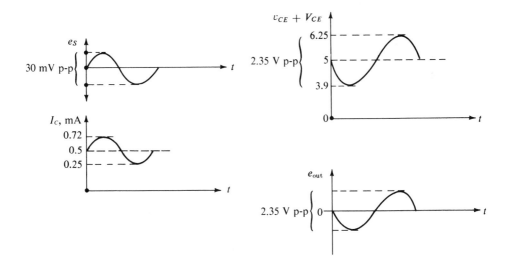

Figure 12.14.   AMPLIFIER WAVEFORMS

tor-emitter voltage contains a DC level (5 V) and an AC component (2.35 V p-p). Only the AC component appears across R_L, producing $e_{out} = 2.35$ V p-p. Thus the overall amplifier gain is

$$G_v = \frac{e_{out}}{e_S} = \frac{2.35 \text{ V p-p}}{30 \text{ mV p-p}} = 78.3$$

From Figure 12.14, an important characteristic of Com.-E amplifiers is observed: the output signal is 180° *out of phase* with the input signal. This is always the case at medium frequencies of operation with a resistive load. At low frequencies the coupling capacitors introduce additional phase shift, while at high frequencies the transistor junction capacitances have the same effect.

Waveform distortion

Waveform clipping will also occur in the Com.-E amplifier as the input signal is increased. It should be apparent from Figure 12.12 that any input signals which would produce a base current variation of more than 20 μA p-p would result in cutoff clipping, while any input signal which would increase the base current to 35 μA or more would result in saturation clipping. For an undistorted (unclipped) output, the maximum base current variation is 20 μA p-p. This gives an output swing from 2.7 V (point C) to 7.2 V (point D), or 4.5 V p-p.

▶ EXAMPLE 12.7

As the input signal is gradually increased, what type of clipping would occur first if the circuit Q point were at point C in Figure 12.12?

The base current at point C is 20 μA. An increase of slightly over 10 μA will cause saturation (point E), while a decrease of 20 μA is needed to reach cutoff. Thus, saturation clipping will occur first. ◀

▶ EXAMPLE 12.8

What is the maximum unclipped output signal which can be obtained from the amplifier under consideration when $R_L = \infty$?

In this case, the AC load line is the same as the DC load line in Figure 12.12. Along this load line, the base current can swing 20 μA p-p before cutoff occurs. This gives a V_{CE} swing from 1 V (point G) to 9.2 V (point F), producing $e_{out} = 8.2$ V p-p. ◀

12.9.   Indicating DC Supply Voltages

Up to now, we have indicated all DC supplies in our circuits by the standard DC battery symbol. In practice, a less cumbersome method of indicating DC supplies is employed whenever one side of the supply is grounded. This is illustrated in Figure 12.15 for the basic Com.-E circuit. Part A shows the complete

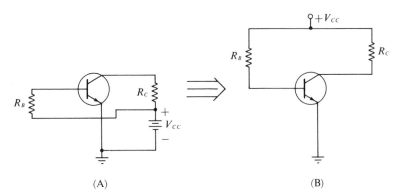

(A)   (B)

Figure 12.15.   EQUIVALENT WAYS TO REPRESENT DC SUPPLIES

battery symbol with its negative terminal connected to ground and its positive terminal connected to one end of R_B and R_C. Part B is an equivalent representation which eliminates the battery symbol. R_B and R_C are shown connected to a point labelled $+V_{CC}$. This $+V_{CC}$ point represents the positive terminal of a power supply of V_{CC} volts whose negative terminal is connected to ground (this ground connection is not shown but is understood to be present.)

In the work to follow, we will occasionally use this latter representation of DC supplies whenever it helps make the circuit diagrams less cluttered. This simpler representation is almost always used in the circuit diagrams of large or complex systems, especially when more than one supply is being used.

12.10.   Common-Emitter Amplifiers: Q-Point Stability

The first consideration in designing transistor amplifiers is the choosing of the quiescent operating point. The DC collector current and voltage must be chosen so that the AC variations caused by the input signal will remain within the linear region of the transistor output characteristics, thus avoiding distortion or clipping of the output waveform. For many applications, the best choice of Q point lies midway along the AC load line in order to allow the maximum undistorted output signal. Any change in the Q point will result in distortion at a lower level of output signal.

Two principal factors can contribute to a change in Q point once it has been set: one is the effect of temperature on the transistor parameters; the other is the change in transistor characteristics which occurs with transistor replacement. Of the three transistor amplifier configurations, the Com.-E is the only one which suffers considerably from shifts in the Q point. Thus our concentration will be centered on the Q-point stability of the Com.-E amplifier for various types of biasing circuits.

The simplest method of setting the Q point in the Com.-E amplifier is to arrange to have a DC base current flow so that when this base current is multiplied by β_{DC} the required value of collector current is obtained. This type of bias is illustrated in Figure 12.16 and is the type which has been used in all of the Com.-E amplifiers we have looked at so far. Since the base current is determined by V_{CC} and R_B and is essentially independent of the transistor param-

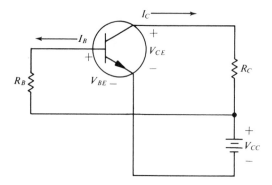

Figure 12.16.   BASE CURRENT BIAS

eters, this type of amplifier bias is called *base current bias*. In this circuit the base current is given by

$$I_B = \frac{V_{CC} - V_{BE}}{R_B} \qquad (12.14)$$

and the collector current then becomes

$$I_C = \beta_{DC}I_B + I_{CEO} \qquad (12.15)$$

This produces a collector-emitter voltage:

$$V_{CE} = V_{CC} - I_C R_C \qquad (12.16)$$

Examination of Equation 12.15 reveals the limitations of this simple method of setting the amplifier Q point. First, with any given type of transistor, the variation of current gain β_{DC} between different transistors is usually between two and three to one. This means that if I_B is chosen to give the desired value of I_C for one transistor, changing the transistor may result in a much different I_C, and the Q point will have shifted from the desired point. Secondly, as we have seen in Chapter 9, the leakage current I_{CEO} and current gain β_{DC} both increase rapidly with temperature. For this reason the value of collector current is very temperature dependent. Thus, if the circuit Q point is set at 25°C, at higher temperature it will have shifted to a new point. As far as the transistor is concerned, junction temperature increases may be due to a rise in ambient temperature or to heat generated as a result of junction power dissipation, or both.

The shifting of circuit operating point can be demonstrated graphically as in Figure 12.17 for a typical set of Com.-E output curves. The solid curves are for a temperature of 25°C. If we assume $I_B = 10\ \mu A$, the load line intersects the 10 μA curve at the Q point, $I_C = 1$ mA and $V_{CE} = 5$ V. A rise in temperature to 35°C causes the output curves to move upward due to the increase in I_{CEO}

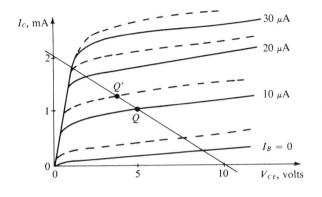

Figure 12.17.   EFFECT OF TEMPERATURE ON Q POINT

and β_{DC} as indicated by the dotted curves. The result is that the Q point has moved up to point Q' where $I_C = 1.2$ mA and $V_{CE} = 4$ V. Further increases in temperature would cause the Q point to move even further up the load line, closer to saturation. This shift in the Q point will cause distortion in the output signal to occur at a lower value of input signal, thus decreasing the possible un-distorted output voltage swing.

From these considerations it can be seen that for stable operation of a Com.-E transistor amplifier some alternative to the base current bias method must be found. This subject is usually referred to as *bias, or Q-point, stabilization*. The problem of Q-point instability does not arise in the Com.-B circuit since changes in α_{DC} are small and the only leakage current flowing is I_{CBO}. Although I_{CBO} increases with temperature, it is small enough to be negligible.

Base voltage bias

One of the most commonly used methods of bias stabilization is called, among other names, *base voltage biasing;* it is illustrated in Figure 12.18A. In this configuration, a fixed DC voltage V_B is applied between base and ground. Note that the emitter is not grounded. The voltage V_B is determined mainly by the voltage divider ratio of R_{B1} and R_{B2} and the supply voltage V_{CC}. This is not exactly true unless the input resistance looking into the base is very large compared to R_{B1} so that it will not affect V_B.

Recall from our discussion of the Com.-C circuit in Chapter 9 that the DC input resistance looking into the base is approximately $\beta_{DC} \times R_E$ when a resistor R_E is connected to the emitter. In Figure 12.18B, this effect is taken into account as the transistor base input is replaced by $\beta_{DC} R_E$. It should be clear that if $\beta_{DC} R_E$ is much greater than R_{B1}, say by a factor of 10 or more, then its effect on the voltage divider will be negligible and we can determine V_B from

$$V_B \simeq \frac{R_{B1}}{R_{B1} + R_{B2}} \times V_{CC} \qquad (12.17)$$

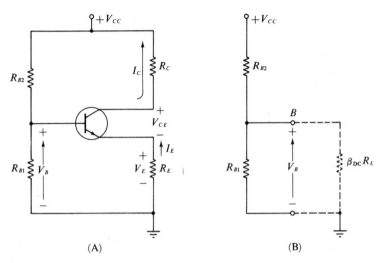

(A)   (B)

Figure 12.18.   (A) BASE VOLTAGE BIAS; (B) EQUIVALENT REPRESENTA-
TION OF BASE CIRCUIT

The voltage V_B, once established, determines the value of V_E in Figure 12.18A. Since

$$V_B = V_{BE} + V_E$$

then

$$V_E = V_B - V_{BE} \qquad (12.18)$$

The voltage V_E is across R_E. Using Ohm's law, we have

$$I_E = \frac{V_E}{R_E} \qquad (12.19)$$

as the emitter current. This value of I_E is fixed by the values of R_{B1}, R_{B2}, V_{CC} and R_E and is not dependent on the transistor parameters except for the small V_{BE} drop of Equation 12.18.

For this fixed value of I_E, the collector current is obtained from

$$I_C = \alpha_{DC} I_E \simeq I_E \qquad (12.20)$$

Once I_C is known, the voltage across R_C is known, and V_{CE} can be determined. Since

$$I_C R_C + V_{CE} + V_E = V_{CC}$$

then

$$V_{CE} = V_{CC} - I_C R_C - V_E \qquad (12.21)$$

Equations 12.18–12.21 essentially determine the Q-point values of the circuit.

▶ EXAMPLE 12.9

Determine I_C and V_{CE} for the circuit of Figure 12.18A for $R_{B1} = R_{B2} = 10$ kΩ, $R_E = 2$ kΩ, $R_C = 1$ kΩ, $V_{CC} = 12$ V and $\beta_{DC} = 100$.

Since $\beta_{DC} \times R_E = 200$ kΩ is much larger than R_{B1}, we can use the voltage divider rule to calculate V_B:

$$V_B \approx \frac{10 \text{ k}\Omega}{10 \text{ k}\Omega + 10 \text{ k}\Omega} \times 12 \text{ V} = 6 \text{ V}$$

If we assume a silicon transistor, then

$$V_E = V_B - 0.7 \text{ V} = 5.3 \text{ V}$$

and

$$I_E = \frac{V_E}{R_E} = \frac{5.3 \text{ V}}{2 \text{ k}\Omega} = 2.65 \text{ mA}$$

Thus

$$I_C \simeq I_E = \textbf{2.65 mA}$$

and

$$V_{CE} = 12 \text{ V} - 2.65 \text{ mA} \times 1 \text{ k}\Omega - \overset{V_E}{5.3 \text{ V}}$$
$$= 4.05 \text{ V} \qquad \blacktriangleleft$$

In this biasing method the Q-point values of I_C and V_{CE} do *not* depend on β_{DC} as long as $\beta_{DC} R_E \gg R_{B1}$. This removal of the dependence of I_C on β_{DC} removes the major cause of Q-point instability. If β_{DC} changes due to temperature effects or transistor replacement, I_C and V_{CE} will be relatively unaffected as long as $\beta_{DC} R_E \gg R_{B1}$. In practice, this condition can usually be met by making R_E no *smaller* than $1/5$ of R_{B1}. In such cases, the Q-point stability is quite adequate. The larger the value of R_E, the more stable will be the Q point. However, R_E cannot be made too large since it reduces I_E. With a lower I_E, h_{ie} would be larger, resulting in a lowering of amplifier voltage gain.

Effect of R_E on voltage gain

Figure 12.19A shows a Com.-E amplifier using base voltage biasing with the input signal capacitively coupled to the base. In this circuit, the input signal e_S does not appear directly across the E-B junction due to the presence of R_E, as is illustrated in part B of the figure. e_S divides up between h_{ie}, the AC resistance between base and emitter, and βR_E, the effective resistance of R_E looking in from the base. This results in a greatly reduced voltage gain since a smaller signal across h_{ie} means a smaller i_B, and, therefore, a smaller i_C signal, which in turn reduces $e_{\text{out}} = i_C R_C$.

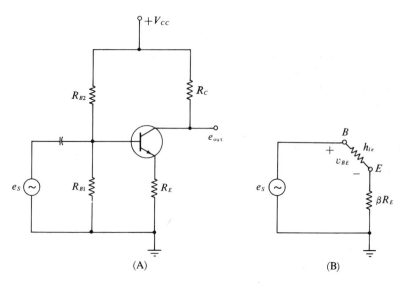

Figure 12.19.   (A) COMMON-EMITTER AMPLIFIER WITH BASE VOLTAGE
                     BIAS; (B) REDUCTION OF *E-B* SIGNAL BY PRESENCE OF R_E

The reduction in voltage gain due to the presence of R_E can be avoided if R_E is shunted by a capacitor C_E as shown in Figure 12.20. The purpose of C_E is essentially to act as an AC short circuit across R_E, thereby reducing its effect on the AC input signal. This means that R_E is shorted out as far as the AC input signal is concerned so that all of e_S appears across h_{ie}. The capacitor C_E, called the *emitter bypass capacitor*, must be chosen large enough so that its reactance $X_C \ll h_{ie}$ at the lowest input frequency, with the effect that most of e_S will be across h_{ie} with very little across the R_E–C_E parallel combination. This will make $v_{be} \approx e_S$.

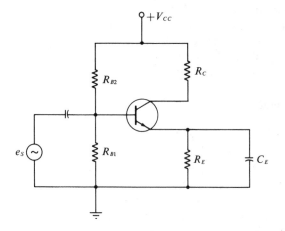

Figure 12.20.   BIAS CAPACITOR C_E SHORTS OUT R_C TO AC

In summary, using the bypass capacitor allows R_E to be used to stabilize the Q point without reducing A_v. In fact, A_v will be exactly the same as if the emitter were grounded, since that is essentially what the capacitor does: it grounds the emitter for AC signals. Of course, C_E has no effect on the DC operation and thus does not affect the Q point.

12.11.   Common-Emitter Amplifier: Summary

The important characteristics of the Com.-E transistor amplifier are summarized in Table 12.2:

TABLE 12.2

COMMON-EMITTER AMPLIFIER

Current gain A_i: β, much greater than 1
Voltage gain A_v: very high; 500 is typical
Power gain A_p: extremely high; 20,000 is typical
Input impedance Z_{ie}:   moderately high; 1 kΩ is typical
Output impedance Z_{oe}:   moderately high; 2 kΩ is typical
Phase shift: 180°

GLOSSARY

β (h_{fe}):   AC forward current gain in common-emitter configuration.

h_{ie}:   transistor AC input resistance in common-emitter configuration.

Z_{ie}:   Com.-E amplifier input impedance.

Z_{oe}:   Com.-E amplifier output impedance.

r_{oe}:   transistor's AC output resistance in common-emitter configuration.

h_{oe}:   transistor's AC output conductance in common-emitter configuration.

Q-point stability: variation of Q point with changes in temperature and with transistor replacement.

Base current bias: method of biasing a common-emitter amplifier with a constant base current.

Base voltage bias: biasing a Com.-E amplifier with a constant base-to-ground voltage.

PROBLEMS

12.1   What transistor parameter measures AC current gain from base to collector?

12.2   The set of data below is for a certain transistor. From this data determine β.

V_{BE}	I_B	I_C	V_{CE}
0.69 V	10 μA	0.51 mA	10 V
0.70 V	14 μA	0.71 mA	10 V

12.3   Define h_{fe}. Determine its value for the transistor of Problem 12.2.

12.4   Define h_{ie}. Determine its value for the transistor of Problem 12.2.

12.5   Refer to the amplifier of Figure 12.2. The circuit uses a silicon transistor with $\beta = \beta_{DC} = 75$. The circuit values are $V_{CC} = 35$ V, $R_B = 1$ MΩ, $R_C = 4$ kΩ and $e_S = 2$ mV p-p.
(a) Determine amplifier Q-point values I_B, I_C, V_{CE}.
(b) Determine the output signal amplitude e_{out}.
(c) Determine A_i, A_v and A_p.
(d) Sketch the total waveforms (DC + AC) of base current, collector current and collector-emitter voltage.

12.6   Indicate what would happen to the amplitude of e_{out} in the amplifier of Problem 12.5 for each of the following changes:
(a) Increase in V_{CC} voltage.
(b) Decrease in R_C.
(c) Increase in R_B.
(d) Increase in h_{fe}.
(e) Connect load to output.

12.7   Determine the input and output impedance of the amplifier of Problem 12.5 if $h_{oe} = 20$ μmhos. Use the general voltage amplifier representation to calculate G_v, G_i and G_p if a 2 mV p-p signal source with $R_S = 500$ Ω is used as input, and the output is capacitively coupled to a 10-kΩ load.

12.8   For each of the following statements indicate whether it refers to Com.-B or Com.-E amplifiers, or both:
(a) Gives a 180° phase shift between e_S and e_{out}.
(b) Has a very low input impedance.
(c) Has bias stability problems.
(d) Has high output impedance (50 kΩ).
(e) Gives high voltage gain (500).
(f) Has highest power gain.
(g) Has low current gain.
(h) Will produce distortion when overdriven.
(i) A_v increases if R_C is increased.
(j) Can be biased from a single DC supply.
(k) Output of one amplifier can easily drive the input of a similar amplifier without significant loss in voltage gain.

12.9 The amplifier in Figure 12.21 uses a germanium transistor with $h_{ie} = 2$ kΩ. The transistor's output curves are shown in Figure 12.12. Using the graphical technique determine:
(a) The complete waveforms of C-E voltage and collector current.
(b) The values of e_{out} and G_v.

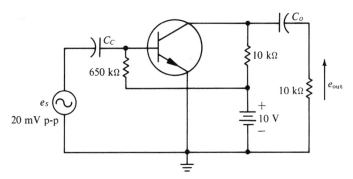

Figure 12.21.

12.10 Repeat Problem 12.9 using $e_S = 40$ mV p-p.

12.11 For the amplifier of Problem 12.9 what is the maximum value of e_S which will produce no clipping?

12.12 (a) Determine I_C and V_{CE} for the circuit in Figure 12.16 if $\beta_{DC} = 50$, $R_B = 100$ kΩ, $R_C = 1$ kΩ, $V_{CC} = 20$ V and the transistor is silicon.
(b) Repeat part (a) if β_{DC} increases to 75 due to an increase in temperature. This should show how unstable the Q point is for base-current biasing.

12.13 Explain how base-voltage biasing improves Q-point stability of the Com.-E amplifier.

12.14 (a) Determine the Q point of the circuit in Figure 12.22 if $\beta_{DC} = 50$.

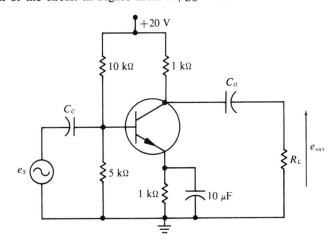

Figure 12.22.

(b) Repeat part (a) if β_{DC} increases to 75. Compare the variation in Q point to that of Problem 12.12.

(c) Explain the purpose of the 10-μF capacitor. What determines how large it should be?

12.15   Consider again the amplifier in Figure 12.22. Indicate the effect each of the following changes will have on the amplitude of e_{out}. In each case assume a change which is large enough to affect e_{out}.

(a) Increase R_E.

(b) Decrease C_E.

(c) Decrease input frequency.

(d) Decrease R_L.

(e) Increase V_{CC}.

(f) Decrease C_o.

(g) Increase β_{DC}.

Common-Collector
Amplifiers and Other Topics

13.1.   Introduction

The Com.-C amplifier possesses characteristics which differ widely from Com.-B and Com.-E amplifiers. Its basic operation will be analyzed in this chapter and these differences will become apparent. In addition, the effects of frequency on amplifier operation will be examined for all three types.

13.2.   Common-Collector Amplifiers: Basic Circuit

The basic Com.-C amplifier circuit is shown in Figure 13.1. Except for the absence of a collector resistor R_C and emitter bypass capacitor C_E, the circuit is identical to the Com.-E circuit with base-voltage bias shown in Figure 12.20. The Com.-C amplifier also utilizes base-voltage bias. However, the output signal is taken from the emitter instead of the collector.

Typical circuit action is as follows. The input signal is coupled through the capacitor into the base of the transistor. This produces a changing base current and consequently a changing emitter current. This changing emitter current produces a signal voltage across R_E, as shown, which is then coupled to the load resistor R_L. We shall find that this output signal will always be slightly less

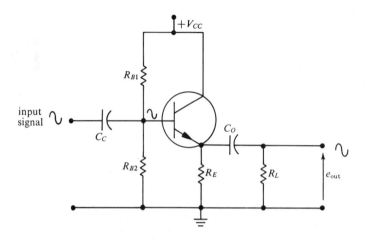

Figure 13.1.   COMMON-COLLECTOR AMPLIFIER

in amplitude than the input signal e_S, resulting in a voltage gain of less than one. A natural question to ask is: Of what use is this circuit if its voltage gain is less than unity? The answer is that this circuit has the advantage of a very high input impedance while at the same time possessing a low output impedance. This characteristic makes it useful as an impedance matching circuit between a high impedance signal source and a low impedance load.

13.3.   Analysis of Operation

To analyze the AC operation of the Com.-C amplifier we will again replace the circuit by its AC equivalent circuit with the DC supply and the coupling capacitors shorted out as in Figure 13.2A. Notice that the collector terminal is grounded for AC since it is tied to $+V_{CC}$. Thus R_{B1} and R_{B2} essentially appear in parallel for AC as shown in the figure. As such, we will refer to this parallel combination as R_B during the rest of this discussion. That is,

$$R_B = \frac{R_{B1} R_{B2}}{R_{B1} + R_{B2}} \tag{13.1}$$

The emitter resistor R_E and load resistor R_L are also in parallel as far as AC is concerned (assuming C_o is an AC short). This parallel combination will be referred to as $(R_L)'$, the effective load. Thus

$$(R_L)' = \frac{R_E R_L}{R_E + R_L} \tag{13.2}$$

The symbols R_B and $(R_L)'$ are used in Figure 13.2B.

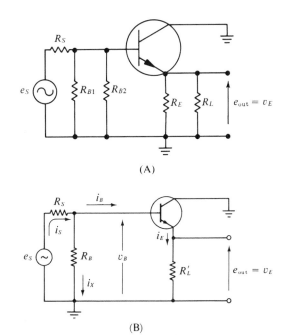

Figure 13.2.   COMMON-COLLECTOR CIRCUIT FOR AC ANALYSIS

The signal source e_S supplies a current i_S to the amplifier. A certain portion of i_S will flow through R_B to ground (i_X), with the remainder becoming base current for the transistor. In most practical circuits the value of i_X will be of the same or greater order of magnitude as i_B and so *cannot* be neglected. In order to find the values of i_X and i_B, we must first determine the impedance seen looking into the base of the transistor since it is in parallel with R_B. Let us call this impedance Z_B. Thus we can write

$$Z_B = \frac{v_B}{i_B} \qquad (13.3)$$

where v_B is the AC voltage from base to ground. Note that v_B is not the same as v_{BE} since the emitter is not at ground potential.

To help us determine Z_B, we can replace the transistor E-B junction by its AC resistance h_{ie}. R_L', when viewed from the base, has an equivalent resistance of approximately $h_{fe} \times R_L'$. This latter fact is derived from our previous discussions of Com.-C circuits and the Com.-E circuit with base-voltage bias. Figure 13.3 shows the circuit with these effects now included. The impedance Z_B between B and ground is now easily seen to be

$$Z_B = h_{ie} + h_{fe}R_L' \qquad (13.4)$$

This Z_B is the total impedance looking directly into the transistor base and is a series combination of h_{ie} (between B and E) and $h_{fe}R_L'$ (between E and ground).

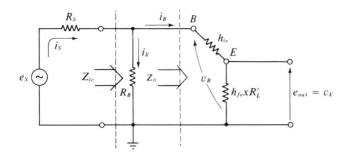

Figure 13.3.   TRANSISTOR REPLACED BY ITS AC EQUIVALENT RESISTANCES

Amplifier input impedance

The total input impedance seen by the source is the parallel combination of R_B and Z_B. That is,

$$Z_{ic} = R_B \parallel Z_B \qquad (13.5)$$

where Z_{ic} is the *Com.-C amplifier input impedance*. It is important to note that Z_{ic} is directly affected by R_L, the load on the amplifier output. This is because R_L is part of R'_L, which in turn is part of Z_B (Equation 13.4) and because Z_{ic} depends on Z_B. The following example illustrates:

▶ EXAMPLE 13.1

(a) Calculate Z_{ic} if $R_{B1} = 50$ kΩ, $R_{B2} = 50$ kΩ, $R_E = 2$ kΩ and $R_L = 2$ kΩ. Assume $h_{fe} = 200$ and $h_{ie} = 2$ kΩ.
(b) Repeat for $R_L = 500$ Ω.

(a)

$$R'_L = R_L \parallel R_E = 1 \text{ k}\Omega$$
$$Z_B = h_{ie} + h_{fe}R'_L = 2 \text{ k}\Omega + 100(1 \text{ k}\Omega) = 102 \text{ k}\Omega$$
$$R_B = 50 \text{ k}\Omega \parallel 50 \text{ k}\Omega = 25 \text{ k}\Omega$$
$$Z_{ic} = R_B \parallel Z_B = 25 \text{ k}\Omega \parallel 102 \text{ k}\Omega$$
$$\simeq 20 \text{ k}\Omega$$

(b)

$$R'_L = 2 \text{ k}\Omega \parallel 500 \text{ }\Omega = 400 \text{ }\Omega$$
$$Z_B = 2 \text{ k}\Omega + 100(400) = 42 \text{ k}\Omega$$
$$Z_{ic} = 25 \text{ k}\Omega \parallel 42 \text{ k}\Omega$$
$$= 15.7 \text{ k}\Omega \qquad \blacktriangleleft$$

It is apparent that the value of R_L has a marked effect on the amplifier input impedance. This effect is peculiar to the Com.-C amplifier but not to the Com.-B or Com.-E types. Values for Z_{ic} are fairly high and range typically from

10 kΩ to 1 MΩ. Thus Com.-C amplifiers usually have a larger input impedance than the other amplifier types.

Calculation of e_{out} and G_v

Once Z_{ic} is determined in Figure 13.3, it is a relatively easy matter to calculate e_{out}. First, it should be seen that the e_S input signal will divide up between R_S and Z_{ic} so that v_B is only a fraction of e_S. Using the voltage divider method,

$$v_B = \left(\frac{Z_{ic}}{Z_{ic} + R_S}\right) \times e_S \qquad (13.6)$$

Next, it can be seen that v_B is divided between h_{ie} and $h_{fe}R_L'$. Therefore,

$$e_{\text{out}} = \left(\frac{h_{fe}R_L'}{h_{fe}R_L' + h_{ie}}\right) \times v_B \qquad (13.7)$$

It should be clear that e_{out} has to be *less than* e_S since only a portion of e_S becomes v_B and only a portion of v_B becomes e_{out}. By combining the above two expressions we can calculate $G_v = e_{\text{out}}/e_S$ as

$$G_v = \frac{e_{\text{out}}}{e_S} = \left(\frac{Z_{ic}}{Z_{ic} + R_S}\right)\left(\frac{h_{fe}R_L'}{h_{fe}R_L' + h_{ie}}\right) \qquad (13.8)$$

The value of G_v has to be less than one, although typically G_v is kept fairly close to one by ensuring that $Z_{ic} \gg R_S$ and $h_{fe}R_L \gg h_{ie}$.

▶ EXAMPLE 13.2

Consider the transistor Com.-C amplifier shown in Figure 13.4. If the transistor parameters are $h_{ie} = 1$ kΩ and $\beta = 100$, calculate the voltage gain, G_v, and e_{out}.

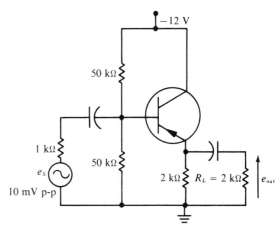

Figure 13.4.

To calculate gain we must determine the amplifier input impedance Z_{ic}, which depends on the transistor input impedance Z_B:

$$Z_B = 1 \text{ k}\Omega + 100(1 \text{ k}\Omega) = 101 \text{ k}\Omega \qquad (13.4)$$

where $(R_L') = 1 \text{ k}\Omega$. The value of R_B is

$$R_B = 50 \text{ k}\Omega \parallel 50 \text{ k}\Omega = 25 \text{ k}\Omega$$

Thus we have

$$Z_{ic} = \frac{101 \text{ k}\Omega \times 25 \text{ k}\Omega}{126 \text{ k}\Omega} \simeq 20 \text{ k}\Omega \qquad (13.5)$$

Now, using Equation 13.8, the voltage gain becomes

$$G_v = \left(\frac{20 \text{ k}\Omega}{20 \text{ k}\Omega + 1 \text{ k}\Omega}\right)\left(\frac{100 \text{ k}\Omega}{100 \text{ k}\Omega + 1 \text{ k}\Omega}\right)$$
$$= \textbf{0.943}$$
$$e_{\text{out}} = G_v \times e_S$$
$$= 0.943 \times 10 \text{ mV}$$
$$= \textbf{9.43 mV p-p} \qquad \blacktriangleleft$$

It should be clear at this point that the output signal of the Com.-C amplifier is *in phase* with the input signal. Referring to Figure 13.3, it can be seen that e_{out} is some fraction (usually close to one) of e_S and in phase with e_S. We can essentially say that the output variations follow almost exactly the input signal variations. For this reason, the Com.-C amplifier is often referred to as an *emitter follower*.

Current gain and power gain

The Com.-C current gain G_i can be calculated by determining i_S and i_{out} as follows:

$$i_S = \frac{e_S}{Z_{ic} + R_S} \qquad \text{(See Figure 13.3.)}$$

$$i_{\text{out}} = \frac{e_{\text{out}}}{R_L}$$

$$G_i = \frac{i_{\text{out}}}{i_S} = \frac{\dfrac{e_{\text{out}}}{R_L}}{\dfrac{e_S}{Z_{ic} + R_S}}$$

$$= \left(\frac{e_{\text{out}}}{e_S}\right)\left(\frac{Z_{ic} + R_S}{R_L}\right)$$

$$G_i = G_v\left(\frac{Z_{ic} + R_S}{R_L}\right) \qquad (13.9)$$

▶ EXAMPLE 13.3

Calculate G_i and G_p for the amplifier of Figure 13.4. From Example 13.2, $Z_{ic} = 20$ kΩ and $G_v = 0.943$. Thus

$$G_i = 0.943\left(\frac{21 \text{ k}\Omega}{2 \text{ k}\Omega}\right)$$

$$\simeq \mathbf{10}$$

$$G_p = G_i \times G_v = 10 \times 0.943$$

$$= \mathbf{9.43} \qquad \blacktriangleleft$$

The above example indicates that the Com.-C current gain G_i can be much less than the transistor current gain h_{fe} (which was 100 for the transistor in the example). There are two reasons for this: first, not all of i_S flows into the base since some signal current flows in the bias resistors R_{B1}, R_{B2}; secondly, the emitter signal current i_E divides up between R_E and R_L. As such, G_i will always be less than h_{fe} except when $R_B \ll Z_B$ and $R_L \gg R_E$. However, G_i will almost always be greater than one.

The Com.-C power gain G_p is usually approximately the same as G_i since G_v is approximately one. Typical values of G_p for Com.-C amplifiers range from 5–500 and are usually not as large as Com.-B and Com.-E power gains.

Output impedance

Let us now consider the output impedance of the Com.-C amplifier. The amplifier's output impedance is what the load resistor sees looking back into the amplifier. Referring to Figure 13.2A, the load resistor R_L looking back into the emitter sees three parallel paths to ground. One of these is the emitter resistor R_E. Another path is back through the emitter and out of the collector; the third path is back through the emitter and out the base. This last path is usually the most significant because it has the least resistance. Any resistance in the base lead appears, when viewed from the emitter, to be *less* than its actual value by approximately the factor h_{fe}. The output impedance will essentially be the impedance seen looking back from the emitter into the base. Denoting the *Com.-C amplifier output impedance* by the symbol Z_{oc}, we have

$$Z_{oc} = \left(h_{ib} + \frac{R_B \parallel R_S}{h_{fe}}\right) \parallel R_E \qquad \textbf{(13.10)}$$

Where h_{ib} is the E-B junction resistance looking from the emitter and the second term is the total resistance in the base circuit ($R_B \parallel R_S$) reduced by the factor h_{fe}. The *value of Z_{oc} is normally very low* and usually is in the range of 10 to 100 ohms.

▶ EXAMPLE 13.4

For the amplifier of Example 13.2 determine the value of output impedance Z_{oc}. Use $h_{ib} = 10 \ \Omega$.

For the transistor, $\beta = 100$, as stipulated in Example 13.2. Using Equation 13.10, we have

$$Z_{oc} = \left(10 \ \Omega + \frac{25 \ \text{k}\Omega \parallel 1 \ \text{k}\Omega}{100}\right) \parallel 2 \ \text{k}\Omega$$
$$= (10 \ \Omega + 9.5 \ \Omega) \parallel 2 \ \text{k}\Omega$$
$$= \textbf{19.0} \ \boldsymbol{\Omega} \qquad \blacktriangleleft$$

Referring to Equation 13.10, it can be seen that the Com.-C output impedance depends on the resistance of the signal source. The larger the source resistance, the higher the output impedance. This effect was neglected in the Com.-B and Com.-E analyses but must be considered in the Com.-C amplifier.

13.4.   Common-Collector Amplifiers: Usefulness

As we have seen, Com.-C amplifiers have no voltage gain and only a moderate current and power gain. However, it was mentioned previously that the high input impedance and low output impedance of a Com.-C amplifier make it suitable for matching a high impedance source to a low impedance load. A practical application would be in preamplifiers that are operated from a high impedance signal source, such as microphones or phonograph pickups. A typical example of such a circuit is shown in Figure 13.5. In this circuit, a fairly high input impedance is provided to the signal source, while the low output impedance of the amplifier can be used to drive a Com.-E amplifier to give the required voltage gain.

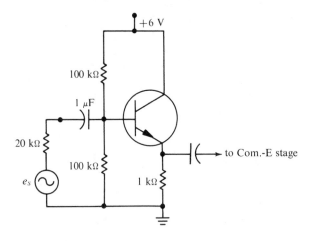

Figure 13.5.   COMMON-COLLECTOR PREAMPLIFIER

A second common application utilizes the very low output impedance of the Com.-C amplifier to drive the base of a Com.-E power amplifier which is normally the last stage of amplification (the output stage). The power amplifier requires a large amount of input current in order to deliver maximum power to the load. Such an arrangement for driving a speaker is shown in Figure 13.6.

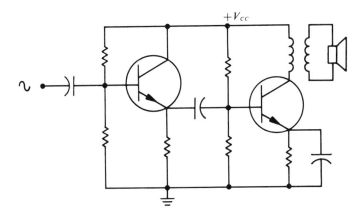

Figure 13.6.   COMMON-COLLECTOR DRIVER

13.5.   Common-Collector Amplifiers: Summary

Table 13.1 summarizes some of the Com.-C characteristics which we have discussed.

TABLE 13.1

COMMON-COLLECTOR AMPLIFIER

Current gain A_i: greater than 1 (as high as h_{fe})
Voltage gain A_v: less than 1; usually close to 1
Power gain A_p: moderately high; 50 is typical
Input impedance Z_{ic}:   high; 50 kΩ is typical
Output impedance Z_{oc}:   very low; 50 Ω is typical
Phase shift: 0°

13.6.   Effects of Frequency on Amplifier Operation

In the previous discussions on transistor amplifiers it was tacitly assumed that the signal frequency was neither low enough nor high enough to affect amplifier operation. In practice, signals of very low or very high frequency are encountered and one must understand how the amplifier responds to these different signals. A typical amplifier will suffer from a loss of gain (current, voltage and power) as the signal frequency increases or decreases beyond certain limits. Figure 13.7 illustrates this in a gain-versus-frequency plot for such an amplifier. It can be seen that over a range of frequencies (sometimes referred to as the

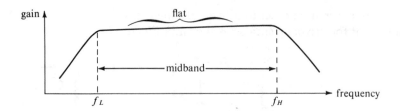

Figure 13.7.   TYPICAL AMPLIFIER GAIN-VS.-FREQUENCY PLOT

mid-band) the amplifier gain is fairly constant but drops off at frequencies below f_L and above f_H. The following discussion deals with the causes of this gain dropoff in a rather qualitative manner. The numerous formulas which usually accompany such a discussion are beyond the scope of this text and so are not included.

Low-frequency effects

If we consider the low-frequency effects first, it can generally be stated that any loss of gain can be attributed to the coupling and bypass capacitors in the circuit. As an example, consider the Com.-E amplifier shown in Figure 13.8 which contains two coupling capacitors and one bypass capacitor. At mid-band frequencies or greater these capacitors will have a very small impedance if chosen correctly. This allows us to assume ideal coupling and bypassing at such frequencies. However, at low frequencies the impedance of these capacitors will eventually increase (recall $X_c = 1/2\pi fC$) to the point where these assumptions are no longer valid. For example, the impedance of C_C will increase to the point where a portion of the input signal e_S will be dropped across it. It follows, then, that a smaller portion of the signal will reach the base of the transistor and become amplified. Similarly, at low frequencies a portion of the output signal will be lost across C_o, resulting in a smaller e_{out}. The function of C_E, as discussed in

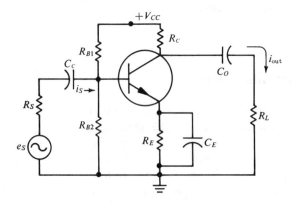

Figure 13.8.   COMMON-EMITTER AMPLIFIER

Chapter 12, is to provide a low-impedance bypass across R_E so that all of the input signal would appear across the *E-B* junction and none would be lost across R_E. Obviously, the bypassing action of C_E at low frequencies becomes less and less effective. All these effects contribute to a smaller output signal and thus lower gain. The individual effects will generally occur at different frequencies. For example, loss of gain due to C_c may begin at 100 Hz, due to C_o at 50 Hz, and due to C_E at 10 Hz. In this case, the most critical effect will be that of C_c because it occurs at the highest frequency. It may be reasoned that elimination of these capacitors entirely would eliminate their effect on low frequency gain. In certain types of amplifiers this is actually done, usually at the expense of lower mid-band gain and problems in biasing and stability. These amplifiers, called *DC (Direct Coupled) amplifiers*, are very important in electronics and should be encountered in later courses. Although the Com.-E amplifier was used in our example, similar low-frequency effects occur in the Com.-B and Com.-C amplifiers.

High-frequency effects

The loss of gain at high frequencies can be attributed to two major factors, one of which is the decrease in transistor current gain α. The decrease in α is related to the transit time of the current carriers as they move from the emitter to the collector. The carriers injected from the emitter into the base will either recombine or diffuse the entire width of the base region to reach the collector. As the signal frequency increases, a frequency will be reached at which the time involved in the transit of the carriers across the base will be longer than the signal period. When this occurs, the change in collector current will not be able to follow exactly the change in emitter current caused by the applied signal across the *E-B* junction. As a result, a build-up of carriers will take place in the base region, thus increasing the amount of recombinations and decreasing α. Of course, the greater the transit time, the lower the frequency at which the decrease in α occurs. The transit time depends on the width of the base region and the diffusion speed of the carriers, which in turn depends on the semiconductor material and carrier polarity. In general, for the same base thickness, germanium will have a shorter transit time than silicon and NPNs will have a shorter transit time than PNPs.

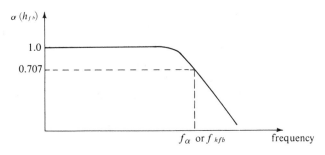

Figure 13.9.   VARIATION OF α WITH FREQUENCY

The variation of α with frequency is shown in Figure 13.9. Manufacturers usually specify the frequency at which α (h_{fb}) drops to 0.707 of its low-frequency value. The symbol f_α or f_{hfb} is given as the frequency at which this occurs. This frequency is referred to as the *common-base cutoff frequency* and is primarily used with the Com.-B configuration since α is the current gain for Com.-B amplifiers.

Since β is related to α by

$$\beta = \frac{\alpha}{1 - \alpha}$$

the value of β will also decrease with frequency but at a much faster rate. The symbol f_β or f_{hfe} is given as the frequency at which β (h_{fe}) drops to 0.707 of its low-frequency value, and is called the *common-emitter cutoff frequency*. This is

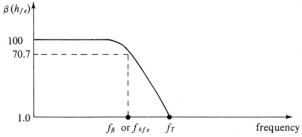

Figure 13.10.   VARIATION OF β WITH FREQUENCY

indicated in Figure 13.10. For a given transistor, this frequency will be much lower than f_{hfb}. In fact, they are related by

$$f_{hfe} \approx \frac{f_{hfb}}{h_{fe}} \tag{13.11}$$

▶ EXAMPLE 13.5

A transistor with a low-frequency $h_{fe} = 50$ has an f_{hfb} of 10 MHz. Determine the value of f_{hfe}.

$$f_{hfe} = 10 \, \frac{\text{MHz}}{50} = 200 \text{ kHz} \qquad \blacktriangleleft$$

This example illustrates the fact that *current gain in the Com.-E and Com.-C configurations will begin dropping at a much lower frequency than the current gain in the Com.-B configuration.* Another frequency parameter of importance for Com.-E operation is given the symbol f_T and is the frequency at which β has decreased to a value of one (shown in Figure 13.10). Such a frequency as f_T is considered because in many Com.-E circuits (not necessarily amplifiers) the value of β need only be greater than one for useful operation. The value of f_T is approximately given by

$$f_T \approx \frac{f_{hfb}}{1.2} \tag{13.12}$$

▶ EXAMPLE 13.6

For the transistor of the previous example determine the value of f_T.

$$f_T = \frac{10 \text{ MHz}}{1.2} = 8.33 \text{ MHz}$$ ◀

The other high-frequency factor is the effect of junction capacitances. An equivalent capacitance exists across both the *E-B* and *C-B* junctions. C_{ob} is the symbol given the output capacitance of the transistor in the Com.-B mode and is essentially the capacitance of the *C-B* junction and any stray or wiring capacitance. It is illustrated in Figure 13.11 and it can be seen that it essentially shunts the load (at high frequencies C_o is an AC short). This indicates that at high frequencies the impedance of C_{ob} will become small enough to shunt signal current away from the load, reducing e_{out}. The value of C_{ob} depends on doping, *C-B* junction geometry and on the DC voltage across the *C-B* junction and is usually around a few pF (10^{-12} farads).

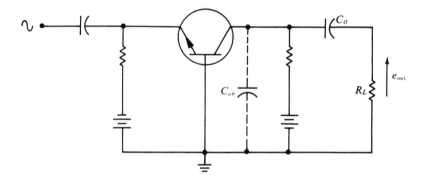

Figure 13.11.   COMMON-BASE OUTPUT CAPACITANCE

C_{oe} is the corresponding symbol for the output capacitance in the Com.-E mode and is a little more complex than C_{ob} since the output is taken from collector to emitter. The capacitances between collector and base and between base and emitter both enter into the output capacity in the Com.-E configuration. There is a relationship between C_{oe} and C_{ob} given below which indicates that the output capacitance in the Com.-E configuration is much greater than in the Com.-B.

$$C_{oe} \approx h_{fe}C_{ob} \tag{13.13}$$

▶ EXAMPLE 13.7

A transistor with $h_{fe} = 100$ has a $C_{ob} = 4$ pF at $V_{CB} = 10$ V. Determine C_{oe}.

$$C_{oe} = 100 \times 4 \text{ pF} = 400 \text{ pF}$$ ◀

Output capacitance in the Com.-C amplifier is relatively insignificant since the output is taken at the emitter, a very low output impedance point. For this reason, Com.-C amplifiers usually have the best high-frequency response of the three amplifier configurations.

Similar effects occur due to transistor junction capacitances at the amplifier input. Here, the capacitances tend to "short out" the input at high frequencies, thereby reducing e_{in} and, consequently, e_{out}.

There are other high-frequency effects, but those just discussed are the most significant. It should be pointed out that many factors go toward determining the high-frequency limit of a transistor amplifier and there are numerous methods of increasing this limit. These methods appropriately belong in a circuit analysis text and will not be covered here.

13.7.   Comparison of Amplifier Configurations

Table 13.2 summarizes the three transistor amplifier configurations and compares their relative characteristics.

TABLE 13.2

Characteristic	Common base	Common emitter	Common collector
Current gain	less than one	high	high
Voltage gain	high	high	less than one
Power gain	moderate	highest	low
Input impedance	very low	moderate	high
Output impedance	high	moderate	very low
Input terminal	emitter	base	base
Output terminal	collector	collector	emitter
Phase shift (mid-band)	0°	180°	0°
Bias stability	good	poor (base current bias)	good
Frequency response	good	fair	best

13.8.   Transistor Design Considerations

In recent years, construction of the transistor has undergone a great many changes and improvements in order to reach its present state of development. A number of manufacturing processes have been devised, have lasted a few years and then have been abandoned as new techniques and processes have come from the laboratory. The design of transistor construction is mainly concerned with the geometry (physical dimensions, shape) and the amounts of doping, both of which can be controlled by the *transistor* designer (not the *circuit* designer). The geometry and doping levels determine to a great extent the behavior of electrical properties such as current gain, frequency response, power-handling capability, switching speed and breakdown voltages.

A thorough discussion of the relation between these electrical characteristics and the transistor physical characteristics would take a prohibitively long time. The important points to be considered have been summarized in Table 13.3. In this table the desired electrical property is followed by the principal physical characteristics which produce it. For example, a high current gain is produced by doping the emitter region heavily, the base region lightly and making the base region as narrow as possible. Looking over this table it can be seen that the transistor designer would have to make several compromises or tradeoffs in determining the best physical characteristics for a given transistor. For example, in considering switching speed a lightly doped collector will reduce *C-B* junction capacitance while, on the other hand, a heavily doped collector will help reduce storage time.

From this table we can briefly sum up the desired attributes of the three transistor regions. First, the emitter region must be heavily doped (producing low resistivity) for high gain. It should have a large area in order to carry high currents but at the same time a small area to increase high-frequency response. Some tradeoff, therefore, may be necessary between frequency response and current-carrying capacity.

Table 13.3

Desired electrical property	Necessary physical characteristics
high current gain	heavy emitter doping; light base doping; thin base region
high frequency response	thin base region; small junction cross-sectional areas, light base doping and light collector doping to reduce junction capacitances
high switching speed	small junction cross-sectional areas, light base doping and light collector doping to reduce junction capacitances; heavy collector doping to reduce storage time
high current and power-handling capability	heavy emitter doping; large junction cross-sectional areas
high *C-B* junction breakdown voltage	light collector doping; wide base region; heavy base doping

The base region should have a low doping level (producing high resistivity) and should be as thin as possible for high gain and good frequency response. A very thin base, however, may result in a relatively low *C-B* junction breakdown voltage unless the collector is also doped very lightly.

Moreover, a high collector resistivity (produced by a low doping level) results in low *C-B* junction capacitance, which is a desirable feature in high-speed and high-frequency transistors. However, this high collector resistivity results in an increased storage time in saturated switching circuits. This apparent dilemma may be resolved by the introduction of gold impurities into the transistor structure which serve to reduce storage time despite high collector resistivity. Thus it is possible to construct transistors which have high breakdown voltage, low junction capacitance and low storage time. The gold-doped transistor, on the other hand, has considerably higher leakage current and lower gain than non-gold devices. For this reason, gold doping is used only for transistors intended for saturated switching, while non-gold-doped devices are used for other purposes.

13.9.   Power Transistors

As we increase the power dissipation demands made on a transistor, the internal heating in the transistor itself becomes a prime consideration. We have seen that transistors are temperature sensitive. As such, they must be prevented from overheating when large amounts of power are being dissipated. *Power transistors*, which are designed to carry high currents and dissipate large amounts of power, must be able to get rid of heat quickly and effectively.

Several methods have been used in the manufacture of power transistors. Some manufacturers have designed their transistor cases with a ribbed or fluted structure to present a large radiating surface area to the environment. This uses the same principle that fins on the cylinder of a motorcycle engine use. In extreme cases, fans or water cooling may be used to increase the rate of cooling. At present, however, the most widely used method employs *heat sinking*. A *heat sink* is essentially a large mass of material which has a good thermal conductivity, such as copper or aluminum. The power transistor is designed in such a way that the heat-producing element—the collector—is placed in contact with the transistor case. The case is then bolted or riveted to a comparatively large piece of metal such as a chassis or subchassis which serves as the heat sink.

Due to the obvious differences in construction, the electrical characteristics of a typical power transistor will be different from those of low power types. In Table 13.4 a comparison between the 2N250A power transistor and the 2N404A general-purpose transistor is made.

First of all, the 2N250A can handle 600 times more power than the 2N404A and 70 times as much collector current. Secondly, the current gain is somewhat lower for the power transistor as is its power gain. This results mainly from the larger base region needed to handle the increased power dissipation. Another result of this is the high leakage current of the power transistor (1 mA). The higher leakage current is not necessarily a great drawback since it must be remembered that the power transistor normally operates in the ampere range when conducting. The larger areas of the power transistor lower its junction resistances and increase its junction capacitances. This manifests itself in the form of lower input impedance and poorer high-frequency response for the 2N250A as compared with its low-power counterpart. The 2N250A is thus relegated to use in the audio frequency range.

TABLE 13.4

Characteristic	2N404A	2N250A (power)
Max. power dissipation	150 mW	90 W
Max. collector-base voltage	40 V	40 V
Max. collector current	100 mA	7 A
Current gain (typical h_{FE})	100	60
Leakage current (I_{CBO})	1 μA	1 mA
f_{hfb}	12 MHz	4 MHz
Power gain (common emitter)	5000	1000
Input impedance (common emitter)	4 kΩ	500 Ω

13.10. Transistor Data Sheets

To analyze or design transistor circuits intelligently, one must be able to use manufacturer's data sheets to obtain information pertinent to the application. Many of the parameters which have been introduced in previous chapters appear on these data sheets, as well as information concerning construction, physical characteristics and electrical characteristic curves. Two sets of typical transistor data sheets are given in Appendix II for the 2N404 (and 2N404A) transistor and the 2N250A (and 2N251A) transistor.

Refer to these data sheets for the following discussion. The first item of information presented on the data sheet is the identification of the transistor according to its JEDEC number (2N404, 2N404A), its type (PNP), its material and construction (germanium alloy transistor) and general area of application (computer and switching circuits). Following a note on environmental tests (which differ according to manufacturer), the device is usually described (TO-5 case); sometimes its outlines and dimensions are given. The lead orientation is also identified here.

The most important set of data, from a user's standpoint, is contained under the heading "absolute maximum ratings at 25°C free-air temperature." Under no circumstances should these ratings be exceeded since alteration of the transistor parameters or ultimate device failure will occur. Of particular importance for the 2N404 is the power dissipation rating (150 mW @ 25°C) which must be derated, according to note 2, by 2.5 mW per degree of ambient temperature above 25°C.

▶ EXAMPLE 13.8

How much power dissipation can the 2N404 handle at $T_A = 50°C$?

$$P_{D(\text{max})} \ (@ \ 50°C) = P_{D(\text{max})} \ (@ \ 25°C) - \frac{2.5 \ \text{mW}}{°C} \times 25°C$$

$$= 150 \ \text{mW} - 62.5 \ \text{mW}$$

$$= \textbf{87.5 mW} \qquad \blacktriangleleft$$

The "electrical characteristics at 25°C free-air temperature" include maximum and minimum values of certain transistor parameters obtained under specified conditions. The parameters listed depend on the intended application. For example, the switching parameters t_d, t_r, t_s and t_f (Chapter 9) are listed for the 2N404 and 2N404A but are not listed for the 2N250A and 2N251A power transistors. It is important to understand that these parameter values are valid only under the specified conditions. For example, a typical 2N404 will have an h_{FE}* of 100 at the DC operating point $V_{CE} = -0.15$ V and $I_C = 12$ mA and measured at 25°C. We can expect that h_{FE} will change with changes in the DC operating point, as well as with changes in temperatures. The same is true of many of the other parameters. For this reason, typical electrical characteristic curves are usually given, among which are curves representing typical parameter variations with changes in certain conditions.

▶ EXAMPLE 13.9

What is a typical value of h_{FE} at $I_C = 12$ mA, $V_{CE} = -0.15$ and ambient temperature $T_A = 50°C$?

We must use the set of curves relating the normalized value of h_{fE} to T_A. In other words, these curves indicate the *relative* value of h_{FE} at different ambient temperatures. Using the appropriate curve (-0.15 V, 12 mA), the value of h_{FE} is seen to increase by 20 percent (1 to 1.2) in going from 25°C to 50°C. Thus, since h_{FE} is typically 100 at 25°C, we can expect its value to increase to 120 at 50°C. ◀

▶ EXAMPLE 13.10

What is the maximum value of I_{CBO} which can be expected at 60°C for the 2N404?

Using the curve relating I_{CBO} to T_A, it can be seen that from 25°C to 60°C, I_{CBO} increases by approximately ten times (1 μA to 10 μA). Since we are looking for the maximum I_{CBO} at 60°C we can assume that if I_{CBO} were at its maximum (5 μA) at 25°C it would increase ten times so that $I_{CBO(\text{max})}$ at 60°C is 50 μA. ◀

---

*h_{FE} is the same as β_{DC}. The uppercase letters *FE* distinguish it from h_{fe}, which is β. Similarly, h_{FB} is the same as α_{DC}, while h_{fb} is the same as α.

It may appear from the list of transistor parameters given for the 2N404 that not enough information is given in case one wished to use the 2N404 in the Com.-B configuration. However, we can easily determine Com.-B parameters from the Com.-E parameters using relationships presented previously.

▶ EXAMPLE 13.11

What are typical values of h_{fb} and h_{ib} for the 2N404 at $V_{CB} = -6$ V and $I_E = 1$ mA?

The values of h_{fe} and h_{ie} are typically 135 and 4 kΩ respectively at $V_{CE} = -6$ V and $I_C = 1$ mA as given on the 2N404 spec sheet. Since

$$\alpha = \frac{\beta}{\beta + 1}$$

then

$$h_{fb} = \frac{h_{fe}}{h_{fe} + 1} = \frac{135}{136} \approx 0.992$$

Also, we know that $h_{ie} \approx h_{fe} \times h_{ib}$ so that

$$h_{ib} \approx \frac{h_{ie}}{h_{fe}} = \frac{4 \text{ k}\Omega}{135} = 29.4 \ \Omega \qquad ◀$$

Of course, as useful as manufacturers' data sheets are, they can only predict typical or worst-case (min and max) transistor operation. In cases where a more accurate prediction is needed, the transistor user must make his/her own measurements to determine the parameter values of interest.

GLOSSARY

Z_{ic}: common-collector amplifier input impedance.

Z_{oc}: common-collector amplifier output impedance.

Emitter follower: another name for common-collector amplifier.

Mid-band frequencies: range of frequencies over which gain is relatively constant.

f_α (f_{hfb}): frequency at which α has dropped to 0.707 of its low-frequency value.

f_β (f_{hfe}): frequency at which β has dropped to 0.707 of its low-frequency value.

f_T: frequency at which β has dropped to one.

C_{ob} (C_{oe}): common-base (-emitter) output capacitance.

Heat sink: large mass of heat-conducting material.

PROBLEMS

13.1   Consider the Com.-C amplifier of Figure 13.12. The transistor has $h_{fe} = 150$.
(a) Determine the Q-point values of I_E, I_C, V_E and V_{CE}. Neglect V_{BE} since it is so low compared to V_{CC}.
(b) Calculate h_{ib} and h_{ie} for the transistor.
(c) Calculate Z_{ic} for $R_L = 10$ kΩ.
(d) Calculate e_{out} and G_v.
(e) Calculate G_i and G_p.
(f) Sketch the *complete* waveform of emitter-to-ground voltage.

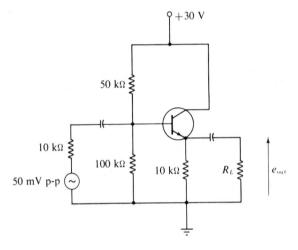

Figure 13.12.

13.2   Calculate the output impedance of the amplifier of Figure 13.12. Indicate how each of the following changes will affect the output impedance:
(a) Decrease of R_L.
(b) Increase of the 50-kΩ resistor to 100 kΩ.
(c) Decrease of R_S.
(d) Decrease in h_{fe}.
(e) Decrease of V_{CC}.

13.3   Will the input impedance of the amplifier in Figure 13.12 increase, decrease or remain the same under the following conditions?
(a) Decrease of R_L.
(b) Increase of R_S.
(c) Increase of the 50-kΩ resistor to 100 kΩ.
(d) Decrease of the 100-kΩ resistor to 82 kΩ.
(e) Increase in transistor h_{fe}.
(f) Increase of V_{CC}.

13.4   Indicate the effects each of the following will have on G_v and G_i for the Com.-C amplifier of Figure 13.12:

(a) Increase of R_L.
(b) Decrease of R_E.
(c) Increase of h_{fe}.
(d) Decrease of the 50-kΩ resistor.
(e) Increase of R_S.

13.5   Indicate which of the three amplifier configurations (Com.-B, Com.-C or Com.-E) pertain to each of the following statements:
(a) Gives 180° phase shift.
(b) Has the highest input impedance.
(c) Has the lowest power gain.
(d) Has high current gain (greater than one).
(e) Has the lowest output impedance.
(f) Has voltage gain less than one.
(g) Has highest output impedance.
(h) Gives no phase shift.
(i) Has bias stability problems.
(j) Uses base-voltage bias.
(k) Input Z depends largely on value of R_L.
(l) Uses bypass capacitor across R_E.

13.6   Explain why transistor-amplifier voltage gain drops at *low* frequencies.

13.7   For the same size input coupling capacitor, what type of amplifier should have the best *low* frequency operation? Explain.

13.8   Explain the factors which cause gain to decrease at *high* frequencies.

13.9   The following data is listed for a particular transistor:

$$h_{fb} = 0.99$$
$$f_{hfb} = 20 \text{ MHz}$$

Determine f_{hfe} and f_T for this transistor.

13.10   For a transistor with a given output capacitance C_{oe}, how will a decrease in R_L affect the high-frequency response? How will it affect mid-band voltage gain?

13.11   With the same value of R_L, what type of amplifier, Com.-B or Com.-E, will have a better high-frequency response?

13.12   If two transistors have the same value of C_{ob} but different values of β, which will have a better high-frequency response in the Com.-E configuration?

13.13   Answer *true* or *false*:
(a) Com.-C amplifiers have a greater power gain than Com.-E amplifiers.
(b) Power transistors can operate at higher frequencies than smaller transistors.
(c) High-speed switching transistors require light base doping and small junction areas.
(d) High current gain requires a narrow base region.
(e) Good high-frequency response requires a fairly large base region.

13.14   What are the chief differences between power transistors and low-power transistors?

13.15   Refer to the transistor data sheets in Appendix II. From the data sheet for
the 2N250A determine the following:
(a) $P_{D(max)}$ at 50°C.
(b) Average I_{CBO} at $V_{CB} = -20$ V and $T_A = 50°C$.
(c) Minimum h_{FE} at $V_{CE} = -1.5$ V and $I_C = 3$ A.

13.16   From the data sheet for the 2N404, determine the following:
(a) Typical total switching time T_t.
(b) Typical h_{FE} at $V_{CE} = -0.15$ V, $I_C = 12$ mA (25°C).
(c) Typical h_{FE} at $V_{CE} = -0.15$ V, $I_C = 100$ mA (25°C).
(d) Typical C_{ob} at $V_{CB} = -10$ V, $f = 1$ mHz.

13.17   Calculate A_v for a Com.-E amplifier using a typical 2N404 biased at $I_C =$
1 mA, $V_{CB} = -6$ V and using $R_C = 10$ kΩ.

13.18   Determine f_{hfe} for a typical 2N404.

13.19   Determine C_{oe} for a typical 2N404.

13.20   Consider the amplifier in Figure 13.13. It is designed to have Q-point values
$V_B = 4$ V and $V_E = 3.3$ V and a voltage gain $G_v = 0.83$. However, when
the circuit is constructed in the lab, the following measurements are made:

$$V_B = 2.8 \text{ V}; \quad V_E = 2.1 \text{ V}; \quad e_S = 10 \text{ mV p-p}$$
$$V_E = e_{out} = 8.3 \text{ mV p-p}$$

The transistor is supposed to have $\beta = \beta_{DC} = 30$.
   Which of the following reasons could be the cause of the discrepancies
between the theoretical and measured values? Explain each choice (there may
be more than one).
(a) Transistor beta is actually less than 30.
(b) Original design calculations are wrong.
(c) C_o is short circuited.
(d) Transistor is shorted between base and emitter.
(e) The 20-kΩ resistor is faulty; its value must be much larger than 20 kΩ.

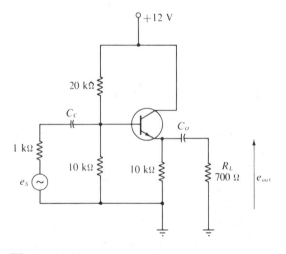

Figure 13.13.

REFERENCES

Foster, J. F., *Semiconductors, Diodes and Transistors*, Vols. 3, 7. Beaverton, Oregon: Programmed Instruction Group, Tetronix, Inc., 1964.

Malvino, A. P., *Transistor Circuit Approximations*. New York: McGraw-Hill Book Company, 1968.

Pierce, J. F., *Transistor Circuit Theory and Design*. Columbus, Ohio: Charles E. Merrill Books, Inc., 1963.

Romanowitz, H. A. and R. E. Puckett, *Introduction to Electronics*. New York: John Wiley & Sons, Inc., 1968.

Transistor Manual. Syracuse, New York: General Electric Company, 1964.

Veatch, H. C., *Transistor Circuit Action*. New York: McGraw-Hill Book Company, 1968.

14

PNPN Devices

14.1.   Introduction

The PNPN class of devices consists of *four* alternately doped semiconductor layers and thus *three* P-N junctions. In contrast to transistors, which are three-layer devices, the PNPNs are utilized solely as semiconductor switches. PNPN devices are often referred to as *thyristors*. The various devices in this class find extensive application in pulse and switching circuitry, power control and motor speed control. In this chapter, the physical principles governing the operation of all PNPN devices will be described, followed by a detailed discussion of the electrical characteristics of some of the more popular PNPN devices. Typical circuit applications will be presented in which the device specifications and ratings will be utilized.

14.2.   PNPNs: General Description

There are two-, three- and four-terminal PNPN devices in use today. Although the various devices have certain distinguishing characteristics, each essentially operates as a switch between the two outer terminals which are called the anode and cathode. The basic four-layer structure is shown in Figure 14.1 with the anode terminal shown coming from the outer P region and the cathode terminal from the outer N region. The operation of a PNPN device resembles that of a mechanical switch between anode and cathode. In other words, it has two operating states: "off," or high resistance, and "on," or low resistance. The various PNPN devices differ mainly in the method used to control or actuate the switching. However, unlike the mechanical switch or the transistor switch, the PNPN switch has the characteristic of *latching*. Latching simply means that once the switch has been closed (the device is "on"), removal of the control or actuating signal does not cause the switch to reopen. This is analogous to manually closing a pushbutton switch and then not having it reopen when the pushbutton is released. Because of this latching characteristic, PNPN devices must be turned "off" by some other means. In the following sections we will investigate this characteristic and how it is used in various applications.

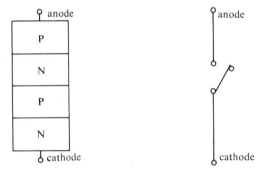

Figure 14.1.   PNPN DEVICES ARE SWITCHES

14.3.   Structure and Basic Operation of PNPNs

The symbolic structure of the basic four-layer PNPN switch is shown in Figure 14.2. Note that the three junctions are labeled *J*1, *J*2 and *J*3 respectively. A

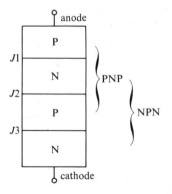

Figure 14.2.   BASIC FOUR-LAYER STRUCTURE

simple, but useful, method of describing the operation of PNPN devices is to consider the PNPN bar to be composed of a PNP structure interconnected with an NPN structure, as shown in Figure 14.3A. Such an interconnection of a PNP and an NPN structure can be represented by two transistors as in Figure 14.3B. We can now examine the operation of PNPN devices using this analogous transistor configuration.

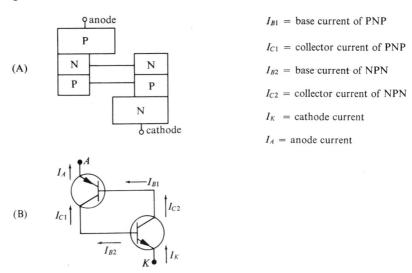

I_{B1} = base current of PNP

I_{C1} = collector current of PNP

I_{B2} = base current of NPN

I_{C2} = collector current of NPN

I_K = cathode current

I_A = anode current

Figure 14.3.   EQUIVALENT REPRESENTATION OF PNPN STRUCTURE

Referring to Figure 14.4A, if a positive voltage is applied from anode to cathode it is apparent that junctions $J1$ and $J3$ will be forward biased and $J2$ will be reverse biased. Since $J2$ will have a much greater resistance than either $J1$ or $J3$ in this case, we can expect that most of the voltage E will appear across $J2$ and very little current will flow from anode to cathode ($I_A = I_K \approx 0$) as long as the value of E is below the breakdown voltage of $J2$.

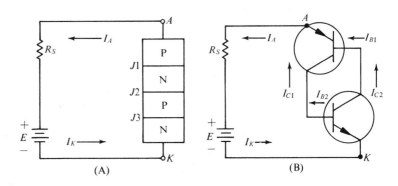

Figure 14.4.   POSITIVE BIAS APPLIED BETWEEN ANODE (A) AND CATHODE (K)

Keeping this in mind, let us now consider the situation using the two-transistor analogy as shown in Figure 14.4B. Both *E-B* junctions are slightly forward biased since *J*1 is the *E-B* junction of the PNP transistor and *J*3 is the *E-B* junction of the NPN transistor. *J*2 is the *C-B* junction of both the PNP and NPN transistors and thus both transistors have a reverse-biased *C-B* junction. Both transistors, then, are biased in the active region. However, since both *E-B* junctions are only barely forward biased, very little current will flow. We can determine an expression for this current by using the basic transistor equations. Consider first the PNP transistor. Its emitter current is the same as I_A, anode current. Thus its base current can be written as

$$I_{B1} = (1 - \alpha_{DC1})I_A - I_{CBO1} \qquad (14.1)$$

where α_{DC1} and I_{CBO1} are the current gain and leakage current, respectively, for the PNP transistor.

If we now consider the NPN transistor, its collector current is given by

$$I_{C2} = \alpha_{DC2}I_K + I_{CBO2} \qquad (14.2)$$

where α_{DC2} and I_{CBO2} are the current gain and leakage current, respectively, for the NPN transistor. The cathode current, I_K, is the same as the emitter current of the NPN device.

We can now utilize the fact that the base current of the PNP serves also as the collector current of the NPN. That is, $I_{B1} = I_{C2}$. Equating the two previous expressions, we have

$$(1 - \alpha_{DC1})I_A - I_{CBO1} = \alpha_{DC2}I_K + I_{CBO2} \qquad (14.3)$$

Noting that $I_A = I_K$, this equation can now be solved for I_A to give

$$I_A = I_K = \frac{I_{CBO1} + I_{CBO2}}{1 - (\alpha_{DC1} + \alpha_{DC2})} \qquad (14.4)$$

The "off" state

This expression for the PNPN current is the basis for explaining the operation of all PNPN devices. In the "off" state the current given by Equation 14.4 will be very low. This may not be readily apparent, but it can be reasoned as follows: Although both transistors are operating in the active region, as mentioned previously, the *E-B* forward bias is very small. Now, for such low values of forward bias, the value of α_{DC} for a transistor can be much lower than one. This is especially true of silicon devices and is the reason why silicon rather than germanium PNPN devices are in use today. For example, if $\alpha_{DC1} = 0.4$ and $\alpha_{DC2} = 0.5$, the value of anode current given by Equation 14.4 is equal to ten times the sum of the leakage currents. Using a typical value of 0.1 μA for I_{CBO1} and I_{CBO2} the anode current becomes 2 μA. This would be the "off" current of the PNPN device. If the value of *E* is 50 volts and the "off" current is 2 μA, this gives a resistance of 50 V/2 μA = 25 megohms between anode and cathode.

The "on" state

Examination of Equation 14.4 will reveal how the device turns "on," which is the essence of PNPN operation. If, through some process, the values of α_{DC1} and α_{DC2} could be increased so that ($\alpha_{DC1} + \alpha_{DC2}$) is equal to one, the device current will increase indefinitely. Actually, the current will be limited by the series resistance in the circuit, such as R_S in Figure 14.4. When this occurs, both transistors reach saturation, in which the junction voltage drops are all small. This may be seen by referring to Figure 14.4B. In the "on" state, anode current will be approximately E/R_S. This is the emitter current of the PNP transistor and produces saturation resulting in a large I_{C1} which becomes I_{B2} and produces saturation in the NPN transistor. Since both transistors are saturated in the "on" state, the anode-to-cathode voltage drop is low, being equal to the sum of one *E-B* forward voltage drop and one $V_{CE(sat)}$, typically resulting in a total of 0.8 V. The "on"-state resistance is very small and for some PNPN devices can go as low as 0.01 Ω. Once the device reaches the "on" state the values of α_{DC1} and α_{DC2} remain very low since both transistors are saturated (I_C is much lower than I_E).

There are several ways of increasing the sum ($\alpha_{DC1} + \alpha_{DC2}$) so that an "off" device may be *triggered* "on." The most common include increasing device temperature, increasing the forward bias on one of the transistor *E-B* junctions and increasing anode-to-cathode voltage. The latter method causes an increase in the *C-B* junction reverse bias of the transistors which, as we know, causes an increase in α_{DC} for each transistor.

Before investigating the various devices which employ one or more of these methods of triggering, a word must be said about what occurs when the voltage at the anode is made negative with respect to the cathode. In this case, both transistors are biased at cutoff (*E-B* junctions *J*1 and *J*3 reverse biased) and only a small reverse leakage current will flow. The entire voltage appears across *J*1 and *J*3, and when this voltage increases sufficiently to cause *J*1 and *J*3 to reach avalanche breakdown, the device behaves in the same way as any P-N diode which has reached breakdown. The device is normally not meant to operate in avalanche breakdown. Biasing with negative anode-to-cathode voltage is called *reverse* bias for the PNPN structure, while the bias used in Figure 14.4 is called *forward* bias. The PNPN structure only behaves as a switch when it is in the forward-bias condition.

14.4. PNPN Four-Layer Diodes: Characteristics and Operation

The *four-layer diode* was invented by W. Shockley, who made a lasting name in the semiconductor field. In fact, it is often referred to as a *Shockley diode*. Its construction is essentially that shown in Figure 14.2 and discussed in the previous section. In the previous section it was determined that the four-layer device will switch from high resistance, "off," to low resistance, "on," if the sum of the alphas is increased to 1. In the four-layer diode this is accomplished by increasing the anode-to-cathode forward voltage until switching occurs. Referring to Figure 14.4, increasing the value of E increases the amount of reverse bias on the center junction, $J2$, which serves as the C-B junction for both transistors. The increase in C-B reverse bias causes the transistor alphas to increase. When the point is reached where $\alpha_{DC1} + \alpha_{DC2} = 1$, switching occurs. This may be visualized more clearly by plotting the I-V characteristic curve of the four-layer diode. This is done in Figure 14.5, where the circuit symbol for the four-layer diode is also shown.

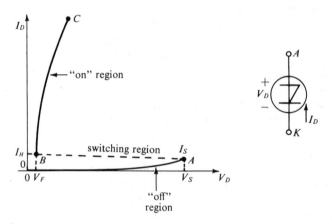

Figure 14.5.   FOUR-LAYER DIODE SYMBOL AND I-V CHARACTERISTIC CURVE

The four-layer diode's operation can be explained with the aid of its characteristic curve as follows:

(a) With $V_D = 0$, the diode does not conduct and $I_D = 0$.

(b) As V_D is gradually increased, the current increases very slowly along the curve O–A. This I_D is a relatively small leakage current and this portion of the curve is called the *"off" region*.

(c) Eventually, as V_D increases, it reaches a value V_S, called the *switching voltage*. At this voltage (point A) the diode will rapidly switch (shown by dotted line) to its "on" region (B–C). In the "on" region, its characteristic is similar to that of a forward-biased P-N diode. In this state the diode's forward voltage drop, V_F, will be very small, typically between 0.5 and 2.0 volts, and the diode current can become very large.

(d) Once the diode has switched to the "on" state, it will remain there as long as its current is kept above the *holding current* I_H which is the value of diode current needed to *hold* the diode "on." If the diode current drops below I_H, the diode rapidly returns to its "off" state and its current decreases to a very low value.

(e) Another diode parameter of importance is I_S, the diode current at the switching point A. I_S is called the *switching current* and is the value of "off" diode current when its voltage is right at its switching voltage value V_S. I_S is always lower than I_H by a factor of ten or more.

Simple diode circuit

A better understanding of four-layer diode operation may be obtained by considering a numerical example. Refer to Figure 14.6 in which a typical four-layer diode is used in a simple circuit. The diode's parameters are also given in the figure. Assume that the voltage E is below 30 V, say 20 V. In this case, the diode will be in its "off" state with only a small leakage current flowing, which is typically 10 μA. We can write, using Kirchhoff's voltage law, that

$$V_D = E - I_D \times 1 \text{ k}\Omega \qquad (14.5)$$

Thus V_D will be approximately 20 V since the voltage drop across the 1-kΩ resistor will be negligibly small (10 mV).

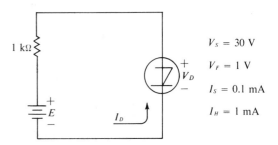

$$V_S = 30 \text{ V}$$
$$V_F = 1 \text{ V}$$
$$I_S = 0.1 \text{ mA}$$
$$I_H = 1 \text{ mA}$$

Figure 14.6.   SIMPLE FOUR-LAYER DIODE CIRCUIT

As E is increased the diode will remain "off" until E becomes large enough to cause the diode voltage to become equal to the switching voltage, 30 V. The value of E which will accomplish this will be equal to 30 V plus the small drop across the series resistor. This resistor drop will be equal to 1 kΩ times the switching current $I_S = 0.1$ mA which flows at the switching point (point A in Figure 14.5). This results in a value of $E = 30.1$ V needed to cause the diode to reach 30 V, at which time it will rapidly switch to its "on" state. In the "on" state the diode voltage drops to approximately 1 volt (V_F). In this state the current will now increase since all but 1 volt of the supply voltage will be dropped across the 1-kΩ resistor. If $E = 30.1$ V, then the current I_D will be $(30.1$ V $- 1$ V)/ 1 k$\Omega = 29.1$ mA. Thus current will begin to flow in the circuit as soon as E is increased to 30.1 V, switching the four-layer diode "on."

Once the diode switches "on" it will remain "on" until the current is made to decrease below the holding current, $I_H = 1$ mA. This means that E can be decreased below 30.1 V and the diode will remain "on." For example, if E is now decreased to 20 V, the current will be $(20$ V $- 1$ V)/1 k$\Omega = 19$ mA, which is higher than the holding current and the diode will remain "on." The smallest value of E which will hold the diode "on" will be equal to that value which produces 1 mA of current. That is,

$$E_{\min} = I_H \times 1 \text{ k}\Omega + V_F$$
$$= 1 \text{ mA} \times 1 \text{ k}\Omega + 1 \text{ V} = 2 \text{ V}$$

Any value of E below this value, say 1.9 V, will cause the current to drop below 1 mA and the device will turn "off." It will then remain in the "off" state until E is again increased to 30.1 V.

This complete sequence of switching from "off" to "on" to "off" is summarized in Table 14.1 for the circuit values used above. Note that when the diode is "off," its voltage is approximately equal to the source voltage.

TABLE 14.1

E	V_D	I_D	Diode state
20 V	20 V	10 μA	"off"
30.1 V	1 V	29.1 mA	"on"
20 V	1 V	19 mA	"on"
2 V	1 V	1 mA	Just "on"
1.9 V	1.9 V	10 μA	"off"

▶ EXAMPLE 14.1

In the circuit of Figure 14.6, if the 1-kΩ resistor is increased to 10 kΩ, will the diode turn "on" at $E = 30.1$ V?

To turn the diode "on" we must have

$$E = 10 \text{ k}\Omega \times I_S + V_S$$
$$= 1 \text{ V} + 30 \text{ V} = 31 \text{ V}$$

Thus $E = 30.1$ V is *not sufficient* to switch the diode "on."   ◀

▶ EXAMPLE 14.2

Using the 10-kΩ resistor, if the diode is switched "on" by increasing E to 31 V, then how low must E be made to turn the diode "off" again?

To turn the diode "off" E must drop below

$$E = 10 \text{ k}\Omega \times I_H + V_F$$
$$= 10 \text{ V} + 1 \text{ V} = 11 \text{ V}$$ ◀

14.5.   Four-Layer Diodes: Parameters and Ratings

Four-layer diodes are currently available with switching voltages in the range from tens of volts to a few hundred volts. The switching current, I_S, is normally, at most, a few hundred microamperes; the holding current, I_H, varies depending on the type, and is in the range from several milliamperes to several hundred milliamperes.

Temperature affects the value of V_S only slightly, causing it to decrease at higher temperatures. Typically, it may drop by 10 percent in going from 25°C to 100°C. The holding current also decreases at higher temperatures, but does so more rapidly. It may change as much as 50 percent over the range from 25°C to 100°C. This means that at higher temperatures the device will remain "on" for lower values of device current.

Limitations on the four-layer diode are mainly those of power dissipation and current. Typical medium power devices have maximum ratings of 150 mW and 150 mA DC at 25°C. However, peak power and current ratings may exceed these values by 50 to 100 times depending on repetition rate and pulse duration.

As far as switching speed is concerned, the four-layer diode will normally turn "on" faster than it will turn "off." Typical values might be 0.2 μs (microsecond) for turn-on time and 0.4 μs turn-off time. These values, however, are highly dependent on the other circuit components.

14.6.   The Rate Effect

The four-layer diode characteristic curve shown in Figure 14.5 is valid only when the applied anode-to-cathode voltage changes slowly. When a rapidly rising voltage is applied to the diode, a *rate effect* occurs that tends to turn the diode "on" at a lower value of voltage.

The cause of this rate effect is the capacitance of the reverse-biased $J2$ junction, which is typically about 100 picofarads (pF). This is illustrated in Figure 14.7. When a rapid change in voltage is applied to the diode, a current will flow

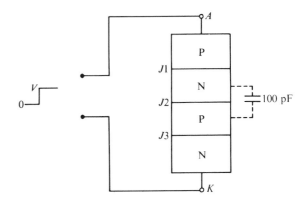

Figure 14.7.   ILLUSTRATION OF RATE EFFECT

through the capacitance which is proportional to the rate of change of the voltage. That is,

$$I_{cap} = C\frac{\Delta V}{\Delta t} \tag{14.6}$$

For example, for a $C = 100$ pF and a voltage which is changing at the rate of 10 volts/μs,

$$I_{cap} = 100 \text{ pF} \times 10 \text{ volts/}\mu\text{s} = 1000 \ \mu\text{A}$$

Because of this capacitive current, the current of the NPN and PNP transistors increases. This will cause the values of α_{DC1} and α_{DC2} to increase. Thus switching will occur at a lower value than V_S. In fact, for a very rapid voltage rise, switching may occur at anode voltages as low as 3 or 4 volts even when $V_S = 30$ V.

This rate effect is very important in the operation of most PNPN devices since it is often the cause of "false triggering" which can occur when a fast rise-time transient (noise voltage) appears at the anode or when a supply voltage is switched into the circuit.

14.7.   Four-Layer Diodes: Applications

Four-layer diodes are suitable for many switching circuit applications where a voltage-sensitive switch is required. In all of its applications, the four-layer diode acts as a very high resistance ("off" state) until its voltage exceeds V_S, after which it becomes a very low resistance similar to a forward-biased P-N diode. In the "on" state its voltage drops to around 1 V. In this respect, it is different from a zener diode, which maintains its breakdown voltage.

Over-voltage indicator

Figure 14.8 shows a four-layer diode used in a circuit as an over-voltage sensor to protect a sensitive load from possible damage due to excessive voltage. The diode has $V_S = 10$ V; as such, as long as the power supply voltage remains

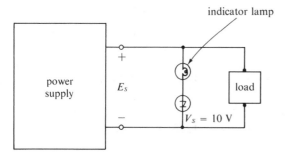

Figure 14.8.   OVER-VOLTAGE INDICATOR

below 10 V, the diode will remain "off" and no current will flow through the indicator lamp. As soon as the power supply voltage exceeds 10 V, the diode switches "on" and its voltage drops to approximately 1 V so that most of the supply voltage appears across the lamp. The indicator lamp thus indicates the over-voltage condition.

The same circuit can be used to remove the load from the circuit when over-voltage occurs by replacing the lamp by a relay. The relay's *normally closed* contacts can then be used to disconnect the load when over-voltage causes the diode to turn "on."

Relaxation oscillator

Consider the circuit shown in Figure 14.9. This particular circuit is called a *relaxation oscillator*. It essentially takes a DC input voltage and produces an oscillating output signal. To operate properly, the supply voltage E must be greater than the diode's switching voltage V_S.

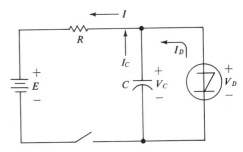

Figure 14.9.   RELAXATION OSCILLATOR

Let us assume that the switch is open and the capacitor is uncharged so that $V_C = V_D = 0$ volts. If the switch is now closed the capacitor will begin to charge up toward the supply voltage through the series resistor R. While this is going on, the diode is in the "off" state so that $I_D \approx 0$. Thus, while the capacitor is charging, the circuit can be redrawn as in Figure 14.10A with the diode re-

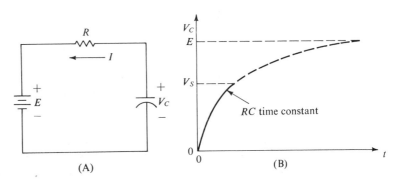

Figure 14.10.   RELAXATION OSCILLATOR REDRAWN DURING CAPACITOR CHARGING INTERVAL

moved. During this time interval the capacitor voltage V_C will be increasing toward E volts with an exponential time constant equal to RC as shown in Figure 14.10B. If it were not for the diode, the capacitor would eventually reach E volts after about five time constants.

The diode, however, will only block current as long as its voltage is below its switching voltage. Since E is made larger than V_S, the capacitor voltage (and thus V_D) will eventually reach the value of V_S. When this occurs the diode will switch to its "on" state, essentially presenting a low resistance across the capacitor. The capacitor will now discharge through the "on" diode. This is illustrated in Figure 14.11A. The discharge time as shown in Figure 14.11B will be very fast since the diode is in its low resistance state. The capacitor will dis-

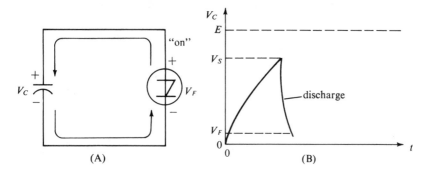

Figure 14.11.   CAPACITOR DISCHARGING THROUGH "ON" DIODE

charge to almost zero volts (actually, to V_F, the diode's "on" voltage) and its discharge current will hold the diode "on." After the capacitor completes its discharge, the current through the resistor R and source voltage E supply current to the diode. If this current is less than the diode's holding current, I_H, the diode will turn "off." The value of this current I will be

$$I = \frac{E - V_F}{R} \tag{14.7}$$

and it should be made less than I_H. Thus

$$\frac{E - V_F}{R} < I_H \tag{14.8}$$

When the diode turns "off," it reverts back to its high resistance state and the situation returns to that in Figure 14.10 where the capacitor again begins charging toward E volts. The above sequence will then repeat itself indefinitely. The waveform of voltage across the capacitor and diode will be that shown in Figure 14.12. It is a repetitive waveform with the capacitor charging and discharging periodically. Equation 14.8 must be satisfied so that the diode turns "off" after the capacitor discharges and the cycle can be repeated.

The frequency of the waveform is essentially determined by the charging time of the capacitor since the discharge time is so short. The charge-up time

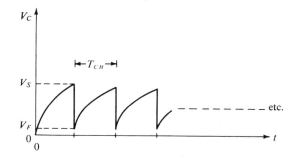

Figure 14.12.   OSCILLATOR OUTPUT WAVEFORM

depends on E, V_S and R and can be obtained from a universal time-constant chart* or from the expression

$$T_{CH} = RC \times \log_e\!\left(\frac{E}{E - V_S}\right) \tag{14.9}$$

This is the period of one oscillation. The frequency is the reciprocal of this period:

$$f = \frac{1}{RC \times \log_e\!\left(\dfrac{E}{E - V_S}\right)} \tag{14.10}$$

It is apparent from Equation 14.10 that the frequency of oscillations is dependent on all the circuit variables. The amplitude of the oscillations, how-ever, depends only on the diode parameter V_S and is essentially constant with frequency since the capacitor must always charge to V_S before discharging.

The most convenient way to vary the frequency of this circuit is to vary the value of R. Increasing R will increase charging time and thus decrease the fre-quency, and vice versa. However, there are limits on the value of R. If R is made too small, then Equation 14.8 will not be satisfied and the diode will never turn "off" once it has been turned "on." On the other hand, if R is made too large the diode will never turn "on" since at its switching point a current I_S must be supplied. To ensure that this does not occur, then,

$$\frac{E - V_S}{R} > I_S \tag{14.11}$$

Or, in other words, the current through R when the diode is at V_S must be greater than I_S.

Equations 14.8 and 14.11 may be arranged to give the minimum and maxi-mum values of R for given values of E, C and diode parameters. The results are

$$R_{\min} = \frac{E - V_F}{I_H} \tag{14.12A}$$

---

*If the time-constant chart is used, simply determine how many τ it takes for the capacitor to charge to V_S. Then $T_{CH} = $ no. of $\tau \times RC$. If Equation 14.9 is used, it may be helpful to know that $\log_e x = \ln x$ (natural log) $= 2.3 \log_{10} x$.

and

$$R_{max} = \frac{E - V_S}{I_S} \tag{14.12B}$$

Keeping the value of R between these values will ensure oscillation. If $R < R_{min}$, the diode will never turn "off" since the diode current will remain above I_H. If $R > R_{max}$, the diode will never turn "on" since the diode current will not reach I_S.

▶ EXAMPLE 14.3

A relaxation oscillator of the type shown in Figure 14.9 uses a voltage supply of 40 V, a 1-μF capacitor, a 100-kΩ resistor and a four-layer diode with $V_S = 20$ V, $I_S = 0.1$ mA, $I_H = 1$ mA and $V_F = 1$ V. Determine the frequency of oscillation, if any.

To determine whether the circuit will oscillate at all, the values of R_{min} and R_{max} must be calculated:

$$R_{min} = \frac{40 \text{ V} - 1 \text{ V}}{1 \text{ mA}} = 39 \text{ k}\Omega$$

$$R_{max} = \frac{40 \text{ V} - 20 \text{ V}}{0.1 \text{ mA}} = 200 \text{ k}\Omega$$

Since our resistor is 100 kΩ, the circuit will oscillate. The frequency will be given by Equation 14.10.*

$$f = \frac{1}{(100 \text{ k}\Omega \times 1 \text{ }\mu\text{F}) \times \log_e 2}$$
$$= 14.5 \text{ Hz} \qquad \blacktriangleleft$$

The frequency of oscillation may also be controlled by varying the power supply voltage. Again, however, Equations 14.8 and 14.11 must be satisfied to ensure oscillation. These equations may be rearranged to give

$$E_{min} = V_S + RI_S \tag{14.13A}$$

and

$$E_{max} = V_F + RI_H \tag{14.13B}$$

for oscillations to occur. If $E < E_{min}$, the diode will not turn "on"; if $E > E_{max}$, the diode will not turn "off" once it switches to the "on" state.

▶ EXAMPLE 14.4

What are the limits on E for the circuit of Example 14.3? What is the range over which frequency may be varied by varying the value of E?

$$E_{min} = 20 \text{ V} + (100 \text{ k}\Omega \times 0.1 \text{ mA}) = 30 \text{ V}$$

---

*Table of natural logarithms, $\log_e x$, is given in Appendix III.

and

$$E_{\max} = 1 \text{ V} + (100 \text{ k}\Omega \times 1 \text{ mA}) = 101 \text{ V}$$

Using $E = E_{\min} = 30$ V, the frequency will be, using Equation 14.10,

$$f_{\min} = 8.7 \text{ Hz}$$

and for $E = E_{\max} = 101$ V, the frequency will be

$$f_{\max} = 50 \text{ Hz} \qquad \blacktriangleleft$$

Time-delay circuit

A similar circuit that is often used in timing applications is shown in Figure 14.13. The function of this circuit is to provide a sharp positive pulse output that occurs a certain time delay after the application of input voltage. The operation of this circuit is similar to the oscillator previously discussed except that a small resistor R_{out} is placed in series with the diode. This means that the capacitor discharge current also flows through R_{out}, producing a pulse across it. The pulse will occur a predetermined amount of time after the application of power. The time delay will be essentially the capacitor charge-up time given by Equation 14.9.

$$T_{\text{delay}} = RC \times \log_e\left(\frac{E}{E - V_S}\right) \qquad \textbf{(14.14)}$$

The value of R_{out} must be small to assure a fast capacitor discharge to provide a sharp pulse. Values between 10 and 100 ohms are typical. The output pulse amplitude will be approximately equal to V_S.

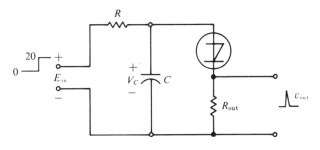

Figure 14.13.   TIME-DELAY CIRCUIT

Usually only one pulse is desired rather than a series of pulses as would occur in an oscillator. For this reason, it is desirable to choose R so that the diode will remain "on" after the first pulse and no further pulses will occur. In other words, R should be less than $R_{\min}$. Once R is chosen, then C can be chosen to give the necessary time delay. Figure 14.14 shows the waveform relationships for this circuit.

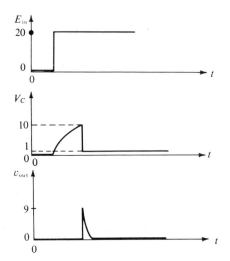

Figure 14.14.   TIME-DELAY CIRCUIT WAVEFORMS

▶ EXAMPLE 14.5

Using a diode with $V_S = 8$ V, $I_S = 0.1$ mA, $V_F = 1$ V and $I_H = 2$ mA, determine values of R and C for a 5-ms time delay if V_{in} goes from 0 V to 20 V.

   The value of R must be small enough so that only *one* pulse occurs and then the diode latches "on." For this,

$$R < R_{min} = \frac{E - V_F}{I_H} = \frac{20 \text{ V} - 1 \text{ V}}{2 \text{ mA}}$$

or

$$R < 9.5 \text{ k}\Omega$$

We can use $R = 8$ kΩ. The value of C will be chosen to give the desired time delay. We have

$$5 \text{ ms} = RC \times \log_e(2.5)$$
$$= 8 \text{ k}\Omega \times C \times 0.92$$

or, solving for C,

$$C = 0.69 \ \mu\text{F} \qquad \blacktriangleleft$$

14.8.   Other Thyristor Diodes

There are several other devices which operate similarly to the four-layer diode. A device called a *silicon unilateral switch* (*SUS*) behaves essentially like a four-layer diode. Its circuit symbol is the same as that for a four-layer diode, although its construction is somewhat different. The SUS is a low-voltage device whose switching voltage is typically less than 10 V. In addition, the switching voltage of the SUS is more accurately controlled and less temperature-dependent than that of a four-layer diode.

A *silicon bilateral switch* (*SBS*) behaves exactly like the four-layer diode and SUS, except that it does so for both polarities of applied voltage. In other words, the SBS will act as a switch in both directions. Its characteristic curve is shown in Figure 14.15 along with its circuit symbol. This *I-V* characteristic shows that the device switches from "off" to "on" for 10 V of either polarity.

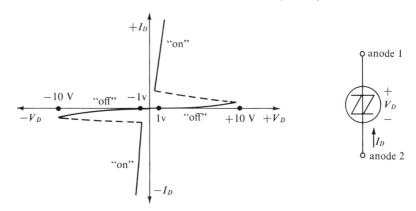

Figure 14.15.   SYMBOL AND *I-V* CHARACTERISTIC OF A SILICON BILATERAL
                  SWITCH (SBS)

The SBS, like the SUS, has switching voltages of 10 V or less and is more accurately controlled and less temperature-dependent than four-layer diodes.

A DIAC is also a *bilateral* switch and is similar to the SBS in operation. The DIAC, however, can operate at higher switching voltages (up to a few hundred) and in that respect is similar to the four-layer diode. Figure 14.16 shows the *I-V* characteristic and symbol for the DIAC. It should be noted that the "on" portion of the curve is not at as low a voltage as the previously discussed devices. This means that when the DIAC switches at $V_S = 30$ V, its "on" voltage drops to only 15 V rather than to 1 V or 2 V.

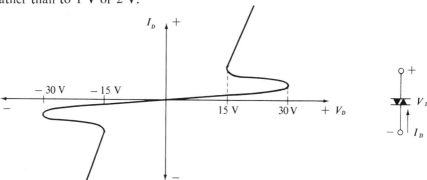

Figure 14.16.   SYMBOL AND *I-V* CURVE OF DIAC

14.9.   The Silicon-Controlled Rectifier: Basic Operation

The *silicon-controlled rectifier (SCR)* is a PNPN device with three terminals: anode, cathode and *gate*. Its gate terminal is attached to the P region closest to the cathode. The SCR construction is illustrated in Figure 14.17 along with the circuit symbol for the SCR. The gate terminal is sometimes referred to as the *cathode gate* to distinguish it from a second gate which is used in certain other PNPN devices.

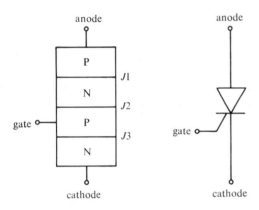

Figure 14.17.   SCR CONSTRUCTION

The two-transistor equivalent for the SCR can be used to explain its basic operation and is shown in Figure 14.18. It is essentially the same as that of the four-layer diode except that the base of the NPN transistor is accessible via the gate terminal. In fact, when the gate is not used (when it is open circuited), the SCR operates the same as the four-layer diode. The gate, however, allows us to turn the device "on" (low resistance between anode and cathode) by a means other than increasing the anode-to-cathode voltage.

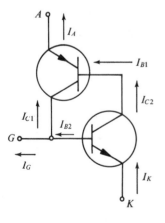

Figure 14.18.   TWO-TRANSISTOR EQUIVALENT

Recall from Section 14.3 that to cause switching in a PNPN device the sum of the transistor alphas must be increased to 1. In the SCR this is accomplished by supplying gate current, in the direction shown, by forward biasing the *E-B* junction ($J3$) of the NPN transistor (gate made positive relative to cathode). This gate current I_G increases the base current I_{B2} of the NPN transistor. This in turn will cause increases in I_{C2} and I_{B1} which, therefore, increase I_{C1}. Thus both transistors will be conducting more current. As a result, the values of their respective alphas will increase. For a given value of anode-to-cathode forward voltage the gate current can be increased to the point where ($\alpha_{DC1} + \alpha_{DC2}$) equals 1. At this point, the SCR turns "on" and the resistance between anode and cathode drops to a low value similar to the four-layer diode.

This may be made clearer upon inspection of a set of typical SCR *I-V* characteristics shown in Figure 14.19. Referring to the figure, it can be seen that with $I_G = 0$ the anode-to-cathode characteristic is similar to that of the four-layer diode. That is, with the anode positive relative to the cathode so that $V_{AK} > 0$, the SCR will block current, making I_A low, until the value of V_{AK} is increased to the *forward breakover voltage*, $V_{(BR)FO}$, at which time the SCR turns "on" and the anode-to-cathode resistance becomes very low.

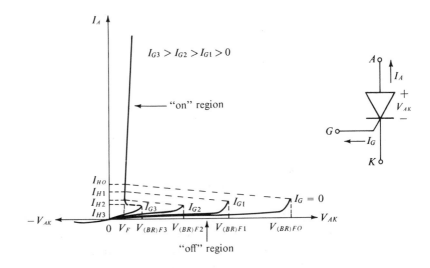

Figure 14.19.   SCR CHARACTERISTIC CURVES

With the application of a gate current, I_{G1}, the operation is essentially the same, except that switching takes place at a lower value of V_{AK}. Increasing the gate current to I_{G2} causes the forward breakover voltage to decrease even further [$V_{(BR)F2}$]. Eventually a value of I_{G3} is reached at which switching will occur at a very low value of voltage [$V_{(BR)F3}$], typically 2–3 volts. Thus the gate current controls the value of anode-to-cathode voltage at which switching will occur; hence the name silicon-*controlled* rectifier.

The reverse characteristic of the SCR ($V_{AK} < 0$) is essentially that of a P-N diode, except that the reverse leakage current will increase if I_G is applied. For this reason, when an SCR is reverse biased, I_G is usually kept at zero to reduce leakage current between anode and cathode.

An important property of the SCR is its latching ability. That is, after the application of a gate current has triggered the SCR "on," the gate current may be removed and the SCR will remain "on" due to the inherent feedback between the two transistors. Thus, only a momentary pulse of gate current is needed to turn the SCR on. This is illustrated in Figure 14.20. Here, the current I_A from cathode to anode will be only a leakage current as long as the gate current is zero. When a pulse of voltage which is positive relative to the cathode is applied at the gate, the resulting gate current triggers the SCR "on." The current through the SCR will then be limited only by the load resistor R_L and will be approximately E/R_L since the SCR anode-cathode voltage will be very low in the "on" state.

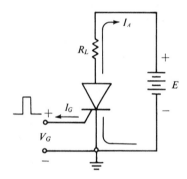

Figure 14.20.   TURNING "ON" SCR WITH GATE PULSE

14.10.   SCR Ratings and Parameters

Circuit designers have available a choice of SCRs whose current capabilities generally range from a few hundred milliamperes to several hundred amperes, and whose voltage capabilities range up to 1000 volts or more. There are SCRs which are designed for fast switching speed, others which need only a low gate current to trigger them "on" and thus have a high gate sensitivity, and still others with characteristics that make them particularly suitable for certain applications. As such, designers and others who select an SCR for a particular application must be familiar with the device's electrical parameters and characteristics and especially its maximum ratings. The following list contains some of the more important SCR ratings and parameters with their symbols and definitions:

(a) *Repetitive peak reverse voltage,* $V_{ROM(\text{rep})}$: maximum allowable instantaneous value of repetitive reverse (anode negative) voltage that may be applied to SCR anode with gate open-circuited ($I_G = 0$). This rating should never be exceeded even though reverse breakdown may not occur.

(b) *Nonrepetitive peak reverse voltage, $V_{ROM(nonrep)}$:* maximum allowable instantaneous value of reverse voltage, of a nonrepetitive nature, that may be applied to the SCR with $I_G = 0$. This rating applies to any transient voltage less than 5 ms in duration, and is always greater than $V_{ROM(rep)}$.

(c) *Peak forward blocking voltage, V_{FOM}:* maximum instantaneous value of forward blocking voltage (anode positive) which will not switch the SCR to the "on" state with the gate circuit open.

(d) *Peak forward blocking voltage, V_{FXM}:* same as V_{FOM} except that a resistor, R_{GK}, is connected between gate and cathode. V_{FXM} is larger than V_{FOM}.

(e) *Forward breakover voltage, $V_{(BR)FO}$:* voltage at which SCR switches into the conductive state with gate circuit open.

(f) *Forward breakover voltage, $V_{(BR)FX}$:* same as $V_{(BR)FO}$ except that a resistor, R_{GK}, is connected between gate and cathode. $V_{(BR)FX}$ is larger than $V_{(BR)FO}$.

(g) *Average forward current, $I_{F(AV)}$:* maximum continuous DC current which may be permitted to flow in the "on" state under stated conditions.

(h) *rms forward current, I_f:* maximum continuous *rms* current which may be allowed to flow in the "on" state under given conditions.

(i) *Peak one-cycle surge forward current, $I_{FM(surge)}$:* highest instantaneous value of forward current that may be permitted to flow through the SCR nonrepetitively. The peak surge current may recur without damage when sufficient time has elapsed to permit the SCR to recover.

(j) *Peak forward (or reverse) gate voltage, V_{GFM} (or V_{GRM}):* maximum allowable peak voltage between gate and cathode when the gate is made positive (or negative) relative to the cathode.

(k) *Peak gate power dissipation, P_{GM}:* maximum instantaneous value of power dissipation permitted between gate and cathode terminals.

(l) *Average gate power dissipation, $P_{G(AV)}$:* maximum value of average power dissipated between gate and cathode.

(m) *Instantaneous "on" voltage, V_F:* voltage drop between anode and cathode in "on" state at a given current level.

(n) *Instantaneous forward blocking current, I_{FO} (I_{FX}):* instantaneous anode current in "off" state at stated conditions of forward voltage and temperature. I_{FO} is with gate open; I_{FX} is with R_{GK} between gate and cathode. I_{FX} is less than I_{FO}.

(o) *Instantaneous reverse blocking current, I_{RO} (I_{RX}):* counterpart of I_{FO} (I_{FX}) under reverse bias.

(p) *Gate trigger current (voltage), I_{GT} (V_{GT}):* gate current (voltage) needed to trigger the SCR with anode terminal at $+6$ volts with respect to cathode terminal at stated temperature conditions.

(q) *Holding current, I_{HO} (I_{HX}):* value of anode current in the "on" state needed to hold the SCR "on" at a given temperature. I_{HO} is used when the gate is open; I_{HX} indicates that R_{GK} is between the gate and the cathode. I_{HO} is lower than I_{HX}.

(r) *Turn-on time, t_{on}:* roughly equal to the time that elapses between the triggering of the SCR and its transition to the "on" state.

(s) *Turn-off time, t_{off}:* time interval required for SCR to go from "on" state to "off" state under stated circuit conditions.

These are the most important SCR characteristics. The meanings of some of these parameters and ratings will become clearer in the following sections.

14.11.   SCR Gate Triggering

The SCR, like the four-layer diode, can be triggered "on" by increasing its anode-to-cathode voltage above the breakover voltage. However, this method is not used in SCR applications since gate triggering offers so many advantages. The main advantage of gate triggering lies in the high power gain which allows a low power signal applied to the gate to control a large power level to a load. In power control circuitry this mode of triggering makes the most use of the SCR.

The conventional SCR is triggered by applying a *positive* gate-cathode voltage to produce gate current. Therefore, the SCR's triggering characteristics can be described in terms of the gate-cathode electrical characteristics at given temperature and anode-cathode bias conditions. The gate-cathode section in the SCR is essentially a P-N junction and its characteristics are similar to those of a P-N junction until the SCR is triggered "on." The SCR will trigger when the gate-cathode junction is sufficiently forward biased and the anode is positive relative to the cathode. It should be pointed out that the gate-cathode junction is somewhat different from ordinary low-current P-N junctions mainly due to the larger physical size needed to handle high currents; the junction area is very large, typically 0.02 in.^{2}.

Trigger requirements

Because of the nature of the manufacturing process, a certain spread of gate-cathode characteristics is likely to be found even between SCRs of the same type. Figure 14.21 shows typical gate characteristics for a 7-A SCR as would normally be supplied on the manufacturer's data sheet. The two curves shown indicate the limits of the possible I_G-V_{GK} curves. That is, any SCR of this type would have a curve somewhere between curves *A* and *B* over the limits of junction operating temperature ($-65°C$ to $125°C$). Note that I_G is on the horizontal axis and V_{GK} on the vertical axis.

A given SCR will trigger from the "off" state to the "on" state for a definite value of gate-cathode voltage. The values of gate-cathode voltage and gate

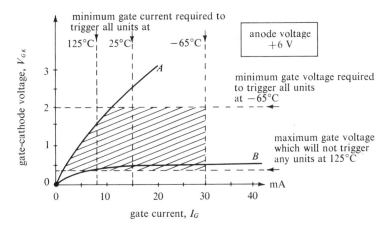

Figure 14.21.   TYPICAL SCR GATE CHARACTERISTICS

current needed to trigger the SCR are called V_{GT} and I_{GT}, respectively. The function of the SCR trigger input (applied to the gate) is to simultaneously supply V_{GT} and I_{GT}. The shaded area shown in Figure 14.21 contains all the possible trigger points (V_{GT}–I_{GT}) for all the SCRs of this type between −65°C and 125°C. In other words, *any* SCR of this type is guaranteed to have a trigger point somewhere in the shaded area.

From the characteristics in Figure 14.21, several important points should be noted:

(a) The minimum gate current required to trigger *all* units at 25°C is 15 mA. This means that a gate current of 15 mA is *guaranteed* to trigger all SCRs of this type at 25°C.

(b) This minimum required gate current decreases to 8 mA at 125° C and increases to 30 mA at −65°C. This means that *an SCR is easier to trigger at higher temperatures*. This is to be expected since the alphas will increase with temperature, thus requiring a lower I_G to bring about triggering.

(c) It is also indicated in the figure that at −65°C the minimum required gate voltage needed to trigger all units is 2 V. This value would be even lower at higher temperatures. What this means is that an applied gate voltage of 2 V or more would be guaranteed to trigger all SCRs of this type over the −65°C to 125°C temperature range.

(d) The figure also indicates that the *maximum* gate voltage which will *not* trigger any units at 125°C is 0.3 V. What this means is that a gate voltage of 0.3 V or less can never trigger any of these SCRs. This is an important quantity in some SCR circuit designs.

Very often the SCR trigger requirements are given in tabular form on SCR data sheets. Table 14.2 gives the same basic information as that in Figure 14.21. However, it is interpreted in a somewhat different manner. For example, at 25°C the maximum value for I_{GT} is 15 mA. This means that for all SCRs of this type, the largest value of gate current ever needed to cause triggering at 25°C is 15 mA. Thus $I_G = 15$ mA will always trigger these SCRs at 25°C.

TABLE 14.2

Gate triggering characteristics ($V_{AK} = +6$ V)			
		I_{GT} (mA)	
Min	Typ.	Max	Conditions
—	10	15	$T_J = 25°C$
—	20	30	$T_J = -65°C$
—	4	8	$T_J = 125°C$
		V_{GT} (volts)	
—	0.7	1	$T_J = 25°C$
—	1	2	$T_J = -65°C$
0.3	0.5	—	$T_J = 125°C$

▶ EXAMPLE 14.6

An SCR with the characteristics of Figure 14.21 and Table 14.2 is used in a circuit where a voltage of 0.8 V is applied to the gate and produces a gate current of 20 mA at 25°C. Will this SCR trigger to the "on" state?

Yes; 15 mA or greater will trigger any SCR of this type at 25°C as indicated by Figure 14.21 and also by Table 14.2.   ◀

14.12.   Gate Triggering Pulse Width

The values of I_{GT} and V_{GT} needed to trigger the SCR as stipulated in the last section are assumed to be either DC values or pulses with sufficient pulse width. It takes a certain amount of time, t_{on}, for the SCR anode current to increase to its final value once a gate trigger is applied. For this reason, a very short pulse at the gate may not turn the SCR "on" even if its amplitude is greater than V_{GT}. The pulse at the gate must be long enough for the SCR anode current to build up to a value called the *latching current*, I_L. When the latching current level is reached the SCR will turn "on" and stay "on" even if the gate pulse terminates at the same instant. The value of I_L can usually be assumed to be about three times the holding current value; that is, $I_L \approx 3I_{HX}$.

The SCR's turn-on time will decrease for a higher amplitude gate trigger. Thus the gate pulse can be made narrower if its amplitude is increased accordingly. Occasionally SCR data sheets present curves indicating the required gate pulse width for various pulse amplitudes.

For example, referring to Figure 14.22, a DC gate voltage signal such as that shown in part A of the figure may be sufficient to trigger the SCR. How-

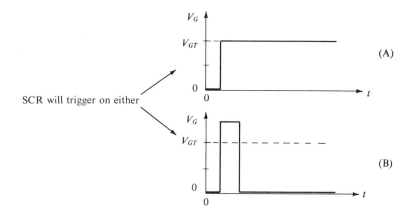

Figure 14.22.   ILLUSTRATION OF EFFECT OF GATE PULSE WIDTH

ever, if a pulse of gate voltage is used, a higher amplitude may be required, as in part B. If the pulse is wide enough, the amplitude need not be any larger than the DC value V_{GT}.

14.13.   SCR Power Gain

It may be beneficial at this time to point out some of the SCR's advantages. In typical operation, the gate current needed to trigger an SCR is less than 1 percent of the anode current that flows when the SCR becomes fully conductive, providing a current gain of greater than 100. Gate voltage needed to trigger is normally around 1 volt, whereas the voltage across the load might be 100 volts once the SCR is "on." Thus voltage gains of 100 or greater are possible. This results in a power gain of $100 \times 100 = 10,000$. The fact that the gate signal may be of much shorter duration than the anode current makes the ratio of load power to gate power even greater. In fact, power gains on the order of one million are not unusual. In view of the inherent large power gain, very low-power-level trigger circuitry can be used for reliable turn-on of SCRs. Typically, a trigger circuit using half-watt resistors, low-voltage capacitors, and a small, switching-type transistor can turn on an SCR controlling 100 kW of power in its anode circuit.

Because of its large power gains and its ability to be triggered by a narrow pulse, the SCR (and related devices) has become the most widely used power control device. In contrast to a typical silicon power transistor, an SCR may supply 5 A of current to a load when it receives a 50-mA gate current pulse for a few microseconds, whereas the transistor might require up to 250 mA of *continuous* base current to achieve the same result. Large SCRs which can handle average currents up to 1000 amperes are available.

14.14.   SCR Turn-Off Considerations

The SCR is a latching device that will remain in its low-resistance "on" state as long as anode current is maintained above the holding current. Unlike the junction transistor, where removal of the base current will bring collector current to zero, removal of SCR gate current will *not* reduce its anode current and turn the device "off." In other words, the gate has no control over the conventional SCR once the device has been triggered "on." There are exceptions to this fact among some of the smaller, low-current SCRs and special gate turn-off devices.

At this point it may appear that by reverse biasing the gate-cathode the SCR could be made to turn "off" since the *E-B* junction of the NPN transistor would seemingly be reversed biased. This would be true if the SCR truly consisted of two conventional junction transistors. However, due to large differences in junction sizes and doping, turning "off" the SCR in this manner will not work except for some of the very low-power devices. For this reason, the gate does not take part in turn-off action and all practical SCR power control circuits most have some provision for effecting anode current turn-off. Some of the turn-off methods will be covered in the following sections.

To turn the SCR "off," the sum of the transistor alphas must decrease below unity. The transistor alphas will decrease as the current through them decreases. Reducing anode current sufficiently, then, will turn the SCR "off." This reduction may be accomplished in various ways, including mechanically opening the anode circuit, reverse biasing the anode to cathode, diverting anode current or using an auxiliary circuit. Regardless of the method used, a certain amount of time, t_{off}, is needed for the SCR to go from "on" to "off" and this should be considered when turning an SCR "off." A typical value of t_{off} is 25 μs and this usually limits the SCR to operating frequencies below 30 or 40 kHz. The value of t_{off} increases with the amount of SCR "on" current.

14.15.   The SCR Used as a DC Switch

The SCR and related PNPN devices are often used in place of mechanical switches or relays. The advantages the SCR offers include smaller size, faster switching speed, silent operation, durability and high sensitivity. SCRs can withstand billions of switching cycles without wearing out, do not suffer from contact bounce or arcing, as relays do, and can survive shock and vibration.

The first SCR circuit we will look at is the simple DC switch shown in Figure 14.23. Here the SCR simply acts like an "on"-"off" switch which controls the DC current to the load. Upon application of the gate signal, the SCR will turn "on." In this state the voltage dropped from anode to cathode will be about 1 volt (V_F). Thus 27 V will appear across R_L. The gate signal could be DC or a pulse of sufficient amplitude and duration. The 1-kΩ resistor connected between gate and cathode is used to stabilize the SCR from turning "on" due to increases in temperature or sudden switching on of the supply voltage (rate effect). This resistor is essentially connected between the base and emitter of the

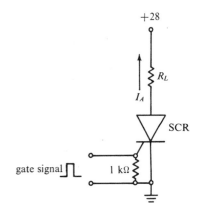

Figure 14.23.   SIMPLE DC SWITCH

NPN transistor (recall the two-transistor equivalent circuit) and shunts some of the current away from the base. This causes the SCR "off" current level, and thus the alphas, to be lower, making the SCR more difficult to turn "on" accidentally.

Once the gate signal triggers the SCR, it will continue conducting as long as the anode current is greater than the holding current and the gate will no longer have any control over the SCR current. To turn the SCR "off" the anode current must be reduced or interrupted. A mechanical switch in series with R_L could be opened momentarily to allow I_A to become zero. Upon reclosing the switch the SCR would stay in the "off" or forward-blocking state until the gate signal appeared again. This process is called *resetting* the SCR.

Inductive loads

The load may be purely resistive, it may be a lamp, a heater, a DC motor field winding or even a relay. In cases where the load is inductive, as in a motor or relay load, special care must be taken due to the amount of time it takes current to build up in an inductance. This case is shown in Figure 14.24. The load is represented by an inductance in series with a resistance. Neglect the resistor R for the moment. If a pulse is to be used to trigger the SCR, its duration must be long enough to allow the current through the load to build up to the SCR latching current, I_L. If the pulse is too narrow the anode current will not have time to reach I_L, thus causing the SCR to turn back "off" when the gate pulse terminates. This can be overcome by using a longer gate pulse. However, in many applications a very narrow gate pulse is to be used or the gate pulse width cannot be controlled accurately. In these cases, the SCR can still be turned "on" by putting a resistor R across the load. Now when the gate is pulsed turning the SCR "on," about 27 volts appear across the parallel combination of the load and R. If R is chosen correctly, the current through it will be greater than I_L, thus providing the SCR with enough current to latch "on" when the gate pulse has terminated. The current through the load will then be able to build up to its final value since the SCR stays "on."

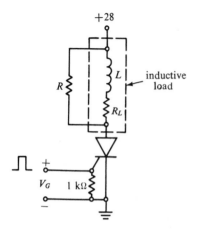

Figure 14.24.   DC SWITCH WITH INDUCTIVE LOAD

▶ EXAMPLE 14.7

The SCR in Figure 14.24 has a holding current I_{HX} of 1 mA and an "on" voltage V_F of 1 volt. What value of R is needed to ensure turn-on with a narrow gate pulse?

To latch "on," the SCR current must reach $I_L = 3I_{HX} = 3$ mA. Thus

$$\frac{27 \text{ V}}{R} \geq 3 \text{ mA}$$

or

$$R \leq 9 \text{ k}\Omega \qquad ◀$$

Another method of turning "off" an SCR in a DC circuit is shown in Figure 14.25, in which a second SCR is used to effectively reverse bias the main SCR.

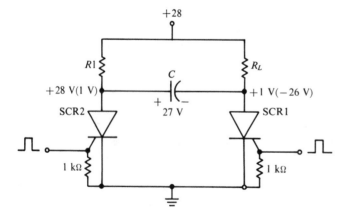

Figure 14.25.   TURNING SCR1 "OFF" USING A SECOND SCR

Assume SCR1 has been triggered "on" by its gate signal and current is supplied to the load. If SCR2 is "off," capacitor C will charge to 27 V, as shown, since its right-hand plate is connected to the anode of SCR1 which is about 1 volt above ground and its left-hand plate is connected to the anode of SCR2 which is at +28 volts. To turn SCR1 "off" a gate pulse is applied to SCR2. This turns SCR2 "on" (since its anode is positive), causing its anode voltage to drop to about 1 volt (shown in parentheses). Since C cannot change its voltage instantaneously, the voltage at the anode of SCR1 will instantaneously become -26 V (shown in parentheses), reverse biasing the anode and turning "off" SCR1. The capacitor C will then charge through R_L to 27 V of the opposite polarity, placing +28 volts at SCR1's anode. When SCR1 is pulsed again, it will turn "on," causing SCR2 to go "off" by the same process.

14.16.   The SCR Used in an AC Circuit

SCRs are often used with an AC voltage supply; in these applications, turn-off is accomplished automatically by the cyclical reversal of the AC supply voltage. Consider the circuit shown in Figure 14.26. Here the anode supply is an AC voltage e_{in}. The circuit operation can be explained with the aid of the voltage waveforms drawn in Figure 14.27.

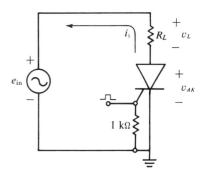

Figure 14.26.   SCR USED IN AN AC CIRCUIT

During the positive half-cycle of e_{in} the anode is positive relative to the cathode. If no gate signal is applied, the SCR will be in the forward blocking state with $v_L \approx 0$. Thus v_{AK} will equal e_{in}, as shown, during the interval from $t = 0$ to $t = t_1$. If at $t = t_1$ a gate pulse is applied, the SCR will be triggered "on" and its voltage will drop to about 1 volt. The load voltage then becomes equal to e_{in} (minus the 1 V drop across the SCR) at $t = t_1$. The SCR will remain "on" with $v_L \approx e_{in}$ until e_{in} decreases to the point where the current through R_L and the SCR drops below I_{HX}, at which point the SCR will turn "off." This occurs at $t = t_2$ when the value of e_{in} is equal to the SCR "on" voltage ($V_F \approx 1$ V) plus the voltage across R_L when $I_L = I_{HX}$.

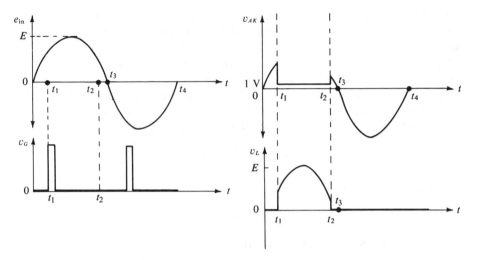

Figure 14.27.   WAVEFORMS FOR CIRCUIT OF FIGURE 14.26

▶ EXAMPLE 14.8

At what value of e_{in} will the SCR turn "off" in the circuit of Figure 14.24 if $R_L = 100 \, \Omega$, $I_{HX} = 10 \, mA$ and $V_F = 1 \, V$?
At turn-off,

$$e_{in} = 100 \, \Omega \times 10 \, mA + 1 \, V$$
$$= 2 \, V$$ ◀

The SCR will remain "off" for the duration of the positive half-cycle with $v_L = 0$ and $v_{AK} = e_{in}$ from t_2 to t_3. At $t = t_3$ the input voltage reverses, causing the SCR to become reverse biased (v_{AK} negative). In this state the SCR remains "off" with $v_L = 0$ and $v_{AK} = e_{in}$ even if a gate pulse occurs. The SCR will not turn "on" again until the gate is pulsed during the next positive half-cycle of e_{in}.

If we examine the load-voltage waveform it can be seen that power is not applied to the load until the gate pulse appears during the positive half-cycle. Thus the average power in the load will depend on the point in the positive half-cycle at which the SCR is triggered "on." That is, the closer t_1 is to 0, the greater the load power will be since it will be conducting for a greater portion of the half-cycle. Various possible load-voltage waveforms are shown in Figure 14.28.

In this figure the phase angle at which the SCR is triggered has been included. In part B of Figure 14.28 the SCR is triggered at the 45° point in the half-cycle. In other words, the *trigger angle*, α, is 45°. From the figure it should be clear that the average load power *decreases* as the trigger angle *increases*. The trigger angle, α, can vary from almost 0° (maximum load power) to almost 180° (minimum load power). This method of controlling the power to the load is called *phase control*, and is used extensively in power controllers, lamp dimmers, motor controllers and in certain types of DC power supplies.

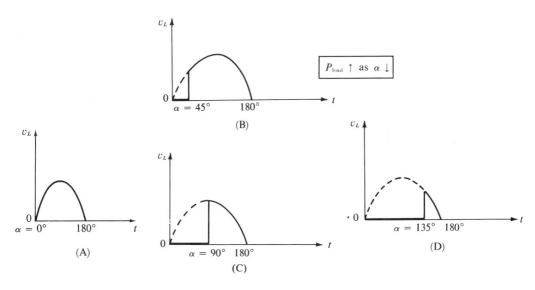

Figure 14.28.   EFFECT OF TRIGGER ANGLE ON LOAD VOLTAGE

Simple phase control circuit

The circuit in Figure 14.29 shows a simple method for varying the trigger angle and therefore the power in the load. Instead of using a gate pulse to trigger the SCR, the gate current is supplied by the AC source voltage e_{in} through $R1$, $R2$ and the series diode. The circuit operates as follows:

(a) As e_{in} goes positive, the SCR is forward biased from anode to cathode; however, it will not conduct ($v_L = 0$) until its gate current exceeds I_{GT}.

(b) The positive e_{in} also forward biases the diode and the SCR's gate-cathode junction; this causes a gate current i_G to flow.

(c) The gate current will increase as e_{in} increases toward its peak value. When i_G reaches a value equal to I_{GT}, the SCR turns "on" and v_L will approximately equal e_{in} (refer to point A on the waveforms in Figure 14.29).

(d) The SCR remains "on" and $v_L \simeq e_{in}$ until e_{in} decreases to the point where the load current is below the SCR holding current. This usually occurs very close to the point where $e_{in} = 0$ and begins to go negative.

(e) The SCR now turns "off" and remains "off" while e_{in} goes negative since its anode-cathode is reverse biased. Since the SCR is now an open switch, the load voltage is zero during this period.

(f) The purpose of the diode in the gate circuit is to prevent the gate-cathode reverse bias from exceeding V_{GRM} during the negative half-cycle of e_{in}. The diode is chosen to have a PRV rating greater than the input amplitude E_{max}.

(g) The same sequence repeats itself when e_{in} again goes positive.

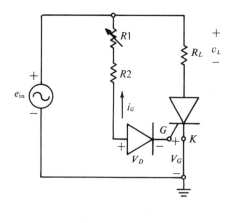

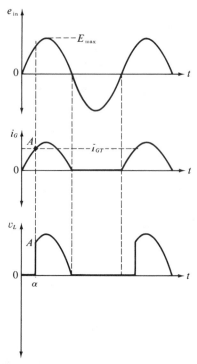

Figure 14.29.   SIMPLE SCR PHASE-CONTROL CIRCUIT

The load-voltage waveform in Figure 14.29 can be controlled by varying $R1$ which varies the resistance in the gate circuit. If $R1$ is increased, the gate current will reach its trigger value I_{GT} at a greater value of e_{in} so that the SCR will trigger at a later point in the e_{in} positive half-cycle. Thus the trigger angle α will increase. The opposite will occur if $R1$ is decreased. Of course, if $R1$ is made large enough the SCR gate current will never reach I_{GT} and the SCR will always remain "off."

▶ EXAMPLE 14.9

The circuit in Figure 14.29 uses an SCR with $I_{GT} = 0.1$ mA, $V_{GT} = 0.5$ V. The diode is silicon and the peak amplitude of the input is 24 volts. Determine the trigger angle α for $R1 = 100$ kΩ and $R2 = 10$ kΩ.

The first step is to determine the instantaneous value of e_{in} at which triggering will occur. At the SCR trigger point, $V_G = V_{GT} = 0.5$ V and $I_G = I_{GT} = 0.1$ mA. Using KVL around the gate circuit, we have

$$e_{in} = I_G(R1 + R2) + V_D + V_G$$

At the trigger point,

$$e_{in(trigger)} = 0.1 \text{ mA}(110 \text{ k}\Omega) + 0.7 \text{ V} + 0.5 \text{ V}$$
$$= 12.2 \text{ V}$$

Since e_{in} is a sine wave, it obeys the expression

$$e_{in} = E_{max}\sin(2\pi\ ft)$$

where $2\pi\ ft$ is the phase angle at any instant of time. For our purposes, this angle is α. Thus, since $E_{max} = 24$ V,

$$e_{in} = 24 \sin \alpha$$

We can find α at $e_{in} = 12.2$ V as follows:

$$12.2 = 24 \sin \alpha$$

$$\sin \alpha = \frac{12.2}{24} = 0.508$$

$$\alpha = 30.6° \qquad \blacktriangleleft$$

The simple circuit of Figure 14.29 suffers from several disadvantages. First, the trigger angle α is greatly dependent on the SCR's I_{GT}, which, as we know, can vary widely even among SCRs of a given type and is also highly temperature-dependent. In addition, the trigger angle can be varied only up to an approximate value of 90° with this circuit. This is because e_{in} is maximum at its 90° point so that the gate current has to reach I_{GT} somewhere between 0°–90° if it will reach it at all. This limitation means that the load-voltage waveform can only be varied from $\alpha = 0°$ (Figure 14.28A) to $\alpha = 90°$ (Figure 14.28C).

More complex phase control circuit

The circuit in Figure 14.30 overcomes the major disadvantages of the circuit just discussed. In this circuit a capacitor $C1$ is used to provide *phase lag* so that triggering of the SCR can be made to occur past the 90° point in the e_{in} cycle. The four-layer diode (or SUS) is used to make the trigger point more well defined and less dependent on the characteristics of the SCR.

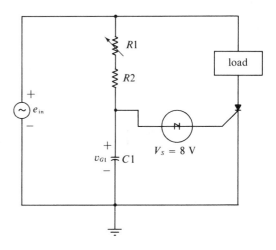

Figure 14.30.   MORE USEFUL PHASE-CONTROL CIRCUIT

The basic operation of this circuit is explained as follows:

(a) The $R1$, $R2$, $C1$ combination forms a low-pass circuit. In this circuit, v_{C1}, the voltage across $C1$, will *lag* behind the e_{in} voltage by a phase angle which depends on the R and C values.

(b) As v_{C1} varies due to e_{in}, it will eventually reach a value which exceeds $V_S = 8$ V, the diode's switching voltage. When it does so, the diode will turn "on," dropping to a low resistance; the capacitor will then discharge through the diode and the SCR gate-cathode junction. The capacitor discharge current supplies the SCR gate trigger current and the SCR turns "on," applying the e_{in} voltage to the load.

(c) The SCR remains "on" through the remainder of the e_{in} positive half-cycle. It stays "off" during the negative half-cycle. When e_{in} goes positive again, the sequence above repeats itself so that the SCR turns "on" at the same trigger angle for each e_{in} cycle.

(d) By varying $R1$, the *phase lag* is varied so that the SCR trigger angle can be varied over a wide range from approximately 0° to typically 130°–140°.

There are several more sophisticated phase-control circuits than the one in Figure 14.30 which allow phase control over the full 0°–180° range. These are discussed fully in the references listed at the end of the chapter.

Inverse parallel operation of SCRs

In the phase-control circuits just discussed, the SCR conducts only during the positive half-cycle of the input voltage so that current flows through the load in only one direction. This is adequate for some DC loads; however, AC loads re-

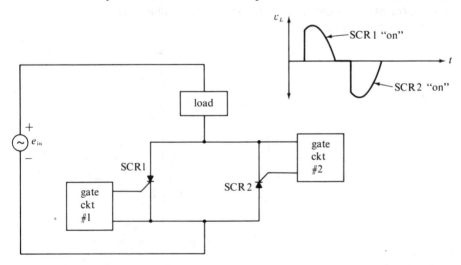

Figure 14.31. INVERSE PARALLEL OPERATION OF SCRs TO PRODUCE AC LOAD VOLTAGE

quire that current flows during both half-cycles. One way to achieve this effect is to use two SCRs in a back-to-back or *inverse parallel* arrangement as shown in Figure 14.31.

Each SCR has its own gate trigger circuit so that the trigger angles are independently controlled, although they are usually made the same so that load current will flow for the same portion of each half-cycle. During the positive half-cycle of e_{in}, SCR1 will be forward biased and will turn "on" when its gate is made positive relative to its cathode. During the negative half-cycle, SCR2 can be triggered "on." The voltage (and current) across the load will have both positive and negative polarities and will have an average value of zero.

The circuit of Figure 14.31 suffers from the fact that two SCRs and two separate gate trigger circuits are required. This drawback is eliminated by the use of a PNPN device called a TRIAC, which acts as an SCR in both directions.

14.17.   The TRIAC

The TRIAC is a semiconductor device that performs the same function as two SCRs connected in inverse parallel and can, therefore, be used to control power to an AC load. At present, TRIACs are limited to relatively low currents (100 A *rms*) and thus cannot replace SCRs for high current applications. Figure 14.32 shows a typical TRIAC circuit.

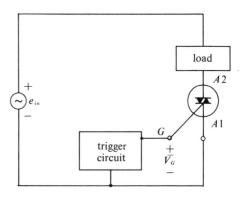

Figure 14.32.   TYPICAL TRIAC CIRCUIT

The symbol for the TRIAC as shown in the figure includes a gate (G) and two anodes ($A1$ and $A2$). The TRIAC will conduct in either direction ($A1$ positive relative to $A2$ and vice versa) when an appropriate trigger is applied to the gate. The major difference between SCR and TRIAC triggering is in the polarity of gate triggering signals. As we know, the SCR is triggered "on" by applying a positive gate voltage while the anode is positive relative to the cathode. The TRIAC, on the other hand, can be triggered "on" under any of the *four* following conditions:

(a) $A2$ positive; V_G positive.
(b) $A2$ positive; V_G negative.
(c) $A2$ negative; V_G positive.
(d) $A2$ negative; V_G negative.

Of these four possible trigger modes, numbers 1 and 4 are most often used since they require the least amount of gate current.

Once the TRIAC is triggered "on" in either direction, it acts similar to the SCR in that its voltage drops to a very low value (typically 1 V) and it will remain "on" until its current drops below its holding current level. Like the SCR, the TRIAC's gate cannot be used to turn the TRIAC "off." Turn-off is accomplished by reduction of the TRIAC current below I_H.

14.18. Other Variations of the SCR

There are other PNPN devices whose operation is similar to the SCR. These devices are somewhat specialized and are used in certain types of applications. The SCR and TRIAC are the workhorses of industrial electronics since they can be used to control load currents of hundreds of amperes and load powers into the kilowatt range. The following PNPN devices are low-current devices by comparison and are used in low-power applications such as sensing, counting and digital logic.

The light-activated SCR (LASCR)

The LASCR is an SCR that can be triggered "on" by the application of light energy. Figure 14.33A shows a simple LASCR circuit. Note that the LASCR

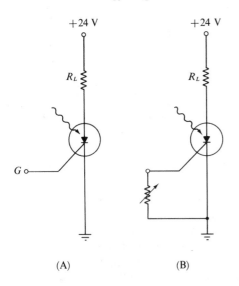

(A)   (B)

Figure 14.33.   (A) SIMPLE LASCR CIRCUIT AT MAXIMUM SENSITIVITY;
(B) SIMPLE LASCR WITH VARIABLE SENSITIVITY

has a gate lead just like a conventional SCR so that it can also be triggered "on" by a positive signal at the gate. Normally, the LASCR is triggered to the "on" state by focussing light energy onto junction $J2$ (see Figure 14.2). Once the LASCR is triggered to the "on" state, it behaves like a normal SCR. The LASCR will stay "on" even if the light disappears. It will turn "off" only if its anode current is decreased below I_H.

The LASCR is most sensitive to light when its gate is open as in Figure 14.33A. This sensitivity can be varied by connecting a variable resistor as in Figure 14.33B. In this way, the level of light at which the LASCR will trigger can be varied.

The gate-controlled switch (GCS)

Another member of the PNPN family is the *gate-controlled switch* (GCS) shown in Figure 14.34. The GCS is very similar to the SCR, except that it can be turned "off" by a *negative* signal at the gate. In other words, a positive gate signal will close the switch, which will remain closed until a negative gate pulse occurs or until the anode current is reduced below I_H.

A major disadvantage of the GCS is that its *turn-off gain* (ratio of controlled anode current to required gate turn-off current) is very low. Typically, a GCS may require a negative gate current of 1 mA to turn "off" 10 mA of anode current. The GCS is limited to low current levels (up to 1 A of anode current).

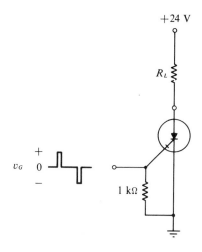

Figure 14.34.   GCS CIRCUIT

The silicon-controlled switch (SCS)

The *silicon-controlled switch* (*SCS*) is a commonly used low-power PNPN device. It is essentially a miniature SCR with leads attached to all *four* of the semiconductor regions. The additional lead is connected to the N region below the anode P region. The SCS is the only four-terminal PNPN device.

The SCS can be triggered "on" or turned "off" by an appropriate signal at either gate. Refer to Figure 14.35A, in which a simple SCS circuit is shown. The cathode gate G_C is the normal gate used with SCRs. The anode gate G_A is the new gate. The SCS can be triggered "on" by either a positive pulse at G_C or a negative pulse at G_A. In the "on" state the SCS behaves like an SCR, namely as a low-resistance with a voltage drop of typically 1 V.

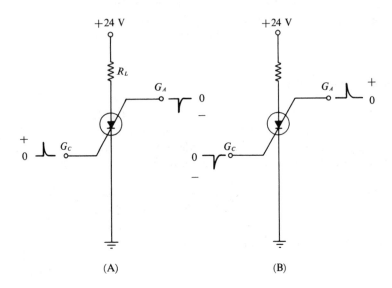

Figure 14.35.   BASIC SCS CIRCUIT: (A) TURN-ON; (B) TURN-OFF

The SCS can be turned off in any of three ways: (1) by reducing its anode current below I_H (same as SCR); (2) by applying a negative pulse at G_C; (3) by applying a positive pulse at G_A. Refer to part B of the figure.

Although the SCS can be turned off at either gate, it suffers from relatively low turn-off gain similar to the gate-controlled switch. However, SCSs are low-current devices designed for low-current applications; therefore, this is not a serious drawback since low anode currents will require low gate currents for turn-off.

The SCS is used mainly in low-power sensing circuits, timing and counting circuits and digital logic circuits. It is extremely versatile since it can be turned "on" and "off" by signals of either polarity.

14.19.   PNPN Devices: Summary

Table 14.3 contains a summary of the various PNPN devices discussed in this chapter.

TABLE 14.3

Device	Turned on when	Turned off when	Comments
Four-layer diode (and SUS)	anode-cathode voltage exceeds V_S.	current drops below I_H.	conducts in one direction (unilateral).
DIAC and SBS	voltage across it exceeds V_S of either polarity.	current drops below I_H.	conducts in both directions (bilateral).
SCR	positive voltage applied to gate; or when anode-cathode voltage exceeds breakover voltage.	current drops below I_H.	unilateral; used mainly for high-current, high-power control circuits.
TRIAC	positive or negative signal applied to gate; or when anode-anode voltage exceeds breakover voltage of either polarity.	current drops below I_H.	bilateral; used for high-power control circuits for AC loads.
LASCR	positive signal applied to gate; or when anode-cathode voltage exceeds breakover voltage; or when light energy is applied.	current drops below I_H.	unilateral; relatively low current device used for light sensing.
GCS	(same as SCR.)	current drops below I_H; or when a *negative* signal is applied to gate.	unilateral; relatively low current device.
SCS	same as SCR *plus* a negative pulse at the anode gate, G_A.	current drops below I_H; or negative pulse at G_C; or positive pulse at G_A.	unilateral; relatively low current device.

GLOSSARY

PNPN devices: four-layer devices of alternate doping.

Latching: characteristic of a PNPN switch which allows it to remain "on" after the excitation has been removed.

Triggering: process of causing a PNPN device to switch from "off" to "on."

Forward switching voltage: anode-to-cathode voltage at which a four-layer diode switches.

Holding current: current required to maintain a PNPN switch in the "on" state.

Rate effect: process by which a PNPN device switches "on" due to a rapidly rising anode-to-cathode voltage.

Forward breakover voltage: anode-to-cathode voltage at which an SCR switches "on."

Latching current: amount of SCR current which must flow at turn-on in order for the SCR to latch on.

Trigger angle: phase angle of input voltage at the instant of SCR triggering.

Phase control: control of power to a load by controlling trigger angle.

PROBLEMS

14.1   How does a PNPN switch differ from a mechanical switch aside from physical characteristics?

14.2   Sketch the PNPN structure, label all junctions and terminals and draw its two-transistor equivalent.

14.3   What is the difference between a four-layer diode and a P-N diode?

14.4   What process produces switching in the four-layer diode?

14.5   Sketch the *I-V* characteristic of the four-layer diode. Label the voltages V_S and V_F and the currents I_S and I_H.

14.6   In the circuit of Figure 14.6, the four-layer diode has the following parameters:

$$V_S = 8 \text{ V}$$
$$I_S = 0.5 \text{ mA}$$
$$I_H = 1.5 \text{ mA}$$
$$V_F = 1 \text{ V}$$

The input voltage E is changing according to the waveform in Figure 14.36. Sketch the waveform of V_D in response to this input.

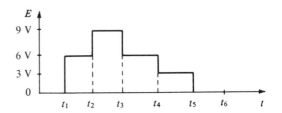

Figure 14.36.

14.7   Repeat Problem 14.6 using a series resistor of 10 kΩ.

14.8   Explain how the "rate effect" can cause a four-layer diode to turn on at lower than V_S.

14.9   Consider the circuit of Figure 14.37. The photocell is a CdS type. The light on the photocell can vary. The circuit is to function as an *over-light* indicator. If the light exceeds a certain threshold level, the alarm is to be activated and must remain activated even if the light level subsequently decreases.
(a) Explain how the circuit operates.
(b) How can the threshold level be easily varied?

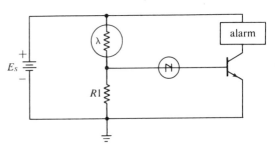

Figure 14.37.

14.10   A relaxation oscillator of the type shown in Figure 14.9 uses the diode with the parameters given in Problem 14.6. If a capacitor of 0.1 μF is used and $E = 12$ V, find $R_{\min}$ and $R_{\max}$ and the resulting limits on the oscillator frequency.

14.11   For the oscillator of Problem 14.10, indicate the effects each of the following changes will have on the frequency and amplitude of the capacitor voltage waveform:
(a) An increase in C.
(b) A decrease in E.
(c) Replacement of the diode with one that has $V_S = 6$ V.
(d) An increase in R.

14.12   Design a time-delay circuit like the one in Figure 14.13 using the diode specified in Problem 14.6 so that a *single* output pulse occurs 10 ms after the input jumps to 20 V.

14.13   What is the approximate amplitude of the output pulse in the delay circuit of the preceding problem?

14.14   Modify your time-delay circuit to produce a periodic train of output pulses at a frequency of 100 Hz (one pulse every 10 ms).

14.15   Explain the differences between the four-layer diode, the SUS, the SBS and the DIAC.

14.16   Consider the circuit in Figure 14.38. It uses an SBS with $V_S = \pm 8$ V. Determine the output-voltage waveform. Assume $V_F = 1$ V and $I_H = 1$ mA.

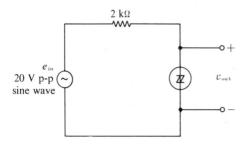

Figure 14.38.

14.17   Draw the circuit symbol for the SCR and label all terminals.

14.18   How does the SCR differ from the four-layer diode?

14.19   Sketch the *I-V* characteristic of an SCR for three values of gate current. Label all important points.

14.20   Can the gate terminal be used to trigger the SCR when its anode is negative relative to cathode?

14.21   Refer to the specification sheet for the GE C30 and C32 SCRs in Appendix II. For the C30A SCR determine the following:
(a) Minimum anode-cathode forward breakover voltage.
(b) Maximum allowable repetitive reverse voltage.
(c) Typical I_{GT} at 25°C.
(d) Typical V_{GT} at 25°C.
(e) Typical holding current at 25°C.
(f) $T_{J(\max)}$.
(g) Maximum allowable reverse gate voltage.
(h) Maximum allowable *rms* forward current.

14.22   Indicate whether the values in answer to 14.21(a), (c), (d), (e), (f) and (h) will increase, decrease or remain the same at higher junction temperatures.

14.23   Will the SCR with the gate characteristics of Figure 14.21 trigger under the following conditions?
(a) $I_G = 5$ mA, $V_G = 0.2$ V, $T_J = 25$°C.
(b) $I_G = 20$ mA, $V_G = 1$ V, $T_J = 25$°C.
(c) $I_G = 20$ mA, $V_G = 2.5$ V, $T_J = -65$°C.

14.24 In Figure 14.39 the SCR has a holding current of 1 mA and a latching current of 3 mA. How wide must the gate pulse be in order for the SCR to turn "on"? Assume $V_F = 1$ V. *Hint:* Find how long it takes the load current to reach 3 mA once the SCR is "on."

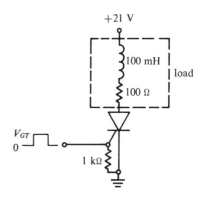

Figure 14.39.

14.25 If a gate pulse with a narrower pulse width than that calculated in Problem 14.24 is to be used, what value of R should be placed across the load to ensure SCR turn-on?

14.26 Why is the SCR useful for supplying large amounts of power to a load?

14.27 Can most SCRs be turned "off" by reverse biasing the gate?

14.28 One method of turning "off" an SCR is shown in Figure 14.40. The transistor is normally held "off" by reverse bias at the *E-B* junction. If the SCR is triggered it will turn "on." When V_{in} is positive the transistor will turn "on" (saturate) and divert all the anode current to the collector. This will turn the SCR "off" and it will remain "off" until its gate is pulsed again when $V_{\text{in}} = 0$. For this circuit determine the value of V_{in} needed to turn the SCR "off" if the transistor is silicon and has $h_{FE} = 30$.

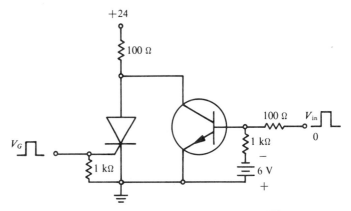

Figure 14.40.

14.29   The SCR in Figure 14.41 has a holding current of 10 mA, $V_F = 1$ V and $V_{GT} = 2$ V. For the input and gate signal waveforms shown, sketch the resulting load-voltage waveform.

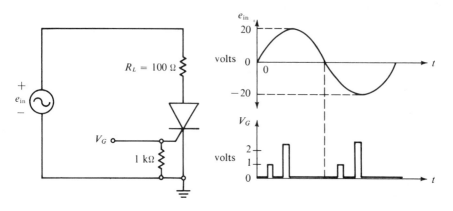

Figure 14.41.

14.30   The circuit of Figure 14.29 utilizes an SCR with $I_{GT} = 0.1$ mA and $V_{GT} = 0.5$ V. If $R_2 = 10$ kΩ and the diode is silicon, determine the value of R_1 needed to cause triggering when e_{in} reaches 3.2 V.

14.31   If e_{in} in Problem 14.30 is a 40 V p-p sine wave, determine the trigger angle. If R_1 is increased, what happens to the trigger angle? To the load power?

14.32   Figure 14.42 shows a TRIAC used in a simple phase control circuit. Assume the TRIAC's gate current needed to trigger is 1 mA (for either polarity). Also assume $I_H \simeq 0$ and $V_G \simeq 0$. Sketch the waveform of load voltage and TRIAC voltage showing trigger angles. Use $R_2 = 40$ kΩ.

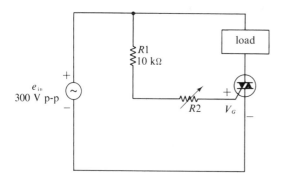

Figure 14.42.

14.33   Figure 14.43 shows an SCR used in a over-heat detector circuit. The *thermistor* used in the circuit is a temperature-sensitive resistor; its resistance *increases* with temperature. The SUS shown in the circuit has $V_S = 8$ V. (a) Explain what happens as the temperature of the thermistor increases.

(b) At what value of R_T will the SCR trigger "on" and activate the indicator light?

(c) How can the temperature at which triggering occurs be varied?

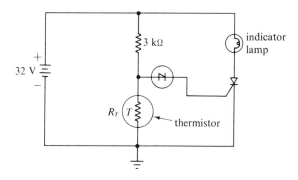

Figure 14.43.

14.34 For the circuit of Figure 14.44 determine how long it takes current to flow through the load once the switch is closed. Assume $V_s = 8$ V.

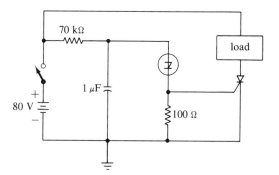

Figure 14.44.

14.35 Consider each of the following descriptions. For each one indicate which PNPN device or devices the statement describes. A given statement can pertain to more than one PNPN device.

(a) A bilateral two-terminal device.

(b) Turned "on" only by exceeding anode-cathode switching voltage.

(c) A bilateral device which is triggered by a gate signal.

(d) A unilateral device that can be turned "off" by a negative gate signal.

(e) Can be turned "on" or "off" at either gate.

(f) A low-current, four-terminal device.

(g) A unilateral two-terminal device with V_S usually ≤ 10 V.

(h) Can be turned "off" by reducing current below I_H.

(i) Can be turned "on" by a *negative* pulse.

(j) Once it is turned "on" it remains in the low-resistance state even after the cause that produced turn-"on" has disappeared or decreased.

(k) Triggered "on" by applying light energy.

REFERENCES

Gentry, F. E., F. W. Gutzwiller, N. Holonyak and E. E. Von Zastrow, *Semiconductor Controlled Rectifiers: Principle and Applications of PNPN Devices*. Englewood Cliffs, N. J.: Prentice-Hall, Inc., 1964.

Heller, S., *Understanding Silicon Controlled Rectifiers*. New York: Hayden Book Company, Inc., 1968.

Romanowitz, H. A. and R. E. Puckett, *Introduction to Electronics*. New York: John Wiley & Sons, Inc., 1968.

Silicon Controlled Rectifier Manual. Auburn, New York: General Electric Company, 1964.

Sowa, W. A. and M. M. Toole, *Special Semiconductor Devices*. New York: Holt, Rinehart and Winston, Inc., 1968.

Unijunction Transistors

15.1. Introduction

The *unijunction transistor*, abbreviated UJT, is a three terminal, single-junction device. The basic UJT and its variations are essentially latching switches whose operation is similar to the four-layer diode; the most significant difference being that the UJT's switching voltage can be easily varied by the circuit designer. Like the four-layer diode, the UJT is always operated as a switch and finds its most frequent applications in oscillators, timing circuits and SCR/ TRIAC trigger circuits.

15.2. Basic Operation

A typical UJT structure, pictured in Figure 15.1, consists of a lightly doped, N-type silicon bar provided with ohmic contacts at each end. The two end connections are called *base-1*, designated $B1$ and *base-2*, $B2$. A small, heavily doped P region is alloyed into one side of the bar closer to $B2$. This P region is the UJT *emitter*, E, and forms a P-N junction with the bar.

An *interbase resistance*, R_{BB}, exists between $B1$ and $B2$. It is typically between 4 kΩ and 10 kΩ and can easily be measured with an ohmmeter with the

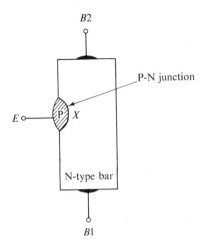

Figure 15.1.   BASIC UJT STRUCTURE

emitter open. R_{BB} is essentially the resistance of the N-type bar. This interbase resistance can be broken up into two resistances, the resistance from $B1$ to emitter, called R_{B1}, and the resistance from $B2$ to the emitter, called R_{B2}. Since the emitter is closer to $B2$, the value of R_{B1} is greater than R_{B2} (typically 4.2 kΩ versus 2.8 kΩ).

The operation of the UJT can better be explained with the aid of an equivalent circuit. The UJT's circuit symbol and its equivalent circuit are shown in Figure 15.2. The diode represents the P-N junction between the emitter and the base bar (point x). The arrow through R_{B1} indicates that it is variable since during normal operation it may typically range from 4 kΩ down to 10 Ω.

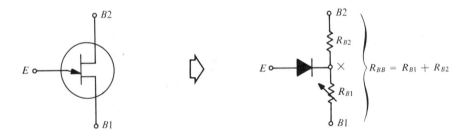

Figure 15.2.   UJT SYMBOL AND EQUIVALENT CIRCUIT

The essence of UJT operation can be stated as follows:

(a) When the emitter diode is reverse biased, only a very small emitter current flows. Under this condition, R_{B1} is at its normal high value (typically 4 kΩ). This is the UJT's "off" state.

(b) When the emitter diode becomes forward biased, R_{B1} drops to a very low value (reason to be explained later) so that the total resistance between E and $B1$ becomes very low, allowing emitter current to flow easily. This is the "on" state.

UJT biasing

Figure 15.3 shows the normal bias arrangement for the UJT. The UJT's equivalent circuit is used to aid in the explanation. In this arrangement, $B2$ is biased positive relative to $B1$ by virtue of the V_{BB} supply. If we neglect the diode momentarily, it can be seen that R_{B2} and R_{B1} act as a voltage divider so that V_x, the voltage at point x relative to ground, is given by

$$V_x = \left(\frac{R_{B1}}{R_{B1} + R_{B2}}\right) \times V_{BB} = \underbrace{\left(\frac{R_{B1}}{R_{BB}}\right)}_{\eta} \times V_{BB}$$

or

$$V_x = \eta V_{BB} \qquad (15.1)$$

where η (the Greek letter "eta") is the voltage divider ratio and is called the *intrinsic standoff ratio*. The value of η is typically between 0.5 and 0.8. For example, if $\eta = 0.65$ and $V_{BB} = 20$ V, then $V_x = 0.65 \times 20 = 13$ V.

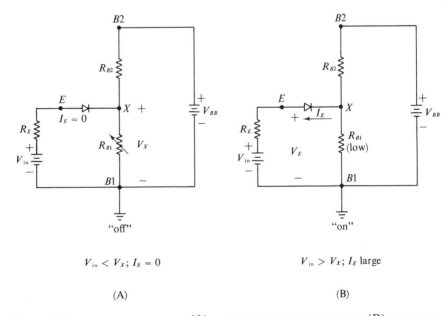

$$V_{in} < V_x; I_E \approx 0 \qquad\qquad\qquad V_{in} > V_x; I_E \text{ large}$$

(A)   (B)

Figure 15.3.   UJT BIASING: (A) EMITTER REVERSE BIASED; (B) EMITTER FORWARD BIASED

The voltage at point x is essentially the voltage on the N side of the P-N junction. The voltage V_{in} is shown applied to the emitter (P side of the P-N junction); thus, if V_{in} is less than V_x the emitter diode will not conduct and I_E

will be approximately zero. This is the "off" state of the UJT (part A of Figure 15.3) where the resistance seen by the input source is very high.

If V_{in} increases to a value slightly higher than V_x, things begin to happen. The diode begins to conduct as it becomes forward biased. Holes from the heavily doped emitter are injected into the N-type bar, specifically into the $B1$ region. The bar, which is lightly doped, offers very little chance for these holes to recombine. As such, the lower half of the bar becomes replete with additional current carriers (holes) and its resistance R_{B1} is drastically lowered (see part B of Figure 15.3). The decrease in R_{B1} causes V_x to drop, which in turn causes the diode to become more forward biased, causing I_E to increase even further. The larger I_E injects more holes into $B1$, further reducing R_{B1}, and so on. When this *regenerative*, or *snowballing*, process ends, R_{B1} is very low (10 Ω) and I_E is very large, limited mainly by resistance R_E.

This turn-"on" process occurs when V_{in} is large enough to turn on the diode. Since the diode is silicon, this occurs when V_{in} is approximately 0.5 V greater than V_x. In other words, the emitter current will jump from zero to a high value when the emitter voltage exceeds the *peak-point voltage* V_P where

$$V_P = V_x + V_D = \eta V_{BB} + V_D \tag{15.2}$$

V_D represents the diode's threshold voltage, usually 0.5 V. V_P is essentially the voltage needed at the UJT emitter in order to switch the UJT from "off" to "on." For example, if $\eta = 0.65$ and $V_{BB} = 20$ V, then V_P will be 13.5 V.

Turning off the UJT

Referring to Figure 15.3B it can be seen that once the UJT is "on," V_x will drop to a very low value and so will V_E, the emitter-to-base-1 voltage. The value of V_E in the "on" state is called the *emitter saturation voltage*, $V_{E(sat)}$, and is typically around 2 V. If V_{in} is increased, I_E will increase through the "on" UJT. If V_{in} is decreased, I_E and V_E will both decrease, but the UJT will remain "on" until I_E is reduced to a value lower than the UJT's *valley current*, I_V. This I_V is essentially the holding current needed to keep the UJT "on." When I_E drops below I_V, the UJT turns "off" and the situation reverts back to that in Figure 15.3B. The emitter voltage at the valley current is V_V, the *valley voltage*. It is typically around 1 V.

▶ EXAMPLE 15.1

A certain UJT has $\eta = 0.6$, and $I_V = 2$ mA. For the circuit of Figure 15.3, assume $V_{BB} = 30$ V and $R_E = 1$ kΩ. (a) Determine the value of V_{in} needed to turn "on" the UJT. (b) Determine the approximate value of I_E when $V_{in} = 25$ V. (c) Determine the value of V_{in} which will cause the UJT to turn "off."

(a) $$V_P = \eta V_{BB} + V_D$$
$$= 0.6(30) + 0.5 = 18.5 \text{ V}$$

Thus for turn-on we need $V_{in} =$ **18.5 V**.

(b) With $V_{in} = 25$ V, the UJT is "on" and we can assume $V_E = V_{E(sat)} \simeq 2$ V. Thus there must be 23 V across R_E so that

$$I_E = 23 \text{ V}/1 \text{ k}\Omega = \textbf{23 mA}$$

(c) At turn-off, $I_E = I_V = 2$ mA and $V_E = V_V = 1$ V. Using KVL,

$$V_{in} = I_E \times R_E + V_E$$
$$= 2 \text{ mA} \times 1 \text{ k}\Omega + 1 \text{ V} = \textbf{3 V}$$

which is the value of V_{in} that will produce turn-off. Any V_{in} lower than 3 V will reduce I_E below I_V and cause the UJT to turn "off."   ◀

15.3.   Comparison of UJT with Four-Layer Diode

From the description in the last section, the student might recognize the similarities between the UJT and four-layer diode operation. Table 15.1 lists the corresponding parameters of each device.

TABLE 15.1

Parameter	UJT	Four-layer diode
Voltage needed to turn-on	V_P	V_S
Current at turn-on point	I_P	I_S
Current needed to hold device on	I_V	I_H
Voltage at holding current point	V_V	V_F

Both devices are switched to the low resistance state by applying a voltage which exceeds a given value. For the four-layer diode (FLD) this value is V_S and it must be applied from anode to cathode; for the UJT this value is V_P (peak-point voltage) and is applied from emitter to base-1. At the turn-on point for each device a small current flows just before the device switches "on." For the FLD this current is I_S; for the UJT it is I_P, called the *peak-point current*.

Once either device has switched to the "on" state it will remain there until its current drops below a given value. For the FLD this value is I_H; for the UJT this value is I_V, the valley current. The "on" voltage at the holding current point is very low for both devices. For the FLD it is V_F and for the UJT it is V_V (valley voltage), both of which are approximately 1 V.

Thus far the basic difference between the UJT and the FLD lies in the variability of the switching voltage. Whereas V_S for the FLD is fixed by its physical construction, V_P for the UJT can be varied by varying the V_{BB} supply voltage. This ability to vary V_P makes the UJT much more versatile and useful than the FLD.

15.4.   UJT Parameters and Ratings

A set of parameters and ratings for a typical UJT are listed in Table 15.2. Some of the entries were defined earlier. Those which require explanation are:

(a) Maximum reverse emitter voltage V_{B2E}. This is the maximum reverse bias which the emitter-base-2 junction can tolerate before breakdown occurs.

(b) Maximum interbase voltage. This limit is caused by the maximum power that the N-type base bar can safely dissipate.

(c) Maximum *peak* emitter current. This represents the maximum allowable value of a *pulse* of emitter current.

(d) Emitter leakage current I_{EO} is the emitter current which flows when V_E is less than V_P and the UJT is "off."

Most of the UJT parameters are temperature-sensitive to some degree. V_V, I_V and I_P will decrease with an increase in temperature. I_{EO} increases slightly with an increase in temperature. The interbase resistance R_{BB} will typically *increase* with temperature at the rate of around 1 kΩ every 25°C. The intrinsic standoff ratio η, however, changes only very slightly with temperature since it is essentially the resistor ratio (R_{B1}/R_{BB}); R_{B1} and R_{BB} have about the same percentage change with temperature; therefore, η stays relatively constant.

TABLE 15.2

UJT CHARACTERISTICS AT $T_J = 25°$C

Max reverse emitter voltage (V_{B2E}):   60 V
Max interbase voltage (V_{BB}):   45 V
Max peak emitter current: 1 A
Max average power dissipation: 500 mW
Interbase resistance (R_{BB}):   4.7 kΩ to 6.8 kΩ
Intrinsic standoff ratio (η):   0.51 to 0.62
Emitter leakage current (I_{EO}):   2 μA (max) at $V_{B2E} = 60$ V
Valley current (I_V):   8 mA (min) at $V_{BB} = 10$ V
Valley voltage (V_V):   1 V (typical) at $V_{BB} = 10$ V
Peak-point current (I_P):   12 μA (max) at $V_{BB} = 10$ V

15.5.   UJT Relaxation Oscillator

The most common UJT circuit in use today is the relaxation oscillator, shown in Figure 15.4. Its operation is similar to the four-layer diode relaxation oscillator. However, its differences are significant enough to warrant a detailed investigation of its operation.

Let us consider the situation in which the capacitor is at zero volts ($V_C = 0$) and the switch is suddenly closed at $t = 0$ applying V_{in} to the circuit. Since

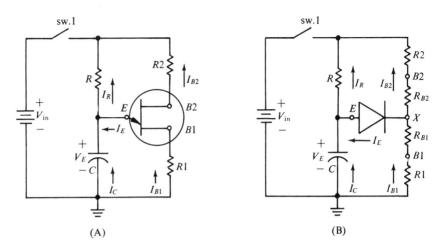

Figure 15.4.   (A) UJT BASIC RELAXATION OSCILLATOR; (B) ITS EQUIV-
ALENT CIRCUIT

$V_E = V_C = 0$, the UJT emitter diode is reverse biased and the UJT is "off."
The amount of reverse bias is V_x volts which can be obtained using the voltage
divider rule

$$V_x = \frac{(R1 + R_{B1})V_{in}}{R1 + R_{B1} + R2 + R_{B2}} \qquad (15.3)$$

In many cases, R1 and R2* are much smaller than R_{B1} and R_{B2} and V_x becomes
approximately equal to ηV_{in} (Equation 15.1).

In this condition the only emitter current flowing will be a small reverse
leakage, I_{EO}. Also, R_{B1} will be at its "off" value (typically 4 kΩ). Thus we can
consider the emitter to be open ($I_E \approx 0$) and the capacitor will begin to charge
toward the input voltage V_{in} through resistor R. The capacitor voltage increases with a time constant of RC as illustrated in Figure 15.5A. It will continue to increase until the voltage at the emitter reaches the peak-point value,
V_P, given by Equation 15.2. At this time, T_1, the emitter diode becomes forward
biased and the UJT turns "on" with R_{B1} dropping to a very low value (typically
10 ohms). Since the diode is now forward biased, the capacitor will discharge
through the low resistance path containing the diode, R_{B1} and R1.

The capacitor discharge time constant is normally very short compared to
its charging time constant (see Figure 15.5B). An analytical expression for the
discharge time constant is difficult to obtain since R_{B1} will continually change
as the current I_E decreases. The discharging capacitor provides the emitter
current needed to keep the UJT "on"; it will remain "on" until I_E drops below
the valley current I_V (note the similarity to I_H), at which time the UJT will turn
"off." This occurs at time T_2 when the capacitor voltage has dropped to the
valley voltage V_V (typically 1 to 2 volts). At this time, R_{B1} returns to its "off"
value, the diode is again reverse biased and $I_E \approx 0$.

---

*The reasons for including R1 and R2 in the circuit will become apparent later.

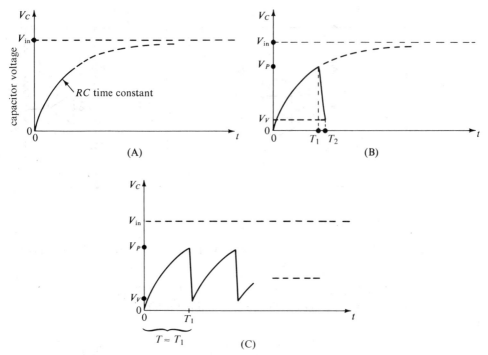

Figure 15.5.

The capacitor will begin charging toward V_{in} once again and the previous chain of events will repeat itself indefinitely as long as power is applied to the circuit. The result is a periodic saw-tooth-type waveform as shown in Figure 15.5C. To calculate the frequency of this waveform we first calculate the period of one cycle. The length of one period, T, is essentially the time it takes for the capacitor to charge to V_P since the discharge time T_2 is usually relatively short. Thus $T \approx T_1$ and is given by

$$T = RC \log_e\left(\frac{V_{in}}{V_{in} - V_P}\right)^*$$ **(15.4)**

In most cases, $V_P \approx \eta V_{in} + V_D$ and the period can be written as

$$T \approx RC \log_e\left[\frac{V_{in}}{V_{in}(1 - \eta) - V_D}\right]$$ **(15.5)**

The small diode drop V_D can often be ignored if $V_{in} > 10$ V, resulting in the more approximate expression

$$T \approx RC \log_e\left(\frac{1}{1 - \eta}\right)$$ **(15.6)**

---

*The value of T can also be obtained using a universal time-constant curve by (a) determining what percentage V_P is of V_{in}; (b) looking on the chart for the number of time constants to reach this percentage; and (c) using T = number of time constants $\times$ RC.

Examination of Equation 15.6 brings out an important point, namely that T is relatively independent of supply voltage V_{in}. This characteristic is important when designing a stable oscillator circuit.

The oscillator frequency is given by $1/T$ and can be obtained by using either of the three previous equations for T.

▶ EXAMPLE 15.2

The circuit of Figure 15.4 uses a UJT with $\eta = 0.6$, $R_{BB} = 10$ kΩ and $V_D = 0.5$ V. If $V_{in} = 20$ V, $R1 = R2 = 100$ Ω, $R = 10$ kΩ and $C = 1$ μF, determine the amplitude of the capacitor waveform (V_P) and the oscillation frequency.

The value of V_P can be calculated using

$$V_P = \eta V_{in} + V_D$$

since $R1$ and $R2$ are $\ll R_{BB}$. Thus we have

$$V_P = 0.6(20 \text{ V}) + 0.5 \text{ V} = \textbf{12.5 V}$$

The period of the waveform is approximately given by Equation 15.4 as

$$T \approx 10 \text{ k}\Omega \times 1 \text{ }\mu\text{F} \times \log_e\left(\frac{20}{20 - 12.5}\right)$$
$$\approx 10 \text{ ms} \times \log_e(2.67)$$
$$\approx 10 \text{ ms} \times 0.98 = 9.8 \text{ ms}$$

This gives a frequency of $1/T = \textbf{102 Hz}$.   ◀

▶ EXAMPLE 15.3

Repeat Example 15.2 for an input supply voltage of 30 V.

$$V_P = 0.6 \times 30 \text{ V} + 0.5 \text{ V} = 18.5 \text{ V}$$

Thus

$$T = 10 \text{ k}\Omega \times 1 \text{ }\mu\text{F} \times \log_e\left(\frac{30}{30 - 18.5}\right)$$
$$= 10 \text{ ms} \times \log_e(2.61)$$
$$= 10 \text{ ms} \times 0.96$$
$$= 9.6 \text{ ms}$$

This gives a frequency of $1/T = \textbf{104 Hz}$.

Notice that V_P increases approximately proportional to V_{in} so that the amplitude of the capacitor saw-tooth waveform is increased. The frequency, however, has changed only slightly; for a 50 percent increase in V_{in} (20 V to 30 V), the frequency increased by only 2 percent (102 Hz to 104 Hz).   ◀

▶ EXAMPLE 15.4

The approximate period of the waveforms of the last two examples could have been calculated using Equation 15.6 since $V_{in} > 10$ V. In each case the result would have been

$$T = 10 \text{ k}\Omega \times 1 \text{ }\mu\text{F} \times \log_e\left(\frac{1}{1 - 0.6}\right)$$

$$= 10 \text{ ms} \times \log_e(2.5)$$

$$= 10 \text{ ms} \times 0.92 = 9.2 \text{ ms}$$

so that $f \simeq 1/T \simeq 109$ Hz. This value is very close to those obtained using the more exact expression of Equation 15.4. The simpler formula of Equation 15.6 is accurate enough for most practical situations.   ◀

The similarity between the operation of the UJT oscillator and the four-layer diode oscillator should now be apparent. The main differences are that the UJT circuit has a *frequency* relatively independent of V_{in} and a capacitor waveform amplitude which changes with V_{in}, while the FLD circuit has a frequency which depends on V_{in} and a capacitor waveform amplitude which is constant at a value equal to V_S.

Pulse outputs

The UJT relaxation oscillator circuit can also supply pulse waveforms. If the output is taken from $B1$, the result is a train of pulses occurring during the discharge of the capacitor through the UJT emitter. The waveform of V_{B1} is illustrated in Figure 15.6A. The amplitude of the $B1$ pulses is always less than V_p, but is greater for larger values of C. The voltage at $B1$ during the UJT "off" time will be very small and is determined by the voltage divider formed by $R1$, R_{BB} and $R2$ (see Figure 15.4B). That is,

$$V_{B1 \text{ (off)}} = \left(\frac{R1}{R1 + R_{BB} + R2}\right)V_{in} \qquad (15.7)$$

This voltage is important in certain applications, as we shall see.

The rise time of the pulses at $B1$ is very short (less than 1 μs), but the fall time depends on the values of C and $R1$. A larger value of C or $R1$ will cause a slower capacitor discharge and a longer fall time.

If the output is taken at $B2$ a waveform of negative-going pulses is obtained as shown in Figure 15.6B. This results from the decrease in R_{B1} when the UJT turns "on." This increases I_{B2} which increases the drop across $R2$ and thus reduces V_{B2}. The amplitude of these pulses is usually about a couple of volts, but can be increased by increasing $R2$.

The pulses at $B1$ are usually the ones of most interest; they are of relatively high amplitude and are not affected by loading since they appear across a low-valued resistor $R1$. These positive pulses are often used to trigger SCRs or other gated PNPN devices. The amplitude of these pulses is to some degree de-

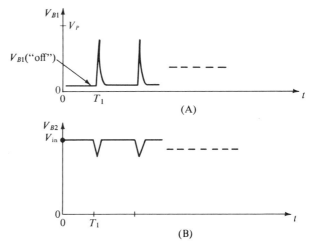

Figure 15.6.   WAVEFORMS AT (A) $B1$ AND (B) $B2$

pendent on the value of C. For values of C of 1 μF or greater, the amplitude of the pulses is approximately equal to V_P (less the 1–2 V UJT drop). As C becomes smaller, the $B1$ pulses decrease in amplitude. The reason for this is that the smaller value for C discharges a significant amount during the time that the UJT is making its *transition* from the "off" to "on" state. Thus, when the UJT finally reaches the "on" state, C has lost some of its voltage (V_P), and less voltage can appear across $R1$ as the capacitor continues its discharge.

Varying the frequency

The frequency of oscillations is normally controlled by varying the charging time constant RC. There are, however, limits on R as there were in the four-layer diode circuit. These limits are dictated by the same considerations and are:

$$R_{\min} = \frac{V_{\text{in}} - V_V}{I_V} \qquad \text{(15.8A)}$$

$$R_{\max} = \frac{V_{\text{in}} - V_P}{I_P} \qquad \text{(15.8B)}$$

Keeping R between these limits will ensure oscillation. If R is greater than $R_{\max}$, the capacitor never reaches V_P since the current through R is not large enough to both charge the capacitor and supply I_P to the UJT. The UJT will stay in the "off" state.

If R is smaller than $R_{\min}$, the capacitor will reach V_P and discharge through the UJT, but the UJT will not turn "off" since the current through R is greater than the I_V needed to hold the UJT "on." In some circuits this is useful where only *one* output pulse is desired after application of power. A time-delay circuit similar to that using the four-layer diode is a good example. The time delay is equal to *one* period (Equation 15.6).

▶ EXAMPLE 15.5

For $C = 1 \ \mu F$, $V_{in} = 20$ V and $\eta = 0.6$, determine the limits on the charging resistor R and the resulting limits on frequency. The UJT has $I_V = 10$ mA, $V_V = 2$ V and $I_P = 5 \ \mu A$.

$$R_{min} = \frac{20 \text{ V} - 2 \text{ V}}{10 \text{ mA}} = \textbf{1.8 k}\Omega$$

and

$$R_{max} = \frac{20 \text{ V} - \overset{V_P}{\overbrace{12.5 \text{ V}}}}{5 \ \mu A} = \textbf{1.5 megohms}$$

Using these values to calculate T in Equation 15.6,

$$T_{min} = 1.8 \text{ k}\Omega \times 1 \ \mu F \times \log_e(2.5) = 1.6 \text{ ms}$$

and

$$T_{max} = 1.5 \text{ M}\Omega \times 1 \ \mu F \times \log_e(2.5) = 1.35 \text{ s}$$

This results in

$$f_{max} = \textbf{625 Hz}$$

and

$$f_{min} = \textbf{0.74 Hz} \qquad ◀$$

Examination of Equation 15.8 indicates that to obtain a greater upper limit on frequency (a lower R_{min}) the value of I_V should be made larger. Similarly, to obtain a smaller lower limit on frequency (a higher R_{max}) the value of I_P should be made smaller. UJTs with I_V as high as 20 mA and I_P as low as 2 μA are presently available, resulting in a possible frequency range of 4000 : 1.

The frequency may also be varied by varying C. The lower limit on C is normally around 0.001 μF, while the upper limit depends on the size of $R1$ (which limits discharge current). In most applications of this circuit the value of C is kept fixed and a variable resistor is used for R.

The temperature stability of the UJT relaxation oscillator frequency is normally very good. This is because η varies only slightly with temperature and the only variation in V_P is due to the small decrease in V_D (2 mV/°C) with temperature. Its stability of frequency with variations in temperature and supply voltage coupled with its simplicity and low cost make the UJT oscillator a popular circuit for timing and pulsing applications.

15.6.   The UJT as an SCR Trigger

The circuit examined in the previous section is often used as the gate trigger source in SCR applications. The basic circuit is shown in Figure 15.7 where the $B1$ pulse output is used to trigger the SCR a predetermined interval of time

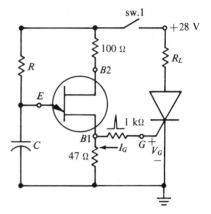

Figure 15.7.   UJT OSCILLATOR AS GATE TRIGGER SOURCE

after the switch is closed. That is, the first $B1$ pulse occurs T seconds after the 28 V is supplied to the UJT circuit. After the SCR has been triggered "on," subsequent pulses at its gate have no effect.

An important design consideration in this type of circuit concerns premature triggering of the SCR. The voltage at $B1$ when the UJT is "off" (Equation 15.7) must be smaller than the voltage needed to trigger the SCR, otherwise the SCR will be triggered immediately upon switch closure. Thus we have the requirement

$$V_{B1\,(\text{off})} < (I_{GT} \times 1\text{ k}\Omega + V_{GT}) \qquad\qquad \textbf{(15.9)}$$

▶ EXAMPLE 15.6

The UJT in Figure 15.7 has $R_{BB} = 6$ kΩ and the SCR has $I_{GT} = 0.1$ mA and $V_{GT} = 0.5$ V. Determine whether the circuit will cause premature SCR triggering.

Using Equation 15.7, we have

$$V_{B1\,(\text{off})} = \frac{47\ \Omega \times 28\text{ V}}{6147\ \Omega} = 0.21\text{ V}$$

The requirement of Equation 15.9 is

$$V_{B1\,(\text{off})} \leq (0.1\text{ mA} \times 1\text{ k}\Omega + 0.5\text{ V}) = 0.6\text{ V}$$

Our value of 0.21 V satisfies this requirement.   ◀

▶ EXAMPLE 15.7

The same UJT of the previous example has $\eta = 0.7$, $V_D = 0.7$ V, $V_V = 2$ V and $I_V = 3$ mA. Choose values for R and C that will give a 10-ms delay in the circuit of Figure 15.7.

The value of R is to be chosen to produce only one $B1$ pulse. That is, we want $R < R_{min}$ as given by Equation 15.8A. We have

$$V_P = 0.7 \times 28 \text{ V} + 0.7 \text{ V}$$
$$= 20.3 \text{ V}$$

Thus

$$R \leq R_{min} = \frac{28 \text{ V} - 2 \text{ V}}{3 \text{ mA}}$$
$$= 8.67 \text{ k}\Omega$$

If we choose $R = 2.2$ kΩ, the value of C is obtained using Equation 15.6:

$$C = \frac{10 \text{ ms}}{2.2 \text{ k}\Omega \times \log_e(3.3)}$$
$$= 3.8 \text{ }\mu\text{F} \qquad \blacktriangleleft$$

The basic UJT oscillator can also be used to trigger an SCR in an AC phase-control circuit such as that shown in Figure 15.8A. The circuit operates as follows:

(a) The AC input is applied to the load in series with the SCR. The load voltage remains at zero until the SCR is switched on by a gate pulse.

(b) e_{in} is also applied to zener diode clipper circuits (R_S and V_Z). The zener diode breaks down at V_Z volts and thus limits or clips the positive peaks of e_{in} (see v_{xy} waveform in the figure). During the negative half-cycle of e_{in} the zener is forward biased and maintains $\simeq 0$ V between x and y.

(c) This clipped sine wave is present as v_{xy} and is the supply voltage for the UJT circuit. During the positive half-cycle of e_{in} while v_{xy} is at $+V_Z$ volts, the capacitor charges through R until it reaches V_P of the UJT. When it does, the UJT turns "on" and discharges C, producing a positive pulse across $R1$ (see v_{B1} waveform). This pulse is fed to the gate of the SCR and turns "on" the SCR.

(d) Once the SCR is "on," the load voltage becomes approximately equal to e_{in} for the duration of the positive half-cycle. (See the v_{load} waveform.)

(e) During the negative half-cycle the SCR stays "off" and v_{load} stays at zero.

(f) The amount of power delivered to the load is controlled by varying the RC time constant, which causes C to charge slower or faster, thereby triggering the UJT and SCR later or earlier in the e_{in} half-cycle. In other words, the RC time constant controls the SCR trigger angle and therefore the load power. As RC is increased, the trigger angle increases and the load power decreases.

There are many other applications of the UJT which take advantage of its simplicity and versatility. Those interested in further examples of UJT circuits should consult the references listed at the end of this chapter.

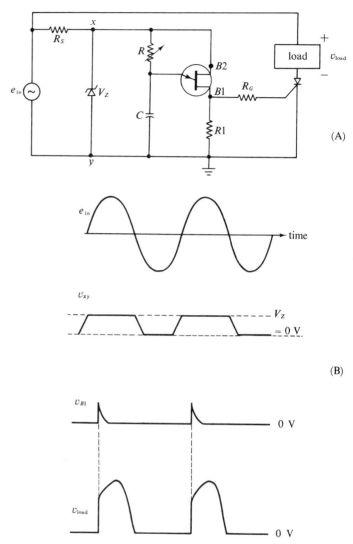

Figure 15.8.   UJT AS AN SCR TRIGGER IN AC POWER CONTROL CIRCUIT

15.7.   The Complementary Unijunction Transistor (CUJT)

In 1967, a new type of UJT was introduced. Although its physical structure is much different than the conventional UJT, its operation is like that of a UJT with its N and P regions interchanged. The *complementary unijunction transistor (CUJT)* is actually fabricated as an integrated circuit containing two junction transistors and two resistors diffused into the same silicon chip. For our purposes we can consider the CUJT to be the opposite-polarity counterpart of the UJT; that is, the CUJT will operate with voltage and current polarities opposite those of the regular UJT. The CUJT circuit symbol is given in Figure 15.9 along with that of the conventional UJT. The only difference is the direction of the emitter arrow; the emitter is N type for the CUJT.

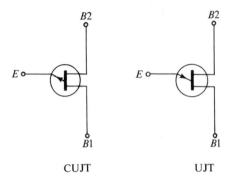

CUJT   UJT

Figure 15.9.   CUJT AND UJT CIRCUITS

The construction of the CUJT produces characteristics which have several important advantages over the conventional UJT. These advantages include:

(a) Has less parameter variation from unit to unit. Typically, the value of η might vary between 0.58 and 0.62 for a CUJT, while it may vary between 0.56 and 0.75 for a UJT.
(b) Operates at approximately one-third the supply voltage (more compatible with integrated circuits).
(c) Has factor of 10 or more reduction in emitter reverse-leakage current.
(d) Is less affected by temperature. The conventional UJT, as was already mentioned, is fairly stable with temperature. Over a temperature range of $-55°C$ to $+150°C$, a conventional UJT oscillator will have a frequency variation of typically 10 percent; over the same temperature range, a CUJT oscillator will have a frequency variation of only 1 percent.

The principle disadvantage of the CUJT is that the maximum interbase voltage and emitter breakdown voltage are limited to 15 V and 8 V, respectively. Although this limits the CUJT to low-voltage applications, it is not a serious drawback.

CUJT oscillator

Operation of the CUJT relaxation oscillator is exactly the same as that of the UJT relaxation oscillator except that all polarities are reversed. Figure 15.10 shows the two circuits side by side. The circuit layout is the same for both; however, V_{BB} and all the waveforms are of opposite polarity for the CUJT circuit.

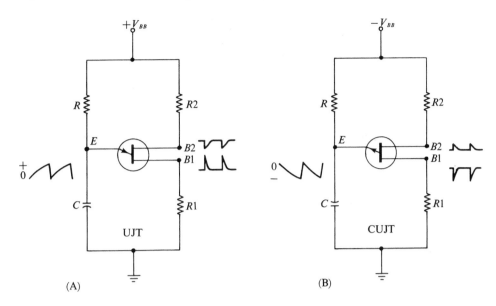

Figure 15.10.   (A) UJT OSCILLATOR; (B) CUJT OSCILLATOR

In practice, the CUJT oscillator is usually operated from a positive V_{BB} supply. This can be done by simply grounding the top terminal of the circuit in Figure 15.10B and connecting $+V_{BB}$ to the bottom terminal. This is the same biasing since the top terminal is still negative relative to the bottom terminal. This situation is shown in Figure 15.11A. It is also shown redrawn upside down in part B of the figure so that the grounded terminal appears on the bottom of the diagram, as is the usual convention.

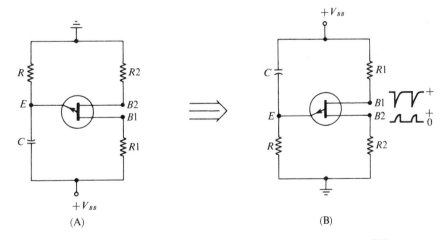

Figure 15.11.   (A) CUJT OSCILLATOR USING POSITIVE SUPPLY; (B) CIRCUIT REDRAWN INVERTED

15.8.   The Programmable Unijunction Transistor (PUT)

Another improved UJT is the programmable UJT (abbreviated PUT). Like the CUJT, the PUT is not constructed like a UJT. In fact, the PUT is a PNPN device whose characteristics can be made similar to the UJT. As we shall see, the PUT actually behaves like a four-layer diode whose switching voltage V_S can be set by the circuit designer via an external voltage divider.

Figure 15.12 shows the PNPN structure and the circuit symbol for the PUT. The anode (A) and cathode (K) are the same as for any PNPN device. The gate (G) is connected to the N region next to the anode. Thus the anode and gate constitute a P-N junction. It is this P-N junction which controls the "on" and "off" states of the PUT. The gate is usually positively biased relative to the cathode by a certain amount, V_G. When the anode voltage is less than V_G, the anode-gate junction is reverse biased and the PNPN device is in the "off" state, acting as an open switch between anode and cathode. When the anode voltage exceeds V_G by about 0.5 V, the anode-gate junction conducts, causing the PNPN device to turn "on" in the same manner as does forward biasing the gate-cathode junction of an SCR (Chapter 14). In the "on" state, the PUT acts like any PNPN device between anode and cathode (low resistance and $V_{AK} \simeq 1$ V).

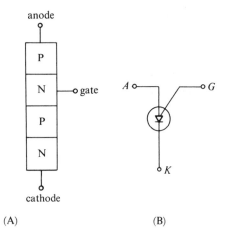

(A)   (B)

Figure 15.12.   (A) PUT STRUCTURE; (B) CIRCUIT SYMBOL

The normal bias arrangement for the PUT is illustrated in Figure 15.13. The voltage divider, $R1$ and $R2$, sets the voltage at the gate V_G. Note that $R1$ and $R2$ are *external* to the device and can therefore be chosen to produce any desired value of V_G. The anode-cathode bias is provided by E_{in}. As long as $E_{in} < V_G$, the device is "off" with $I_A = 0$ and all of E_{in} present across the anode-cathode ($V_{AK} = E_{in}$). The "off" state is summarized in part A of the figure.

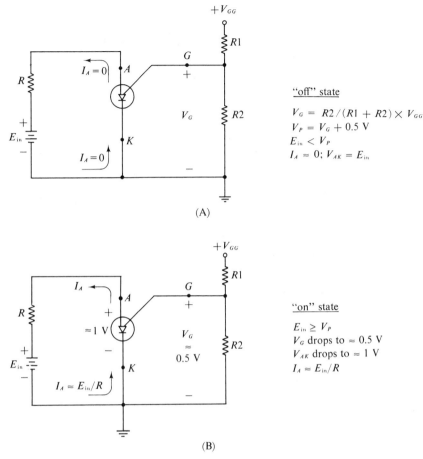

Figure 15.13.   (A) circuit with put in "off" state; (B) put in "on" state

If E_{in} is increased to about 0.5 V greater than the V_G bias value, the device turns "on." In other words, the *peak-point voltage* V_P for the PUT is given by

$$V_P = V_G + 0.5 \text{ V} \tag{15.10}$$

In the "on" state, the anode-cathode voltage, V_{AK}, drops to $\simeq 1$ V and the anode current, I_A, is essentially equal to E_{in}/R, being limited by R. In addition, V_G drops to a very low value ($\simeq 0.5$ V) since $R2$ is now shunted by the "on" PNPN structure. The PUT will remain in the "on" state until the anode current is decreased below the valley current, I_V. The "on" state is summarized in part B of the figure.

PUT parameters

The important PUT parameters are the same as those for the UJT, namely V_P, I_P, I_V and V_V. The value of V_P, the PUT firing voltage, is determined by the external bias V_G. The valley voltage, V_V, is essentially the PUT's "on" voltage of approximately 1 V. I_P, the peak-point current (current needed to turn-on), and I_V, the valley current (same as holding current), are usually lower for the PUT than for the UJT. In fact, both I_P and I_V are very much dependent on the resistances in the V_G voltage divider. Figure 15.14 shows how I_P and I_V typically vary for different values of R_G, where R_G is the *parallel* combination of $R1$ and $R2$ (the Thevenin equivalent resistance between the gate and ground). As the graphs show, I_P and I_V both decrease for higher values of R_G. Especially important is the fact that I_P can be reduced to very low values using large values of R_G. This characteristic is very useful in long time-delay circuits and will be discussed again later.

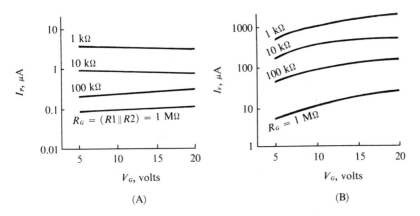

Figure 15.14. (A) PUT I_P VERSUS V_G CHARACTERISTIC AT VARIOUS VALUES OF R_G; (B) I_V VERSUS V_G CHARACTERISTIC AT VARIOUS VALUES OF R_G

15.9.   PUT Relaxation Oscillator

Figure 15.15A shows the PUT relaxation oscillator, whose operation is very similar to the UJT oscillator. The various circuit waveforms are shown in part B of the figure. These waveforms reveal the following important points:

(a) The capacitor voltage charges toward the 20-V supply voltage through the 10-kΩ resistor until it reaches $V_P = 10.5$ V. At that point the PUT turns "on" and the capacitor discharges rapidly through the PUT and the 100-Ω resistor. The trigger voltage of 10.5 V is set by the voltage divider consisting of the two 20-kΩ resistors which bias V_G at +10 V.

(b) The voltage at G remains at 10 V, while the capacitor charges and the PUT is "off." When the PUT turns "on," v_G drops to approximately zero. After the capacitor discharges, the PUT turns "off" (assuming current through 10 kΩ is less than I_V) and v_G returns to 10 V. This results in a negative-going pulse at G.

(c) A positive pulse is produced across the 100-Ω resistor as the capacitor discharges. The amplitude of this pulse is slightly lower than the capacitor peak voltage due to the anode-cathode "on" voltage of $\simeq 1$ V.

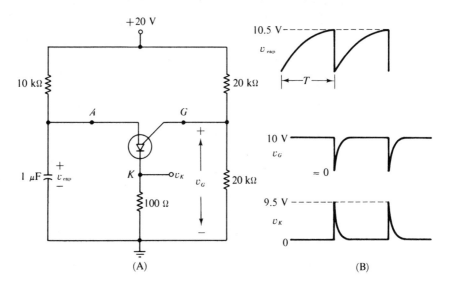

Figure 15.15.   (A) PUT OSCILLATOR; (B) CIRCUIT WAVEFORMS

The period of the oscillator waveforms can be calculated from

$$T = RC \log_e\left(\frac{E_{in}}{E_{in} - V_P}\right) \tag{15.11}$$

The frequency is simply $1/T$. Note that the above expression and the expression for the conventional UJT oscillator (Equation 15.4) are exactly the same.

▶ EXAMPLE 15.8

Calculate the period and frequency of the oscillator in Figure 15.15.
    For this circuit, $R = 10$ kΩ, $C = 1$ μF and $V_G = 10$ V. Thus $V_P = 10.5$ V. Using Equation 15.11, we have

$$T = 10 \text{ k}\Omega \times 1 \text{ μF} \times \log_e\left(\frac{20}{20 - 10.5}\right)$$

$$= 10 \text{ ms} \times \log_e(2.1)$$

$$= 10 \text{ ms} \times 0.7 = \textbf{70 ms}$$

Thus

$$f = \frac{1}{T} \simeq \textbf{143 Hz}$$ ◀

▶ EXAMPLE 15.9

Indicate how each of the following changes will affect the frequency of the oscillator of Figure 15.15: (a) decreasing R or C; (b) doubling the supply voltage; (c) increasing the 20-kΩ resistor in the top leg of the voltage divider.

(a) If either R or C is decreased, the charging time constant decreases. Thus the period will decrease proportionally. This effect is also evident from Equation 15.11. The frequency will consequently increase.

(b) Doubling the supply voltage will have two counteracting effects. The capacitor will charge at twice the rate, but it will have to reach a PUT switching voltage which is twice as large since the V_G bias will also double. The net effect, then, is that the capacitor will take the same amount of time to reach the turn-"on" point of the PUT. (This characteristic can also be verified using Equation 15.11 and is left for the student to verify.) The period and frequency are therefore unaffected by changing the supply voltage. However, the amplitudes of the various waveforms will be affected proportionally.

(c) Increasing the upper 20-kΩ resistor will change the gate bias V_G; in fact, V_G will have to decrease. Thus the capacitor will reach the lower switching voltage in a shorter time, causing T to be less, and frequency greater. ◀

15.10.   Advantages of the PUT

The PUT operation, though similar to the conventional UJT, has several important advantages over its predecessor. First, the switching voltage is easily varied by changing V_G through the voltage divider ratio. Second, the PUT shares the CUJT advantage of low operating voltage capability (as low as 4 V).

Probably the most important advantage of the PUT is its low peak-point current, I_P. Recall from our earlier discussion of the UJT oscillator that the maximum charging resistor which could be used depended on I_P (Equation 15.8B). The PUT can be made to have a very low I_P (0.1 μA) by using large resistors for the gate bias voltage divider. With a lower I_P it is possible to use a much larger charging resistor. This is a distinct advantage in long time-delay applications since a larger R would reduce the required value of C. The following example illustrates.

Figure 15.16 is a PUT time-delay circuit similar to its UJT counterpart. Note that the voltage divider resistors are very large so that the PUT's I_P will be very low. Since $R_G = R1 \parallel R2 \simeq 1$ MΩ, according to Figure 15.14A, we can expect an I_P of around 0.1 μA. The value of V_G is easily seen to be

$$V_G = \frac{3 \text{ M}\Omega}{5 \text{ M}\Omega} \times 10 \text{ V} = 6 \text{ V}$$

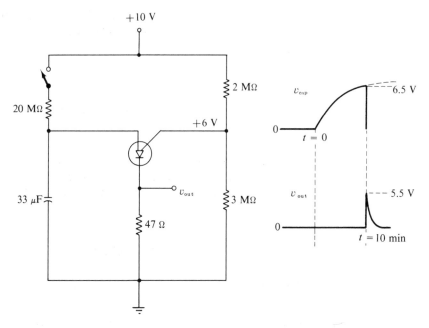

Figure 15.16.   PUT 10-MINUTE TIME-DELAY CIRCUIT

Thus V_P = 6.5 V. When the switch is closed the capacitor will charge toward 10 V. When it reaches 6.5 V it will discharge through the PUT, producing a pulse across the 47-Ω resistor. The time delay between closing the switch and the occurrence of the output pulse is to be designed for approximately 10 minutes.

To achieve such a long time delay requires a very large RC time constant. Equation 15.11 can be used to determine the required RC:

$$T = RC \log_e\left(\frac{E_{in}}{E_{in} - V_P}\right)$$

$$600 \text{ s} = RC \log_e\left(\frac{10}{10 - 6}\right) = RC \log_e(2.86) = RC\,(1.05)$$

so that

$$RC = \frac{600 \text{ s}}{1.05} = 571 \text{ s}$$

It is usually desirable to keep the value of C as low as possible. Large values of C usually require electrolytic capacitors which suffer from high leakage and are physically quite large. Thus it is necessary to make R very large. The largest R that can be used can be calculated using Equation 15.8B:

$$R_{max} = \frac{E_{in} - V_P}{I_P} = \frac{10 \text{ V} - 6 \text{ V}}{0.1 \text{ } \mu\text{A}} = 40 \text{ M}\Omega$$

To be conservative, R should be kept well below R_{max} to ensure proper operation. The value of 20 MΩ is chosen. The value of C can now be found by satisfying $RC = 571$ s:

$$C = \frac{571}{R}$$

$$= \frac{571}{20 \times 10^6} = 28.6 \times 10^{-6} \text{ F}$$

$$= 28.6 \ \mu\text{F}$$

Thus a 30-μF capacitor can be used.

If this same circuit were designed using a conventional UJT, the values of R and C might be 1 MΩ and 600 μF. The higher I_P of the UJT results in a lower R_{max}.

GLOSSARY

Unijunction transistor: switching device consisting of a silicon N-type bar and one alloyed P-N junction.

Base-1, base-2:   terminals at either end of the N-type bar of the UJT.

Emitter: terminal at alloyed P region.

Interbase resistance (R_{BB}):   resistance between base-1 and base-2 ("off" state).

Intrinsic standoff ratio (n):   UJT's internal voltage divider ratio.

Peak-point voltage (V_P):   voltage needed at the emitter terminal to turn UJT "on.'

Peak-point current (I_P):   emitter current at peak-point voltage, V_P.

Valley voltage (V_V):   emitter voltage in the "on" state.

Valley current (I_V):   emitter current needed to hold UJT in the "on" state.

Complementary UJT: device with characteristics of a negative-polarity UJT.

Programmable UJT: PNPN device which operates like a UJT whose V_P can be externally programmed.

PROBLEMS

15.1   In the circuit of Figure 15.3, if $V_{BB} = 18$ V and the UJT has $R_{B1} = R_{B2} = 5$ kΩ, calculate the peak-point voltage, V_P. What value of V_{in} is needed to turn the UJT "on"?

15.2   After the UJT is turned "on" what happens to the value of R_{B1}?

15.3   How can the value of V_P be varied for a given UJT?

15.4   How is the UJT turned "off"?

15.5   A UJT with the following parameters is used in the circuit of Figure 15.3:

$$\eta = 0.66$$
$$V_D = 0.7 \text{ V}$$
$$I_V = 4 \text{ mA}$$
$$V_V = 1 \text{ V}$$
$$I_P = 10 \ \mu\text{A}$$

(a) Assume that the UJT is initially "off." To what value must V_{in} be raised to turn "on" the UJT? Use $V_{BB} = 20$ V.

(b) If $R_E = 1$ kΩ, to what value must V_{in} be reduced before the UJT turns "off"?

15.6   The UJT of Problem 15.5 is used in the oscillator circuit of Figure 15.4. The circuit values are $R1 = 100 \ \Omega$, $R2 = 50 \ \Omega$, $R = 10$ kΩ, $C = 2 \ \mu$F and $V_{\text{in}} = 24$ V.

(a) Determine V_P.

(b) Determine whether the circuit will oscillate. If it oscillates, determine its frequency.

(c) Accurately sketch and label the capacitor waveform.

15.7   For the oscillator of Problem 15.6 determine the range of frequencies which can be obtained by varying R. How may this range of frequencies be changed without changing the UJT?

15.8   Consider the circuit of Figure 15.17. The oscillator is supposed to produce pulses at a rate of 10 kHz. When power is applied to the circuit, however, the circuit fails to oscillate. The technician testing the circuit notices that when the power is turned off the circuit temporarily oscillates as the power supply voltage drops to zero. Explain these observations and determine what is wrong with the circuit. How should the circuit be modified for proper operation?

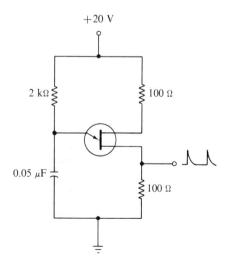

Figure 15.17.

15.9   In Figure 15.18, how long after the switch is closed will the motor be ener-
            gized?

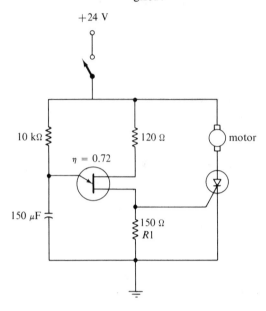

Figure 15.18.

15.10   Suppose when the switch is closed in Figure 15.18 that the motor goes
            on instantly (no time delay). Which of the following reasons could be cause
            for the malfunction? (There may be more than one possible answer.) Ex-
            plain each choice.
            (a) The capacitor is shorted.
            (b) $R1$ is too large.
            (c) The supply voltage is too large.
            (d) The SCR is too sensitive.
            (e) The UJT is shorted from $B2$ to $B1$.

15.11   The UJT in Figure 15.18 has $R_{BB} = 4.7$ kΩ. The SCR has the characteristics
            listed in Table 14.2 (Chapter 14). What is the maximum value of $R1$ that
            can be used if the SCR is not to be prematurely triggered? (Assume room
            temperature operation.)

15.12   Consider the circuit of Figure 15.8. e_{in} has a peak amplitude of 160 volts.
            The zener diode has $V_z = 10$ V; therefore, it can be assumed that the voltage
            applied to the UJT circuit is essentially 10 V for the duration of the positive
            half-cycle of e_{in}. If $R = 6.8$ kΩ, $C = 1$ μF and $\eta = 0.7$, determine the trigger
            angle of the SCR. Sketch the v_{load} waveform showing its value at the trigger
            point. Input frequency is 60 Hz.

15.13   What is the difference between the CUJT and the UJT?

15.14   What are the advantages of the CUJT over the UJT?

15.15   Consider the CUJT oscillator in Figure 15.11B. If $V_{BB} = 20$ V and $\eta = 0.6$,
            sketch the waveform of voltage at the emitter (relative to ground). Note that
            the CUJT turns "on" when the E-$B1$ diode becomes forward biased.

15.16 Explain how the operation of the PUT differs from that of the UJT.

15.17 Determine the period and frequency of the PUT oscillator of Figure 15.19. Sketch the waveforms of v_{cap}, v_G and v_K, showing approximate amplitudes.

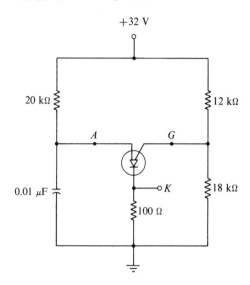

Figure 15.19.

15.18 Indicate *three* ways to decrease the frequency of the oscillator of Problem 15.18.

15.19 Consider the PUT time-delay circuit of Figure 15.20. The PUT has the characteristics of Figure 15.14. If C is to be limited to a maximum of 10 μF, what is the maximum delay which this circuit can produce?

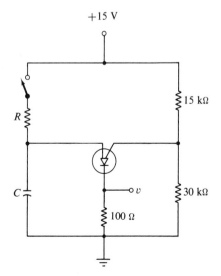

Figure 15.20.

REFERENCES

Burrows, C. N. and G. J. Deboo, *Integrated Circuit and Semiconductor Devices: Theory and Application*. New York: McGraw-Hill Book Company, 1971.

Davis, C. A., *Industrial Electronics*. Columbus, Ohio: Charles E. Merrill Publishing Company, 1973.

16

Field-Effect Transistors

16.1.   Introduction

The field-effect transistor is a semiconductor device whose operation consists of controlling the flow of current through a semiconductor channel by application of an electric field (voltage) perpendicular to the conduction path. There are two major categories of field-effect transistors, the *junction field-effect transistor* (abbreviated JFET) and the *insulated-gate field-effect transistor* (IGFET), more commonly called the *metal-oxide-semiconductor field-effect transistor* (MOSFET). In this chapter we will examine the operation of these devices as well as the differences in their characteristics.

Field-effect transistors have several important differences when compared to junction transistors (NPN and PNP), including the following:

(a) FET operation depends only on the flow of majority carriers, holes for P-channel FETs and electrons for N-channel FETs. They are therefore called *unipolar* devices. Junction transistors, as we have seen, depend on both minority and majority current carriers and are therefore called *bipolar transistors.*

(b) FETs are much easier to fabricate and are particularly suitable for integrated circuits because they occupy less space than bipolar transistors.

(c) FETs exhibit a much higher input resistance, typically 100 megohms or greater.

(d) FETs are normally less sensitive to temperature.

(e) When operated as amplifiers, FETs have less voltage gain and a poorer high-frequency response.

(f) FETs are less *noisy* than bipolar transistors. *Noise* is the term for random electrical fluctuations caused by movement of electrons inside the semiconductor structure. Noise usually appears as an unwanted electrical signal.

16.2.   The Junction Field-Effect Transistor (JFET)

In its simplest form an N-channel JFET starts as a bar of N-type silicon (Figure 16.1A) with metal contacts at either end, referred to as the *source* and *drain*. By doping P-type regions on either side of the bar, the structure of Figure 16.1B results. Metal contacts are made to these P regions and are called *gates*. Normally, the gate leads are connected together as shown in Figure 16.1C. The region between source and drain is referred to as the *N channel*. As the diagram shows, there are two P-N junctions formed by the P-type gate regions and the N-channel.

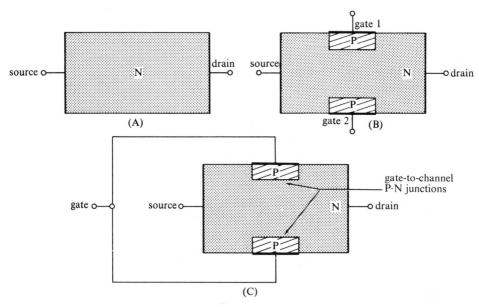

Figure 16.1.   BASIC N-CHANNEL JFET STRUCTURE

Unlike the bipolar transistor, current is not made to flow across these P-N junctions during normal FET operation. Instead, current flow takes place only through the N channel from source to drain. This is shown in Figure 16.2

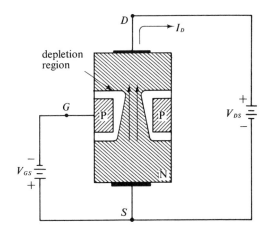

Figure 16.2.   BIAS FOR N-CHANNEL JFET

where the normal bias arrangement for the N-channel FET is also illustrated. The drain (D) is biased *positive* relative to the source (S) by the V_{DS} source. This causes electrons to flow from source to drain through the channel. This current is called the drain current, I_D.

The gate (G) is shown biased *negative* relative to the source by the V_{GS} source. (Only one gate lead is shown for clarity; it is assumed that the two gate leads are connected together.) Thus the gate-to-channel P-N junctions are reversed biased so that essentially *zero* current flows in the gate lead. During normal JFET operation the gate-source voltage is never allowed to forward bias the gate-source P-N junction sufficiently to cause gate current to flow. This effect allows the input resistance at the gate to remain very high (typically 100 MΩ).

Depletion regions

The reverse-biased gate-channel junctions give rise to depletion regions at the junctions. Recall that these depletion regions are areas where almost no charge carrier exist. These depletion regions penetrate into the channel (see white areas in Figure 16.2) and thereby reduce the conducting portion of the channel. In other words, the presence of these depletion regions *increases* the resistance of the channel by making it narrower.

The depletion regions are wedge-shaped, being larger at the drain end and narrower at the source end. This is because the gate-to-channel reverse bias is greater at the drain end since the drain is biased positive. In fact, the reverse bias on the drain end of the P-N junction is equal to the absolute sum of the two bias voltages. For example, if $V_{DS} = 20$ V and $V_{GS} = -5$ V, then this reverse bias is 25 V. The depth of penetration of the depletion regions will depend on this reverse bias and will vary as either V_{GS} or V_{DS} is varied. Thus the channel resistance and therefore the channel current are functions of both V_{GS} and V_{DS}.

16.3.   JFET Drain Curves

The operation of the JFET can best be explained by studying a typical set of *drain curves*, such as those shown in Figure 16.3. Each curve shows how the drain current, I_D, varies with V_{DS} for a given value of gate-source voltage, V_{GS}. Note that V_{GS} is either 0 V or negative in value so that the gate junction never conducts. Actually, V_{GS} can be allowed to go slightly positive (less than +0.5 V) before any gate current flows. Each drain current curve corresponds to a different value of V_{GS}. This implies that the JFET is a *voltage-controlled* device, whereas the bipolar transistor is a *current-controlled* device.

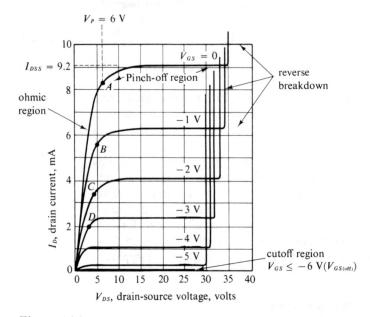

Figure 16.3.   TYPICAL DRAIN FAMILY OF CURVES

$V_{GS} = 0$ case

First, let us consider the case where $V_{GS} = 0$ V. As V_{DS} is increased from 0 V, the drain current, I_D, increases almost linearly until $V_{DS} = 6$ V (point A). In this region the channel resistance remains fairly constant (obeys Ohm's Law), so it is called the *ohmic region*. In this region the depletion regions have not penetrated the channel enough to significantly affect the channel resistance.

As V_{DS} is increased further the drain current does not increase proportionally, because two opposite-acting effects are taking place. First, the increase in V_{DS} tends to accelerate electrons through the channel so as to increase I_D; second, the increase in V_{DS} causes the depletion regions to spread into the channel, thereby increasing the channel resistance. The first effect initially dominates so that I_D does increase, but it does not dominate as strongly as it would if the channel resistance had not increased.

However, as V_{DS} continues to increase, eventually a point is reached where these two effects cancel each other out so that I_D does not increase with V_{DS}. As the curve shows, above $V_{DS} = 15$ V the drain current levels off at 9.2 mA (until breakdown). The region between point A and breakdown is called the *pinchoff* region. In this region the depletion regions have essentially *pinched off* the channel so that I_D cannot increase any further. The value of V_{DS} at point A where the curve begins to level off is called the FET's *pinchoff voltage* V_P. For this case, $V_P = 6$ V. *The value of V_P is the amount of gate-to-drain reverse bias which causes pinchoff to occur.* Every FET has a certain value of V_P (just as every junction transistor has a certain β).

If V_{DS} is increased to around 35 V, the gate-to-drain reverse bias exceeds the junction's reverse breakdown rating and the drain current increases drastically. The breakdown region is usually avoided during normal operation.

$V_{GS} = -1$ V case

If the gate is biased 1 V negative relative to the source ($V_{GS} = -1$ V) the FET behavior is essentially the same, except that the drain current begins to level off at a lower value of V_{DS}. Inspection of the $V_{GS} = -1$ V curve shows that I_D begins levelling off at $V_{DS} = 5$ V (point B) and eventually levels off at 6.3 mA. In other words, the pinchoff region starts at $V_{DS} = 5$ V rather than at $V_{DS} = 6$ V, as it did for the $V_{GS} = 0$ case, because, with the 1-V reverse bias on the gate, only a 5-V drain voltage is needed before the gate-to-drain reverse bias needed for pinchoff ($V_P = 6$ V) is reached. Since pinchoff begins at a lower value of V_{DS}, the I_D will necessarily level off at a lower value.

Other values of V_{GS}

Examination of the $V_{GS} = -2$ V curve shows that I_D begins to level off at $V_{DS} = 4$ V (point C); for $V_{GS} = -3$ V, I_D begins to level off at $V_{DS} = 3$ V (point D), and so on. The I_D curves become progressively lower as V_{GS} is made more negative, until for $V_{GS} = -6$ V the I_D curve is essentially flat at $I_D = 0$. In fact, for V_{GS} values of -6 V or more, I_D will be zero for all values of V_{DS} (until breakdown), because the gate reverse bias is sufficient to cause the depletion regions to extend completely across the channel, thereby making channel resistance very large and cutting off *all* channel current. This region of the FET characteristics is called the *cutoff* region.

The value of V_{GS} which produces cutoff ($I_D = 0$) is given the symbol $V_{GS(off)}$. For this case, $V_{GS(off)} = -6$ V. The value of $V_{GS(off)}$ will *always* have the same magnitude as V_P, the pinchoff voltage. That is,

$$|V_{GS(off)}| = |V_P| \qquad (16.1)$$

This relationship is true of all JFETs. On most JFET specifications sheets only one of the values V_P or $V_{GS(off)}$ is given since the other one can be easily found using Equation 16.1.

More on pinchoff

The difference between *pinchoff* and *cutoff* is often confusing to the student. Perhaps the confusion can be cleared up with the help of Table 16.1:

TABLE 16.1

Gate-source bias V_{GS}	V_{DS} at start of pinchoff	Pinchoff current I_{DP} at 15 V
0 V	6 V $= V_P$	9.2 mA (I_{DSS})
−1 V	5 V	6.3 mA
−2 V	4 V	4.1 mA
−3 V	3 V	2.4 mA
−4 V	2 V	1.2 mA
−5 V	1 V	0.3 mA
−6 V $= V_{GS(off)}$	0 V	0 mA

This table is simply a summary of the pinchoff conditions for various values of V_{GS} reverse bias. For example, at $V_{GS} = 0$ V the pinchoff region begins at $V_{DS} = 6$ V (I_D begins levelling off). The value of I_D well into the pinchoff region (I_{DP}) levels off at 9.2 mA. Similarly, for $V_{GS} = -1$ V the pinchoff region begins at $V_{DS} = 5$ V and $I_{DP} = 6.3$ mA. The table shows that as V_{GS} is made more negative, the value of V_{DS} required for pinchoff decreases. As stated earlier, it takes a certain gate-to-drain reverse bias to produce pinchoff (this is V_P). Thus we can say that pinchoff will occur when

$$|V_{GS}| + |V_{DS}| = V_P \tag{16.2}$$

For our example JFET, $V_P = 6$ V. Examination of Table 16.1 shows that Equation 16.2 is satisfied for each case.

When $|V_{GS}| \geq V_P$ by itself, then the channel current is completely *cut off* independent of V_{DS}. When V_{GS} is equal to or more negative than -6 V for our JFET, the channel current is cut off.

▶ EXAMPLE 16.1

A certain JFET is specified to have $V_{GS(off)} = -4$ V. (a) Determine the drain current for $V_{GS} = -6$ V; (b) At what value of V_{DS} will pinchoff begin for $V_{GS} = -1.5$ V?

    (a) The drain current will be zero for $V_{GS} \leq -4$ V. Thus, for $V_{GS} = -6$ V, we must have $I_D = \mathbf{0}$.

    (b) $|V_P| = |V_{GS}|$ so that $V_P = 4$ V for this FET. This means that pinchoff will begin whenever the combined V_{GS} and V_{DS} voltages produce a gate-to-drain reverse bias of 4 V. Using Equation 16.2 at $V_{GS} = -1.5$ V,

$$1.5 \text{ V} + V_{DS} = 4 \text{ V}$$

or

$$V_{DS} = 2.5 \text{ V}$$

at the start of the pinchoff region.   ◀

Definition of I_{DSS}

An important JFET parameter is the value of I_{DP} when $V_{GS} = 0$ V. This value is given the symbol I_{DSS} and represents the *drain-to-source pinchoff current when the gate is shorted to the source ($V_{GS} = 0$).* The drain curves in Figure 16.3 indicate a value of 9.2 mA for I_{DSS}. The value of I_{DSS} will vary from one JFET to another even if they are of the same type. Variations of 2 to 1 and even 3 to 1 are not uncommon. I_{DSS} represents the maximum JFET current which will flow during normal operation.

Transfer characteristics

Table 16.1 gives values of I_{DP} for each value of V_{GS}. These are the values of drain current well into the pinchoff region (at $V_{DS} = 15$ V) after I_D has almost completely levelled off. These values may be plotted in graphical form as shown in Figure 16.4. This curve, referred to as the FET's *transfer characteristic*, has the approximate shape of a parabola whose equation is

$$I_{DP} \simeq I_{DSS}\left(1 - \frac{|V_{GS}|}{|V_P|}\right)^2 \qquad\qquad \textbf{(16.3)}$$

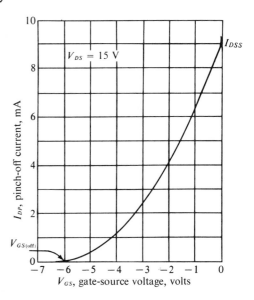

Figure 16.4.   PINCHOFF CURRENT (I_{DP}) VARIATION WITH V_{GS} (TRANSFER CHARACTERISTIC)

Once I_{DSS} and V_P or $V_{GS(off)}$ are known, I_{DP} can be found for any value of V_{GS} by using this relationship. For example, with $I_{DSS} = 9.2$ mA and $V_P = 6$ V we can calculate I_{DP} for $V_{GS} = -3$ V as follows:

$$I_{DP} \simeq 9.2 \text{ mA}\left(1 - \frac{3}{6}\right)^2 = 9.2\left(\frac{1}{2}\right)^2$$
$$\simeq 2.3 \text{ mA}$$

This value agrees very closely with the I_{DP} value obtained from Figure 16.3, Table 16.1 or Figure 16.4. Note that only the *magnitude* of V_{GS} was used in the calculation.

P-channel JFET

Up to now our discussion has concerned only an N-channel JFET. P-channel JFETs are also used; they operate in basically the same manner, except that all polarities are reversed. For the P-channel JFET the gate is N type so that V_{GS} is biased positive and V_{DS} is biased negative. A typical set of P-channel drain curves can be obtained by changing the V_{DS} and V_{GS} polarities in Figure 16.3.

The circuit symbols for the N-channel and P-channel JFETs are shown in Figure 16.5 along with their proper bias polarities relative to the source terminal. Note that the arrow on the gate lead points iN for the N-channel JFET and out for the P-channel JFET.

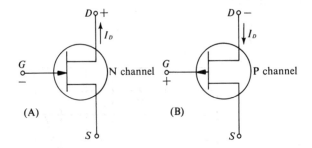

Figure 16.5.   JFET CIRCUIT SYMBOLS: (A) N-CHANNEL; (B) P-CHANNEL

▶ EXAMPLE 16.2

A certain P-channel JFET has $V_P = -2.5$ V and $I_{DSS} = 4$ mA. What is I_{DP} at $V_{GS} = +1$ V?

Equation 16.3 is equally applicable to P-channel devices. Thus

$$I_{DP} \simeq 4 \text{ mA}\left(1 - \frac{1}{2.5}\right)^2 = 4 \text{ mA}(0.6)^2$$
$$\simeq 1.44 \text{ mA}$$

Note that only the magnitude of V_P was used in the calculation. For this FET, $V_{GS(off)} = +2.5$ V. If it had been given instead of V_P it could have been used with the same result since $|V_P| = |V_{GS(off)}|$.   ◀

16.4.   JFET Ratings and Parameters

The most important JFET rating which must be considered by a circuit designer is gate-to-channel reverse breakdown voltage. As mentioned in the last section, breakdown can cause the drain current to increase drastically so that the FET's power dissipation rating is exceeded.

BV_{GSS}

The most common way of expressing the FET breakdown rating is through the quantity BV_{GSS}, which is the *gate-to-source breakdown voltage with drain shorted to source*. In essence, this voltage is the amount needed to cause breakdown between gate and source. However, since source and drain are normally interchangeable, this voltage also specifies the gate-to-drain breakdown voltage. The value of BV_{GSS} might typically be 35 V, which indicates that the gate should never be more than 35 V reverse biased relative to the source or to the drain.

If we examine the drain characteristics of Figure 16.3 we see that, for $V_{GS} = 0$, breakdown occurs when the *drain* is $+35$ V relative to the *gate*. As such, for $V_{GS} = -1$ V, breakdown occurs at $V_{DS} = +34$ V, and so on. This explains why the breakdown voltage is different for each curve.

There are many JFET parameters which are specified on a manufacturer's data sheet. The most important of these parameters will be defined here. We have already discussed V_P, $V_{GS(\text{off})}$ and I_{DSS} as they pertain to the JFET drain characteristic curves.

I_{GSS}

Since the gate is normally reverse biased, there will be a leakage current that will exist in the gate lead. Most FET data sheets specify I_{GSS}, the *gate-to-source leakage current with drain shorted to source*. I_{GSS} is usually measured at a V_{GS} value that is at least half of BV_{GSS}. A typical JFET will have $I_{GSS} = 0.1$ nA at 25°C. This gate leakage current is, of course, temperature sensitive and doubles for approximately every 11-degree increase in temperature. At 175°C it will typically have increased to 30 nA.

In most applications the gate serves as a JFET input terminal. The very low gate leakage current results in very high input impedances which are normally 100 MΩ or greater. Of course, this situation exists only if the gate is not forward biased significantly since forward biasing will produce a heavy current flow and very low input resistance.

JFET capacitances

Like any P-N junction device, the JFET possesses junction capacitances which can affect its behavior at high frequencies. The gate-source capacitance, C_{GS}, and the gate-drain capacitance, C_{GD}, are typically only a few picofarads. However, due to an effect called the *Miller effect* these relatively small capacitances can result in an FET input capacitance which is many times larger.

Typical values of C_{GS} and C_{GD} might be 5 pF each (they are usually equal). On most FET data sheets C_{GS} and C_{GD} are not specified directly. Rather, a quantity called C_{iss}, *input capacitance with drain shorted to source*, is specified. It can normally be assumed that $C_{iss} \simeq C_{GS} + C_{GD}$ so that $C_{GS} = C_{GD} = C_{iss}/2$.

g_{fs}, transconductance

We have seen that application of different gate-to-source voltages produces different drain curves. In other words, V_{GS} controls the flow of drain current for a given value of V_{DS}. In the pinchoff region this control is graphically represented by the transfer characteristic of Figure 16.4. A useful JFET parameter which is a measure of the gate's control over I_D can be obtained from the transfer characteristic. It is called the *forward transconductance* and is given the symbol g_{fs}.* The value of g_{fs} indicates how much the drain current will change (ΔI_{DP}) for a change in V_{GS} (called ΔV_{GS}) at a constant value of V_{DS}. Stated mathematically,

$$g_{fs} \equiv \frac{\Delta I_{DP}}{\Delta V_{GS}}\bigg|_{V_{DS}} = \text{constant} \tag{16.4}$$

The units for g_{fs} are amperes per volt, which are *mhos*, the units used for *conductance*. This, along with the fact that g_{fs} indicates how the gate voltage signal is *transferred* to a drain current signal, is why g_{fs} is called transconductance. Because of the small currents involved, units of micromhos (μmho) rather than mhos are commonly used.

Values for g_{fs} can be found from either the drain curves or the transfer characteristic and are specified for certain values of V_{DS} and V_{GS}. In order to find g_{fs} at a certain value of V_{GS}, it is necessary to change the value of V_{GS} slightly to see how much I_{DP} changes. The following example illustrates:

▶ EXAMPLE 16.3

From Figure 16.4 determine approximate values of g_{fs} near $V_{GS} = 0$ and near $V_{GS} = -2.5$.

The following values are obtained from the transfer characteristics:

	I_{DP}	V_{GS}	
$\Delta I_{DP}\begin{cases} \\ \\ \end{cases}$	9.2 mA	0 volts	$\bigg\}\Delta V_{GS}$
	7.7	−0.5	
	4.1	−2	
	2.4	−3	

---

*Other commonly used symbols for transconductance include y_{fs} and g_m.

Using the first two entries in Equation 16.4,

$$g_{fs} \text{ (at } V_{GS} = 0) \simeq \frac{1.5 \text{ mA}}{0.5 \text{ V}} = 3000 \text{ μmhos}$$

Using the last two entries,

$$g_{fs} \text{ (at } V_{GS} = -2.5 \text{ V}) \simeq \frac{1.7 \text{ mA}}{1 \text{ V}} = 1700 \text{ μmhos} \quad \blacktriangleleft$$

This example illustrates the fact that g_{fs} decreases as V_{GS} reverse bias increases. In other words, at larger values of V_{GS} the gate voltage has less effect on drain current. This effect is also apparent from the transfer curve (Figure 16.4), which is steepest at $V_{GS} = 0$ and gradually levels off as V_{GS} approaches $V_{GS(off)}$. The value of g_{fs} is greatest at $V_{GS} = 0$ and can typically be as high as 10,000 μmhos.

Values of g_{fs} can also be calculated using the following approximate relationship:

$$g_{fs} \simeq \frac{2I_{DSS}}{V_P}\left(1 - \frac{|V_{GS}|}{|V_P|}\right) \tag{16.5}$$

For example, for the JFET we have been using we have $I_{DSS} = 9.2$ mA and $V_P = 6$ V. The value of g_{fs} at $V_{GS} = -2.5$ V can be calculated using this expression

$$g_{fs} \text{ (at } V_{GS} = -2.5 \text{ V}) \simeq \frac{2 \times 9.2 \text{ mA}}{6 \text{ V}}\left(1 - \frac{2.5}{6}\right)$$
$$\simeq 1789 \text{ μmhos}$$

This value agrees closely with that calculated in Example 16.3.

For the special case of $V_{GS} = 0$, Equation 16.5 can be simplified to

$$g_{fs}(0) \simeq \frac{2I_{DSS}}{V_P} \tag{16.6}$$

where $g_{fs}(0)$ is the transconductance at $V_{GS} = 0$. For our JFET we thus have

$$g_{fs}(0) \simeq \frac{2 \times 9.2 \text{ mA}}{6 \text{ V}} = 3067 \text{ μmhos}$$

which agrees closely with the result of Example 16.3.

Output resistance r_{DS}

The parameter which is analogous to a junction transistor's output resistance (r_{ob} and r_{oe}) is the JFET's *output drain-source resistance* r_{DS}. The value of r_{DS} is simply the AC resistance from source to drain for a given value of V_{GS} gate voltage. That is,

$$r_{DS} = \frac{\Delta V_{DS}}{\Delta I_D}\bigg|_{V_{GS} = \text{constant}} \tag{16.7}$$

The value of r_{DS} is defined in both the ohmic and pinchoff regions. Referring to the drain curves in Figure 16.3, it can be seen that r_{DS} will be much lower in the ohmic region (typically 1 kΩ) where V_{DS} strongly affects I_D than in the pinchoff region (typically 500 kΩ) where V_{DS} only weakly affects I_D. Like the junction transistor, the JFET is normally operated in the region where its curves are flat and r_{DS} is high when used as an amplifier.

When the JFET is used as a switch it normally operates between $V_{GS} = 0$ (JFET "on") and $V_{GS} = V_{GS(off)}$. For such operation, the value of drain-source resistance in the ohmic region at $V_{GS} = 0$ is significant. This value is called $r_{DS(on)}$ and is typically a few hundred ohms. $r_{DS(on)}$ is essentially the "on" resistance of the JFET when $V_{GS} = 0$ and V_{DS} is below pinchoff.

16.5.   JFETs: Amplification

The basis for JFET amplification lies in the gate voltage's control over the drain current. A variation in the gate-to-source voltage produces a variation in the drain current. This variation in drain current can produce a variation in the voltage across a load resistor placed in series with the drain (see Figure 16.6). If R_L is large enough, the load voltage signal will be greater than the input gate signal so that a voltage gain A_v is realized.

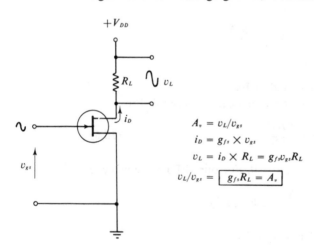

Figure 16.6.   JFET AMPLIFIES GATE VOLTAGE SIGNAL

It should be apparent that a JFET with a greater transconductance g_{fs} will produce a greater voltage gain because larger drain current variations will be produced for the same gate voltage signal. In fact, when g_{fs} is known, the approximate value of A_v can be determined from

$$A_v \simeq g_{fs} \times R_L \qquad (16.8)$$

This formula (derived in Figure 16.6) is valid whenever R_L is kept much *lower* (by factor of ten or more) than the JFET's output resistance r_{DS}. The JFET's voltage gain then can be increased by using large values of R_L and JFET's with large values of g_{fs}.

16.6. Basic Common-Source Amplifier

Figure 16.7A shows the basic N-channel JFET common-source amplifier. The input signal is applied to the gate and the output signal is taken at the drain. A DC bias voltage is connected to the gate. This bias voltage is needed to establish the amplifier Q point at the desired point.

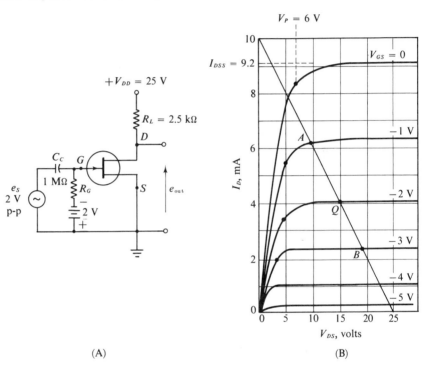

(A)   (B)

Figure 16.7.   (A) BASIC COMMON-SOURCE AMPLIFIER; (B) LOAD LINE

Large-signal analysis

The operation of this amplifier for a large input signal (2 V p-p) can be analyzed using a load-line technique similar to that used for the junction transistor amplifiers. The JFET drain characteristics are shown in Figure 16.7B. The load line is determined by V_{DD}, the drain supply, and R_L. The two points needed to plot the load line are:

$$I_D = 0, \quad V_{DS} = V_{DD} = 25 \text{ V} \quad \text{and} \quad I_D = \frac{V_{DD}}{R_L} = 10 \text{ mA}, \quad V_{DS} = 0$$

The constructed load line is shown in the figure connected between these two end points.

The DC operation of the amplifier can be determined by ignoring the capacitively coupled input signal e_S. The 2-V battery reverse biases the gate-source junction. The very low gate leakage current produces only a negligible voltage drop across the 1-MΩ gate resistor (i.e., 1 nA $\times$ 1 MΩ = 1 millivolt). Thus the DC gate-source voltage is $V_{GS} = -2$ V.

The amplifier Q point is obtained as the intersection of the load line and the drain curve corresponding to $V_{GS} = -2$ V. From Figure 16.7B, the circuit values at the Q point are $I_{DQ} = 4.1$ mA and $V_{DSQ} = 14.5$ V. Notice that the Q point is well into the pinchoff region for the same reason that junction transistors are biased in the active region, namely to minimize distortion of the output signal.

The input signal e_S will cause the gate-source voltage to vary around its Q-point value of -2 V. Since $e_S = 2$ V p-p, the gate-source voltage will vary between -1 V and -3 V. As a result, the amplifier operating point will move along the load line between points A and B. When e_S is at its *positive* peak, the net gate-source voltage is -1 V and the amplifier is at point A where $V_{DS} = 9$ V (down from its Q-point value of 14.5 V). When e_S is at its *negative* peak, the net gate-source voltage is -3 V and the amplifier is at point B where $V_{DS} = 19$ V. The waveforms of e_S and v_{DS} are shown in Figure 16.8.

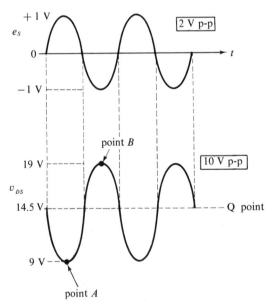

Figure 16.8.   INPUT AND OUTPUT WAVEFORMS

Examination of these waveforms reveals *three* important points:

(a) The v_{DS} output signal is 180° out-of-phase with the e_S input signal. This is a characteristic of the common-source amplifier.

(b) The output-signal-to-input-signal ratio is 10 V/2 V = 5, so that the voltage gain is $A_v = 5$.

(c) The output signal variation is not symmetrical around its Q-point value. The positive peak is 4.5 V above the DC bias value and the negative peak swings 5.5 V below the DC bias value. This *distortion* in the output signal is a result of the unequal spacing of the drain curves. The spacing is greater as V_{GS} approaches zero. This output distortion normally is not desirable since it means that the shape of the input signal is not faithfully reproduced in the amplifier output. For this reason, JFET amplifiers are not normally used to amplify large input signals. For very small input signals (millivolt range) the amount of output distortion is within acceptable limits because the output stays very close to the Q point. As such, the JFET is normally used in small-signal amplifiers.

Small-signal analysis

When the input signal is very small, the operation stays near the Q point so that the JFET's transconductance g_{fs} remains relatively constant over the complete output-signal excursion. For small signals the approximate voltage gain can be found using Equation 16.8. That is,

$$A_v \simeq g_{fs} \times R_L \tag{16.8}$$

using the value of g_{fs} at the Q point. The formula can also be used to give a rough estimate of A_v for large signals. To illustrate, we determined in Example 16.3 that $g_{fs} = 1700$ μmhos at $V_{GS} = -2.5$ V for the JFET curves of Figure 16.7. With $R_L = 2.5$ kΩ the value of expected A_v is

$$A_v \simeq 1700 \times 10^{-6} \times 2.5 \text{ k}\Omega$$
$$\simeq 4.25$$

as compared to $A_v = 5$, which was found using the load-line method.

For maximum voltage gain the JFET should be biased at or close to $V_{GS} = 0$, where g_{fs} is greatest. This is easily accomplished simply by removing the gate bias source in Figure 16.7. Of course, biasing at $V_{GS} = 0$ means that e_S must not be allowed to increase the gate voltage above approximately +0.5 V or it will become forward biased. Thus e_S is limited to 1 V p-p or less.

▶ EXAMPLE 16.4

Calculate A_v for the JFET amplifier of Figure 16.7 if it is biased at $V_{GS} = 0$.

The value of $g_{fs}(0)$ was previously calculated in Example 16.3 as 3000 μmhos. Thus

$$A_v \simeq 3000 \times 10^{-6} \times 2.5 \times 10^3$$
$$\simeq \textbf{7.5}$$

Compare this to $A_v = 5$ obtained when $V_{GS} = -2$ V.   ◀

Input and output impedance

The input impedance of the amplifier in Figure 16.7A is equal to the parallel combination of $R_G = 1$ MΩ and the gate input resistance of the JFET. This latter quantity is usually around 100 MΩ or greater so that $Z_{in} \approx R_G$. Values of R_G are kept large (1–100 MΩ) to provide high input impedance. This high input impedance is one of the main advantages of the JFET, making it useful in amplifying signals from high impedance sources (microphones, cartridge pickups).

The output impedance of the common-source amplifier is equal to the parallel combination of R_L and r_{DS}, the output resistance of the JFET. As stated earlier, r_{DS} in the pinchoff region is very large (greater than 500 kΩ) so that for normal values of R_L we can assume $Z_{out} \approx R_L$. The value of R_L is usually in the 2 kΩ–10 kΩ range so that Z_{out} for the common-source amplifier is in the same range as for the common-emitter amplifier.

In summary, the JFET amplifier provides a tremendous impedance level transformation from input to output. This makes it extremely valuable as an impedance matching circuit between a high impedance signal source and a low impedance load.

16.7.   JFET Bias Stability

Like bipolar transistors, JFETs also suffer from wide variations in parameters for units of the same type. For example, the JFET characterized by the curves in Figure 16.7B has $I_{DSS} = 9.2$ mA, $V_P = 6$ V and $g_{fs}(0) = 3000$ μmhos. This figure contains the curves for one particular JFET of that type. In actuality, this JFET might have parameters that can be anywhere in the following ranges: $I_{DSS} = 5$ mA to 15 mA; $V_P = 4$ V to 8 V; $g_{fs}(0) = 1500$ μmhos to 4500 μmhos.

With parameter variations like these, it should be apparent that the Q point of the amplifier in Figure 16.7 could vary widely from one circuit to another even though they all use the same type JFET. This uncertainty in the Q point is not a critical problem if the amplifier is being used for very small input signals. Since the JFET amplifier voltage gain is usually less than 10, the output signal will also be very small so that operation will not move very far from the Q point. Thus, as long as the Q point does not move out of the pinchoff region, there will be little danger of signal distortion.

The type of bias used in the amplifier of Figure 16.7A is called *gate bias* since the bias voltage source is applied to the gate. This type of bias is very unstable due to the above-mentioned JFET parameter variations. It also has the disadvantage that a second power supply is needed if biasing is at anything other than $V_{GS} = 0$. Another form of JFET biasing, shown in the circuit of Figure 16.9, overcomes these disadvantages to a certain extent and, in addition, provides a more stable voltage gain than the amplifier in Figure 16.7.

In this circuit, a resistor R_S is connected between source terminal and ground. Any drain current flowing through the JFET will also flow through

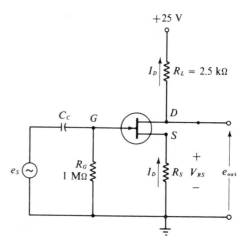

Figure 16.9.   COMMON-SOURCE AMPLIFIER USING SELF-BIAS

this resistor. The voltage V_{RS} across this source resistor is simply equal to $I_D \times R_S$, with the source *positive* relative to ground. Since the gate terminal is at ground potential ($I_G = 0$ so $V_{RG} = 0$), then the gate is negative relative to the source by the value of V_{RS}. That is,

$$V_{GS} = -V_{RS}$$

The gate-to-source reverse bias is thus provided by the voltage drop across R_S due to I_D. This is called *self-bias*. The value of R_S is chosen to give the desired Q point. The following example illustrates:

▶ EXAMPLE 16.5

The FET in Figure 16.9 has the drain characteristics of Figure 16.7B. Determine the required value of R_S if it is desired to bias the amplifier at a drain voltage of 15 V relative to ground ($V_D = 15$ V).

Since $V_{DD} = 25$ V, then 10 V must be dropped across R_L if V_D is to equal 15 V. Thus $I_D = 10$ V/2.5 kΩ = 4 mA is the Q-point drain current. Referring to the FET curves, we can see that a V_{GS} of -2 V will give approximately 4 mA for I_D. Thus the drop across R_S must be 2 V so that

$$R_S = \frac{2 \text{ V}}{4 \text{ mA}} = 500 \text{ } \Omega \qquad \blacktriangleleft$$

Because of the presence of R_S, the Q point of the self-bias amplifier is less affected by FET parameter variations than is the gate-biased amplifier. For example, suppose the FET in Figure 16.9 were replaced by one with a higher I_{DSS}. With the new FET in the circuit, the drain current will tend to be higher. However, the larger I_D produces a larger voltage across R_S and therefore a greater gate-source reverse bias. This increase in V_{GS} will tend to decrease I_D and therefore counteract some of the I_D increase caused by the FET's characteristics. The net result is that I_D will increase less than the amount it would if R_S were not present.

The effect of R_S on the Q-point value of I_D, then, is to counteract any changes (up or down) which come about due to FET parameter variation. The bias stability will improve as R_S is made larger. Of course, R_S has to be chosen as in Example 16.5 to produce the required V_{GS} used to produce the desired I_D.

Effect of R_S on voltage gain

The presence of R_S will also affect the changes in drain current produced by the input signal voltage. For example, as e_S goes positive it reduces the gate-source reverse voltage so that the drain current increases; but the increase in I_D also increases the voltage across R_S, which then increases the gate-source reverse bias, so that I_D does not increase by as much. The net effect is that the drain current signal is smaller, so that less signal voltage will be developed across R_L. This results in a lower voltage gain. If this is not desirable, R_S can be *bypassed* with a capacitor in the same manner used to bypass the emitter resistor in the base-voltage bias common-emitter amplifier. The capacitor effectively shorts out R_S for the AC signal, while R_S is still allowed to affect the DC operating point.

16.8.   The Common-Drain Amplifier

Another useful JFET amplifier configuration is shown in Figure 16.10. Actually, it is the same as the self-bias common-source amplifier of Figure 16.9, except that R_L has been removed and the output signal is taken at the source.

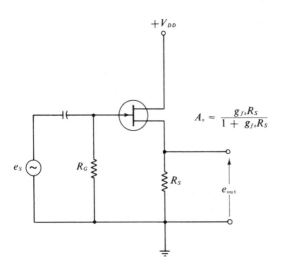

$$A_v \approx \frac{g_{fs}R_S}{1 + g_{fs}R_S}$$

Figure 16.10.   COMMON-DRAIN AMPLIFIER

Thus it is called a *common-drain* amplifier. Its operation is very similar to that of the common-collector amplifier (Chapter 13) in that it has a high Z_{IN}, a low Z_{OUT} and a voltage gain less than unity.

In this circuit R_S has a dual purpose. It is used to produce the self-bias to establish the Q point; it is also used to develop the AC output signal. This output signal is always less than the e_s input signal. It can never equal e_s because, if it did, then there would be no signal voltage across the gate-source junction and therefore no drain current signal to produce e_{out}. The voltage gain can be determined from the following relation:

$$A_v = \frac{e_{out}}{e_S} = \frac{g_{fs}R_S}{1 + g_{fs}R_S} \qquad (16.9)$$

where g_{fs} is the value of transconductance at the Q point. As this formula shows, A_v can never equal one; however, it can approach unity when $g_{fs}R_S \gg 1$.

Another characteristic of the common-drain amplifier is that there is no phase shift between input and output at normal frequencies. Thus e_{out} essentially *follows* the variations of the input signal. For this reason, this amplifier is also referred to as a *source follower*.

The input impedance of the source follower is approximately equal to R_G, which can be made very high, as is done for the common-source amplifier. The output impedance of the source follower is not what it might appear to be at first glance. It is not equal to R_S, but can actually be much lower than R_S. The formula for Z_{OUT} is:

$$Z_{OUT} = \frac{R_S}{1 + g_{fs}R_S} \qquad (16.10)$$

which shows that $Z_{OUT} < R_S$.

▶ EXAMPLE 16.6

A source follower has $R_S = 1 \text{ k}\Omega$ and is biased at a Q point such that $g_{fs} = 4000 \ \mu\text{mhos}$. Calculate the amplifier's voltage gain and output impedance.

$$A_v = \frac{g_{fs}R_S}{1 + g_{fs}R_S} = \frac{4000 \times 10^{-6} \times 1000}{1 + 4000 \times 10^{-6} \times 1000} = \frac{4}{5}$$

$$= \mathbf{0.8}$$

$$Z_{OUT} = \frac{R_S}{1 + g_{fs}R_S} = \frac{1 \text{ k}\Omega}{1 + 4}$$

$$= \mathbf{200 \ \Omega} \qquad ◀$$

Because of its high impedance ratio, the common-drain amplifier is useful in acting as an impedance matching circuit between a high impedance signal source and a low impedance load. It reproduces the input signal with only a slight loss in amplitude and with no shift in phase.

16.9. JFETs: General Comments

Although we have concerned ourselves with the use of JFETs in amplifier circuits, it should be noted that the applications of these devices extend into many other areas. JFETs are commonly used in chopper circuits (converting a DC signal to a time-varying signal), switching circuits and in a communication circuit called a *mixer*. In this latter application, the JFET offers an advantage over the junction transistor due to its nonlinear transfer curve.

In amplifier applications the junction transistor is still the most often used device due to its lower cost and its normally higher gain. The JFET is most useful when high Z_{IN} is required and when the input signal is very small so that biasing at $V_{GS} = 0$ is feasible. For this reason, the JFET amplifier is usually found on the front end of amplifier systems, followed by junction transistor amplifiers.

Large power applications of JFETs are not currently possible since power FETs are still in the developmental stage. Units with power capabilities of 10 W should be available in the near future, but much work remains to be done in this area.

16.10. MOSFETS: Introduction

The metal oxide semiconductor FET (MOSFET) has several characteristics in common with the JFET. The MOSFET is a low-power device with extremely high input impedance. Like the JFET, the MOSFET has a source, drain and gate, and a conducting channel whose resistance is controlled by the gate voltage. The principle difference between the two devices is the structure of the gate. The gate-source path in the JFET is a P-N junction which is normally kept in reverse bias. In the MOSFET there is no gate-source P-N junction; instead, a thin layer of an insulating material (silicon dioxide) is placed between gate and channel. This insulator has an extremely high resistance (typically 10^{14} Ω); therefore, the gate leakage current is even lower than for the JFET (typically 10^{-14} A).

Because there is no P-N junction between gate and channel in the MOSFET, there is no limitation on the polarity of the gate voltage. The insulator between gate and channel maintains its very high resistance for either polarity of bias. Because the gate is insulated from the channel, a MOSFET is also referred to as an *insulated-gate* FET (IGFET). However, MOSFET is the more commonly used term.

16.11. Operation of the MOSFET

In the JFET a conducting channel is present between source and drain, and this channel is made narrower by *depletion* regions caused by gate-source reverse bias. This is the only mode of operation of JFETs and is called the *depletion* mode of operation. MOSFETs can also operate in this mode. In addition, MOSFETs can operate in another manner, in which the width of the conducting channel is *increased* by applying the proper polarity of gate voltage. This mode of operation is called the *enhancement* mode since the channel current is enhanced by application of gate voltage. MOSFETs are thus subdivided into two categories: *enhancement-MOSFETs* (hereafter abbreviated E-MOSFETs) and *depletion-MOSFETs* (hereafter abbreviated D-MOSFETs).

The enhancement-MOSFET

Figure 16.11A shows the structure of an N-channel E-MOSFET. The structure begins with a high-resistivity P-type substrate; two low-resistivity N-type regions are diffused into the substrate as shown. Then the surface of the structure is covered with a layer of insulating silicon dioxide. Holes are cut into the oxide layer, allowing contact to the N regions (source and drain). Then a metal contact area is placed over the oxide, covering the entire channel from source to drain. The contact to this metal area is the gate terminal. Note that there is no physical contact between the gate and P substrate due to the insulation afforded by the silicon dioxide.

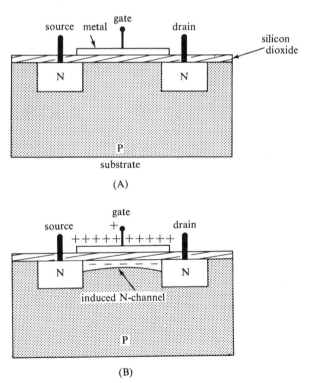

Figure 16.11.   (A) STRUCTURE OF N-CHANNEL E-MOSFET; (B) POSITIVE GATE VOLTAGE INDUCES N CHANNEL

Since the source and drain are separated by the P-type substrate, the source-to-drain current will be extremely low because there are essentially two P-N junctions connected back to back. However, the gate can be used to produce a conductive channel from source to drain. The *metal* area of the gate, the silicon di*oxide* layer, and the *semiconductor* channel form a capacitor. The gate area is the top plate and the P substrate forms the bottom plate, while the silicon dioxide is the dielectric. In other words, this is a metal-oxide-semiconductor (MOS) capacitor. When a positive voltage is placed on the gate (see Figure 16.11B), the metal plate is charged positive. This positive charge will induce a negative charge on the semiconductor plate. As the positive voltage at the gate is increased, holes in the P-type semiconductor are repelled until the region beneath the oxide becomes an N-type semiconductor region. That is, an N channel has been formed between source and drain. Current can now be made to flow from source to drain through this induced N channel by biasing the drain positive relative to source, as was done with the N-channel JFET.

Thus the device current flow is *enhanced* by the application of positive gate voltage. In fact, making the gate voltage more positive will widen the induced N channel, thereby reducing the channel resistance even further. Of course, a zero or negative gate voltage will produce no enhancement effect and the device will not conduct. Because it does not conduct for zero gate voltage, the E-MOSFET is often referred to as a "normally off" MOSFET.

A P-channel E-MOSFET is constructed just like the N-channel unit, except that all P and N regions are interchanged. For the P-channel device, a negative gate voltage induces a P-type channel and enhances current flow through the channel.

E-MOSFET drain curves

A typical set of drain characteristic curves is shown in Figure 16.12. The pinch-off and ohmic regions correspond to those of the JFET. It is significant to note that drain current does not flow until V_{GS} is made at least 2 volts positive. This voltage is called the *threshold* voltage $V_{GS(th)}$ and indicates the minimum positive gate voltage required to induce a significant N channel. It is also significant

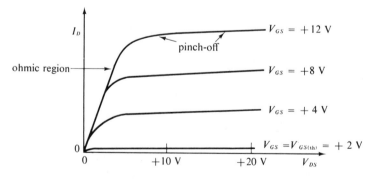

Figure 16.12.   TYPICAL DRAIN CURVE FOR N-CHANNEL E-MOSFET

that the gate-source voltage and drain-source voltage are of the same polarity. This makes biasing MOSFET amplifiers relatively easy.

The threshold voltage $V_{GS(th)}$ depends on the particular type of MOSFET (doping, geometry, etc.) and is typically in the 1–5 V range, although it can go below 1 V in certain E-MOSFETs. The threshold voltage is an important E-MOSFET parameter since it specifies the minimum gate voltage which will allow drain current. For $V_{GS} < V_{GS(th)}$, the E-MOSFET is in *cutoff*.

Typical drain curves for a P-channel E-MOSFET can be obtained from the curves in Figure 16.12 simply by changing the polarities of V_{GS} and V_{DS}. Obviously, for the P-channel device, $V_{GS(th)}$ will be negative so that cutoff occurs whenever V_{GS} is less negative than $V_{GS(th)}$.

E-MOSFET symbols

The use of different MOSFET symbols by various manufacturers has created somewhat of a problem in the interpretation of circuit diagrams. The Institute of Electrical and Electronic Engineers (IEEE) has adopted standard symbols which we will use here. Eventually these symbols should be adopted by all MOSFET manufacturers and users. Figure 16.13 shows the symbols for the N- and P-channel E-MOSFETs along with their normal bias polarities.

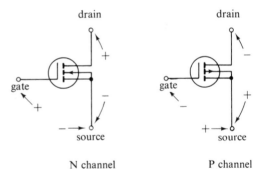

Figure 16.13.   E-MOSFET SYMBOLS

In these symbols the gate is shown separated from the rest of the MOSFET, indicating that the gate is electrically insulated from the channel. As for the JFET, the N-channel device has the arrow pointing *in*, while the P-channel device has the arrow pointing *out*. Also note the broken vertical line joining the drain and source; this indicates a "normally off" device (channel acts like an open circuit).

In both of the symbols in Figure 16.13 the substrate is indicated by the arrow and is connected to the source. Very often this connection is made internally by the manufacturer. Occasionally, however, MOSFETs are supplied as four-terminal devices with the substrate brought out as a separate lead. In those cases the MOSFET symbol is modified accordingly.

The depletion-MOSFET

The D-MOSFET is constructed similarly to the E-MOSFET, except for one important difference: in the D-MOSFET a conducting channel is present between source and drain even without gate voltage. Figure 16.14 shows the structure of an N-channel D-MOSFET which illustrates the presence of a moderate-resistivity N channel. This initial N channel allows current to flow from source to drain even without gate voltage. If a negative voltage is applied to the gate, however, holes from the P subtrate are attracted into the N channel and begin to neutralize the free electrons so that the effective N channel is depleted (reduced). This effect increases the channel resistance and reduces the current. The operation with negative gate voltage is the *depletion* mode.

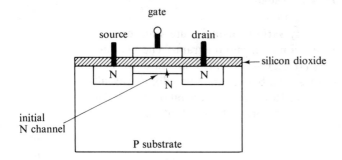

Figure 16.14.   STRUCTURE OF N-CHANNEL D-MOSFET

The N-channel D-MOSFET can also be operated in the *enhancement* mode by making the gate positive. The process is the same as in the N-channel E-MOSFET. The positive gate voltage produces a widening of the initial N channel, causing an increase in channel current. Thus the D-MOSFET is more accurately a depletion-enhancement MOSFET since, depending on the gate polarity, it can operate in both modes.

Since the D-MOSFET will conduct a significant current when $V_{GS} = 0$, it is often referred to as a "normally on" MOSFET. Recall that the E-MOSFET is a "normally off" MOSFET since it requires $|V_{GS}| \geq |V_{GS(\text{th})}|$ in order to conduct.

D-MOSFET drain curves

The set of typical drain characteristic curves for an N-channel D-MOSFET shown in Figure 16.15 illustrates this device's overall operation. The curves are the same shape as those of the other FET devices with different curves for different V_{GS} values. The $V_{GS} = 0$ condition results in a pinchoff current value which is given the symbol I_{DSS} (same as for JFETs). For positive values of V_{GS} the channel is enhanced and the drain current values are greater than I_{DSS}. This is the enhancement region. For negative values of V_{GS} the channel is depleted and the drain current values are lower than I_{DSS}. Eventually, as V_{GS} becomes

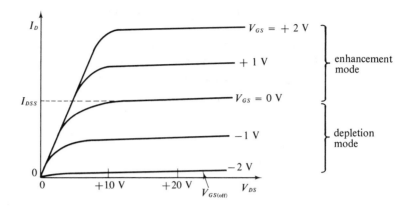

Figure 16.15.   TYPICAL DRAIN CURVES FOR N-CHANNEL D-MOSFET

more negative the channel current is cut off almost completely. This happens at $V_{GS} = V_{GS(off)} = -2$ V for the curves in the figure. $V_{GS(off)}$ is the same parameter discussed in conjunction with the JFET. Negative values of V_{GS} constitute the depletion region of operation.

A typical set of drain curves for a P-channel D-MOSFET can be obtained from those in Figure 16.15 simply by reversing the voltage polarities. P-channel D-MOSFETs are not as common as the N-channel type due to problems in the manufacturing process which are peculiar to the fabrication of P-channel D-MOSFETs.

D-MOSFET symbols

The IEEE standard circuit symbols for D-MOSFETs are given in Figure 16.16 along with the normal drain-source bias polarities. The gate-source polarity is not shown since it can be of either polarity. Once again, note that the arrow points inward for the N-channel device and outward for the P-channel device. Also note that the drain-to-source path is shown as an *unbroken* line to indicate that the channel is normally conducting for D-MOSFETs.

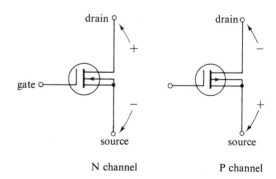

Figure 16.16.   D-MOSFET CIRCUIT SYMBOLS

16.12.   Summary of FET Types

At this point it would be helpful to pause and summarize the basic FET types. Table 16.2 lists the various FETs and the corresponding information concerning mode of operation and bias polarities.

TABLE 16.2

FET type	Operating mode	V_{DS}	V_{GS}	Comments
N-channel JFET	depletion	+	−	V_{GS} can go slightly positive; cut off when $\|V_{GS}\| < \|V_{GS(off)}\|$.
P-channel JFET	depletion	−	+	V_{GS} can go slightly negative; cut off when $\|V_{GS}\| < \|V_{GS(off)}\|$.
N-channel E-MOSFET	enhancement	+	+	conducts when $\|V_{GS}\| > \|V_{GS(th)}\|$.
P-channel E-MOSFET	enhancement	−	−	conducts when $\|V_{GS}\| > \|V_{GS(th)}\|$.
N-channel D-MOSFET	depletion; enhancement	+ +	− +	"normally-on" MOSFET; cut off when V_{GS} more negative than $\|V_{GS(off)}\|$.
P-channel D-MOSFET	Depletion; enhancement	− −	+ −	"normally-on" MOSFET; cut off when V_{GS} more positive than $\|V_{GS(off)}\|$.

From this table it can be concluded that *all* N-channel FETs are operated with *positive* values of V_{DS}, while *all* P-channel FETs use *negative* values of V_{DS}. In addition, it can be seen that *any* FET operating in the depletion mode uses gate-source bias V_{GS} opposite in polarity to V_{DS}. On the other hand, operation in the enhancement mode requires V_{GS} and V_{DS} of the same polarity.

16.13.   Gate Breakdown in MOSFETs

Special care must be taken when handling and using MOSFETs. The silicon dioxide insulating layer between gate and channel is sufficiently thin that if too large a voltage is applied between gate and channel the insulator will break down and rupture. Typically, 100 V is capable of ruining a MOSFET. One hundred volts can easily be applied to the gate accidently by static charges during handling on a dry day or by soldering irons which are poorly insulated from the AC power line. To prevent accidental damage, some manufacturers now supply MOSFETs (enhancement type only) with a built-in zener diode connected from gate to source. This zener is designed to break down at 50 V, thereby keeping V_{GS} well below the value which would cause destructive breakdown of the silicon dioxide layer. Unfortunately, this technique increases the MOSFET gate leakage current to values similar to that of JFETs since there is now a reverse-biased P-N junction (the zener) across the input.

In some applications, the input leakage current must be kept as low as possible; therefore, unprotected MOSFETs must be used. In such cases, precautions must be taken to avoid damaging the device. The gate should be kept shorted to either the source or drain until placed in the circuit and any soldering iron used should have its tip properly grounded.

16.14.   MOSFETs: General Comments

MOSFETs, like JFETs, can be used as signal amplifiers. The previous discussion and formulas for the JFET amplifiers are equally applicable to MOSFETs. MOSFETs used in the depletion mode are biased in the same manner as the JFET. Enhancement MOSFETs are biased differently since V_{GS} and V_{DS} have to be of the same polarity. The use of E-MOSFETs as amplifiers makes it possible to connect the output of one amplifier stage directly to the input of another amplifier stage without the need for coupling capacitors. The DC bias voltage at the output of the first amplifier simply serves as the gate bias for the amplifier which follows. Elimination of the coupling capacitors simplifies the circuit and usually improves the low-frequency response.

The MOSFET parasitic capacitances between the gate and drain (C_{GD}) and the gate and source (C_{GS}) are normally of lower values than those of the JFET. As such, the MOSFET typically has a somewhat better high-frequency response than the JFET.

Temperature effects are minimal with the MOSFET. Its input leakage current does not flow across a reverse-biased P-N junction as input leakage does in a JFET, and is therefore not as temperature-sensitive. MOSFETs are one of the least temperature-dependent semiconductor devices available today.

At present, the most important application of MOSFETs is in the broad area of digital switching circuits. Previously, this area was completely dominated by bipolar transistor circuits. Although the MOSFET does not as yet measure up to the bipolar transistor as far as switching speed is concerned, it does possess many desirable characteristics that give it certain advantages, especially in digital integrated circuitry. MOSFETs are much smaller than bipolar transistors and require fewer processing steps in their fabrication. In addition, power dissipation in MOSFET switching circuits is extremely low. Thus extremely small complex arrays of MOSFET switching circuits can be integrated with fewer complications than comparable bipolar arrays. MOSFET integrated circuits are extensively used in airborne avionic systems, earth satellites, medium-speed computers and electronic calculators. Further information on MOSFETs in integrated circuits will be presented in the next chapter.

GLOSSARY

Field effect: effect of varying the conductivity of a current path by the application of a perpendicular electric field.

Junction field-effect transistor (JFET):   field-effect device utilizing reverse-biased P-N junction gate.

Channel: semiconductor conductive path between source and drain.

Source: source of majority carriers.

Drain: collector of majority carriers.

Gate: control terminal in FETs which produces the electric field which, in turn, varies channel conductivity.

Pinchoff: condition in FETs in which drain current stops increasing with increases in V_{DS}.

Pinchoff voltage (V_P):   total gate-to-drain reverse bias needed to produce pinchoff.

Cutoff: condition in FETs in which gate-to-source voltage causes channel current to be essentially zero.

$V_{GS(off)}$:   gate-to-source voltage needed for cutoff.

Transconductance (g_{fs}):   FET parameter which is a measure of gate's control over drain current.

MOSFET: Field-effect device utilizing an insulated capacitor gate.

Depletion mode: operation of an FET with an *increase* in gate-to-source voltage in order to *decrease* drain current.

Enhancement mode: operation of an FET with an *increase* in gate-to-source voltage in order to *increase* drain current.

Threshold voltage $V_{GS(th)}$:   gate-to-source voltage needed to produce significant drain current in an enchancement-MOSFET.

PROBLEMS

16.1   What is the principal current carrier in the N-channel JFET? The P-channel JFET?

16.2   Sketch the symbol for the N-channel JFET showing the proper bias polarities. Do the same for the P-channel JFET.

16.3 How does an increase in gate-to-source reverse bias affect the drain current in a JFET?

16.4 Why is the input resistance of the JFET so high?

16.5 Sketch a typical set of characteristics for a P-channel JFET. Label the ohmic and pinchoff regions.

16.6 A certain N-channel JFET has $V_P = 5$ V. At what value of V_{DS} will the drain current level off (reach pinchoff) with $V_{GS} = -2$ V? With $V_{GS} = 0$ V? At what value of V_{GS} will the JFET be cut off?

16.7 The FET of the previous question had $I_{DSS} = 10$ mA. Using Equation 16.3, calculate I_{DP} at $V_{GS} = 0, -1$ V, -2 V, -3 V, -4 V and -5 V. Sketch the device's transfer characteristic.

16.8 Using the data sheets for Motorola 2N4220 and 2N4222 N-channel JFETs in Appendix II, determine the following for the 2N4220:
(a) Maximum allowable V_{DS}.
(b) Maximum allowable I_D.
(c) Maximum allowable P_D.
(d) Maximum gate leakage current at 25°C; at 150°C.
(e) Minimum and maximum I_{DSS}.
(f) Typical g_{fs} at $V_{GS} = 0$.
(g) Typical r_{ds} ("on").

16.9 If the 2N4220 were biased at $V_{GS} = 0$, what would be a typical voltage gain using $R_L = 20$ kΩ in the basic JFET amplifier?

16.10 From the transfer characteristic of Figure 16.4 determine the approximate value of g_{fs} at $V_{GS} = -3.5$ V. Then determine, both graphically and by using Equation 16.8, the voltage gain of the amplifier of Figure 16.7 when it is biased at $V_{GS} = -3.5$ V.

16.11 Why is it advantageous to bias a JFET amplifier close to $V_{GS} = 0$? What are the restrictions when this is done?

16.12 It is difficult experimentally to measure V_P for a JFET because the beginning of the pinchoff region occurs too gradually. A more accurate method utilizes the relationship of Equation 16.6 to determine V_P indirectly. Describe an appropriate experimental procedure for determining V_P.

16.13 A particular JFET has $I_{DSS} = 16$ mA and $V_P = 8$ V. Sketch a set of drain curves for this JFET for V_{GS} values of 0 V, -2 V, -4 V, -6 V and -8 V.

16.14 Consider the amplifier in Figure 16.17. The JFET in the circuit has the drain characteristic curves of Figure 16.7(B). For this amplifier determine approximate values for
(a) I_D and V_{DS} at Q point.
(b) voltage gain.
(c) Z_{IN} and Z_{OUT}.
Sketch the *complete* e_S and e_{out} waveforms.

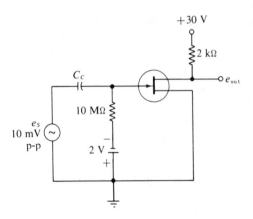

Figure 16.17.

16.15 Remove the 2-V bias source in Figure 16.17. Insert a resistor R_S for self-bias. Determine the value of R_S needed for $V_{GS} = -2$ V. Explain how R_S will affect the voltage gain.

16.16 The JFET with the characteristics of Problem 16.13 is used as a source follower with $R_S = 240 \ \Omega$. Calculate Z_{OUT} and A_V if it is biased at $V_{GS} = -2$V.

16.17 Indicate which of the following descriptions refer to the common-source amplifier and which refer to the common-drain amplifier:
(a) A_V usually close to unity.
(b) 180° phase shift.
(c) A_V increases as JFET is biased closer to $V_{GS} = 0$.
(d) Very high Z_{IN}.
(e) Low Z_{OUT}.
(f) e_s should not exceed 1 V p-p when biased at $V_{GS} = 0$.
(g) Can use self-bias.
(h) R_S can be bypassed to increased A_V.

16.18 Why is the input resistance of MOSFETs so high?

16.19 Explain the difference between enhancement- and depletion-MOSFETs.

16.20 What type of FET can be biased at either polarity of gate voltage?

16.21 Identify the FET types shown in Figure 16.18 and label the normal bias polarities.

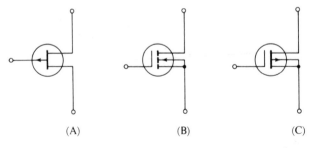

(A)   (B)   (C)

Figure 16.18.

16.22 Indicate which of the following phrases refer to JFETs and which refer to MOSFETs:
(a) Has lower gate leakage current.
(b) Has higher input resistance.
(c) Has higher input capacitance.
(d) Has better frequency response.
(e) Can operate in the enhancement mode.
(f) Is widely used in digital integrated circuits.
(g) Has lower A_V than bipolar transistors.
(h) Has pinchoff region.

REFERENCES

Deboo, G. J. and C. N. Burrows, *Integrated Circuits and Semiconductor Devices.* New York: McGraw-Hill Book Co., 1971.

Malvino, A. P., *Electronic Principles.* New York: McGraw-Hill Book Co., 1973.

Miller, G. M., *Modern Linear Circuits for Electronics Technology.* Englewood Cliffs, N.J.: Prentice-Hall, Inc., 1973.

Richman, Paul, *Characteristics and Operation of MOS Field-Effect Devices.* New York: McGraw-Hill Book Co., 1967.

17

Introduction to Integrated Circuits

17.1.   Introduction

A *discrete* circuit is the kind you build when you connect separate components together in a certain arrangement. Each component that is added to the circuit is discrete; that is, distinct from the others. An *integrated* circuit is a circuit that has been fabricated as an inseparable assembly of electronic components in a single structure. Integrated circuits (ICs) completely eliminate the use of individual components such as resistors, diodes, transistors and capacitors as the building blocks of an electronic circuit or system. Despite its rapid development in the past few years, integrated circuit technology is still considered to be in its infancy. However, even in this early stage ICs have demonstrated significant improvements over discrete circuitry. Basically, these advantages include greatly reduced size and weight, lower cost and improved reliability.

Weight and size reduction are obvious. These savings are particularly important in military and space applications where the increased sophistication of electronic control systems has caused a dramatic increase in the number of electronic circuits and components. In space applications weight is of primary importance since every extra ounce requires several more pounds of thrust to maintain it. In consumer applications weight and size are normally not primary considerations.

The cost reductions afforded by ICs are a result of many factors. A major reduction comes with the fact that a very complex circuit can be fabricated in the same number of process steps as used for a conventional transistor. Also, with complete IC circuits serving as basic components, shipping charges, purchasing costs, inventory costs, incoming inspection and testing costs, and the costs of almost all the factors involved in equipment manufacturing are greatly reduced.

Perhaps the most important advantage of ICs lies in the area of reliability. Increased reliability is due to several factors, the most significant of which is the need for fewer interconnections. In earlier discrete component circuits, up to 50 percent of all circuit failures were caused by interconnections between components. Another factor concerns the lower power operation of ICs. Due to their small size ICs are more suited to low power operation. The closeness of the components within a silicon wafer reduces the chance of stray electrical pickup, allowing very small signal operation. This low power operation means lower internal temperature rises and, consequently, improved reliability. Thus the IC offers a more reliable approach to circuit assemblies than is available using discrete components. This is by far its most important advantage, from both military and consumer application standpoints.

ICs are divided into two major categories: *monolithic* ICs and *hybrid* ICs.

17.2. Monolithic Integrated Circuits

The term *monolithic* describes an IC structure in which all elements of the circuit are built in a single crystal of semiconductor material. In the present state of technology, this material is silicon. The technology of monolothic ICs is based on the silicon diffused planar process, in which all process steps are performed on one surface of the silicon slice and all the contacts to the various components are made at the same surface. Interconnection between the components is made by depositing a metallic wiring pattern on the oxide-covered surface of the silicon wafer.

The geometries of individual components (transistors, diodes, resistors and capacitors) of a given IC are designed so that they can all be formed simultaneously. In fact, all the elements of a complete circuit can be diffused into a single wafer, using exactly the same processes and in almost the same amount of time required to make a single transistor.

The basic structure of a monolithic IC is illustrated in Figure 17.1. It consists of three layers of different materials. The relatively thick bottom layer is of P-type silicon, onto which is grown a thin epitaxial (high-resistivity) layer of N-type silicon. This P-N structure is topped with a thin layer of silicon dioxide.

In *all-diffused monolithic* structures, all the component parts are formed within the thin N-type region. The P region is not an active part of the circuit. Its primary function is to serve as a substrate in order to give the structure mechanical ruggedness. It serves an additional purpose by providing a simple means of electrically isolating (insulating) the various diffused components.

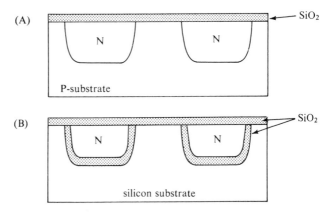

Figure 17.1.   BASIC MONOLITHIC IC STRUCTURE

The silicon dioxide layer has two specific functions. It protects the semi-conductor surface against contamination by external impurities and it provides the means for selectively diffusing the various components into the wafer beneath.

Since each component is formed within an N region, it is necessary to isolate the various N regions from each other in order to isolate each component. Two methods of providing this isolation are illustrated in Figure 17.2. The first method, called P-N *diode isolation*, is shown in Figure 17.2A and uses the high resistance of a reverse-biased P-N junction. Current will not flow from one N region to the other since they are separated by the P-region substrate, which essentially forms two back-to-back diodes. In practice, the substrate is tied to the most negative voltage in the circuit so that each N region is reverse biased relative to the substrate. In this way, each N region is insulated from the substrate and from the other N regions. This method does have drawbacks due to the imperfect isolation afforded by a reverse-biased diode and especially due to the junction capacitances across each isolation diode. These capacitances limit the circuit operating speed.

The second method of isolation is shown in Figure 17.2B and is called *silicon-dioxide isolation*. A layer of silicon dioxide around each N region produces the desired isolation. The substrate is undoped silicon. This type of isola-

Figure 17.2.   ISOLATION   TECHNIQUES: (A) P-N   DIODE   ISOLATION;
                (B) SILICON-DIOXIDE ISOLATION

tion has several advantages over the first type. First of all, the isolation is much more perfect because the silicon dioxide is a better insulator than a reverse-biased P-N junction. In addition, there is no capacitance between the N regions and the substrate. This improves high-frequency operation. Another advantage lies in the greater versatility of the types of circuits which can be manufactured.

Counteracting these advantages is the added complexity of the fabrication process. Even when developed for mass production, the additional process steps add to the manufacturing costs. These costs, of course, are passed on to the IC user.

The components normally associated with a monolothic IC are transistors, diodes, resistors and capacitors. The transistor represents the most complicated component and all the other parts can be fabricated in conjunction with one or more of the transistor processes. This is illustrated in Figure 17.3 where a typical cross section of the basic monolithic components is shown.

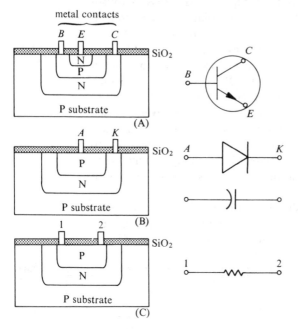

Figure 17.3.   TYPICAL IC COMPONENTS: (A) TRANSISTOR; (B) DIODE
AND CAPACITOR; (C) RESISTOR

The NPN transistor structure is formed by successive impurity diffusions into the top of the substrate. The silicon dioxide layer is appropriately etched to allow diffusion into the proper portion of the substrate. After the diffusion is complete the top surface is covered with silicon dioxide for protection, and metal contacts are made to the various transistor regions.

Integrated circuit diodes are formed by simply utilizing the P-N junctions of the integrated transistor. In Figure 17.3B, the *C-B* junction of the transistor is used as the diode. The anode of the diode is formed during the transistor-base diffusion. The cathode of the diode is the collector region. Contacts are made to these two regions and brought to the top surface. This type of diode is used for general applications. When high-speed switching is required, the *E-B* junction of the transistor is used as a diode.

There are two basic methods of forming monolithic capacitors. The first method, shown in Figure 17.3B, utilizes the junction capacitance of a reverse-biased P-N diode. In this method, the capacitance is limited to about 100 pF and is dependent on the reverse voltage across the junction. Despite these limitations, it has the advantage that it can be produced by diffusion simultaneously with the other components.

The other type of capacitor utilizes the metal-oxide-semiconductor (MOS) structure similar to that used as the gate of MOSFETs. MOS capacitors provide somewhat higher values of capacitance but are still limited to several hundred picofarads. They offer the principal advantages of being nonpolar (voltages of either polarity may be used) and independent of voltage. In addition, they have lower leakage and can normally operate at higher voltages than the diffused capacitor.

Integrated-circuit resistors utilize the resistivity of the doped silicon. A resistor is formed by diffusing an impurity into the silicon wafer. By controlling the concentration of the impurity and the depth of diffusion the resistance values can be controlled. Most resistors are formed during the diffusion of the base (see Figure 17.3C) since this is the highest resistivity region. For very low values of resistance the emitter region is used since it has a much lower resistivity. For reasons associated with the diffusion process, it is difficult to reproduce resistors with tolerances closer than ± 10 percent. However, the ratio of resistor values made on the same wafer can be reproduced to within ± 1 percent tolerances. For this reason, IC designers must use resistance ratios rather than absolute values of resistance as the controlling factor in their circuit designs.

17.3. Formation of a Complete Monolithic IC

To help illustrate the formation of a complete IC, a simple circuit will be investigated. Refer to Figure 17.4. The first step in the formation of the IC is the isolation of the various N regions as shown in Figure 17.4A. In general, they will be different sizes, depending on the components for which each region is used. One N region is needed for each component. Step two in the process (Figure 17.4B) is the diffusion of the P-type base region. This is accomplished by etching away the appropriate areas of silicon dioxide and allowing P-type impurities to diffuse into each N region. As far as the resistor and diode are concerned the process is complete. The transistor, however, must still be provided with its emitter region. This is accomplished in the third step (Figure 17.4C) of the process, in which N-type material is diffused into the P region (base) to form the emitter. In this step the diode and resistor are not affected because the silicon dioxide layer masks these regions from any impurity diffusion.

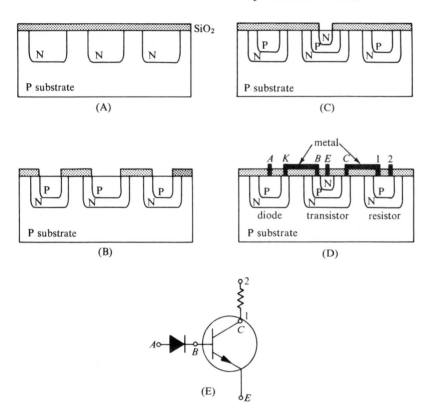

Figure 17.4.   IC PROCESS STEPS: (A) ISOLATION OF N REGIONS; (B) P-TYPE BASE DIFFUSION; (C) N-TYPE EMITTER DIFFUSION; (D) CONTACT FORMATION AND INTERCONNECTION; (E) CIRCUIT SCHEMATIC

The last step, as shown in Figure 17.4D, forms the contacts of the various components and a metallization pattern is then deposited to connect the components according to their circuit arrangement. The final circuit is shown schematically in Figure 17.4E.

From this example it can be seen that the number of process steps needed to produce an integrated circuit is the same as required to produce a transistor (except for the isolation step). This is true no matter how complicated the circuit is. A complete circuit may total 50 components and can fit into an area of 50 mils-square (0.05 in. $\times$ 0.05 in.). A circular silicon wafer of 1.25 inches diameter can contain nearly 350 of these circuits.

Thus a large number of circuits can be formed on one silicon wafer using the same process steps shown in Figure 17.4. After the circuits have been formed they are usually tested right on the wafer before being separated into individual chips. The testing usually consists of DC probe measurements. After testing, the individual circuits on the wafer are separated into individual circuit wafers; this is usually done by scribing and breaking the wafer.

The individual circuit wafer is then prepared for assembly in a package. The wafer is mounted to either a metal or glass base and electrical connections are made from the wafer contacts to the external package leads. This process is usually thermal compression bonding with gold wire of 1-mil diameter. Finally, the package is sealed by molding in plastic or by welding on a metal lid or can.

17.4.   IC Terminology

Like all new technologies, IC technology has developed a terminology of its own. Some of the more common terms are defined as follows:

  (a) *Bonding:* attachment of wires to an IC, or the mounting of an IC to a substrate.
  (b) *Chip:* part of a wafer of silicon upon which a component or an IC is fabricated.
  (c) *Die:* same as chip.
  (d) *Diffusion:* process whereby small quantities of material are allowed to "seep" into a silicon crystal in order to modify the electrical characteristics of the crystal.
  (e) *Epitaxial growth:* process of depositing layers of silicon material on the substrate.
  (f) *Etching:* removal of surface material from a chip by chemical means. In the monolithic process the etching is selective since there is no removal of material from regions covered by the photoresist material.
  (g) *Isolation:* means of electrically isolating the various components of an IC from each other.
  (h) *LSI (large-scale integration):* refers to ICs which contain over one hundred components per chip.
  (i) *MSI (medium-scale integration):* refers to ICs which contain from 12 to 100 integrated components per chip.
  (j) *SSI (small-scale integration):* refers to ICs with fewer than 12 components per chip.
  (k) *Substrate:* insulating semiconductor area upon which an IC is fabricated or mounted.
  (l) *Yield:* percentage of acceptable integrated circuits produced in a production run.

17.5.   MOS Integrated Circuits

The monolithic ICs discussed in the preceding sections would be more accurately referred to as bipolar monolithic ICs since they are based on the bipolar transistor fabrication and employ the bipolar transistor as their major component. A second family of monolithic ICs based on the MOSFET structure is also available and, in fact, is currently being widely used in many applications, especially in the digital field.

The processing of MOS integrated-circuit wafers uses essentially the same technology required for monolithic bipolar ICs. However, the MOS process is inherently less expensive than the bipolar process since it requires only one diffusion step to form both the source and drain regions, as compared with anywhere from two to four diffusion steps for bipolar ICs. In addition, there is no need for isolation techniques for enhancement-MOS devices since each source and drain region is isolated from the others by P-N junctions formed with the substrate. This characteristic is illustrated in Figure 17.5, where *two* P-channel E-MOSFETs are shown fabricated on the same N-type substrate.

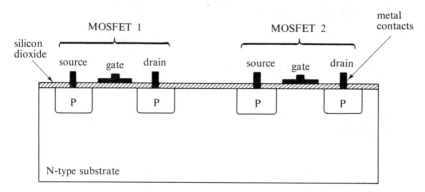

Figure 17.5.   EXAMPLE OF MOS IC STRUCTURE

The MOS IC packing density can be very high, in which case more devices can be fabricated in a given area than with bipolar ICs (typically ten times more). One reason for this is that isolation regions are not needed. Also, MOS-type resistors occupy less than 1 percent of the area of a conventional diffused resistor. This high packing density makes MOSs particularly suited to large-scale integration (LSI).

Most MOS ICs utilize three basic components: MOSFETs, MOS capacitors and MOS resistors. A MOS resistor is formed simply by using an E-MOSFET whose gate is tied to its drain such that the device is always biased "on." The enhanced channel serves as the desired resistor and the resistance value is controlled by various factors, such as geometries and doping levels.

The principle disadvantage of MOS ICs is their slower operating speed when compared to bipolar ICs. As such, MOS ICs do not compete with bipolar ICs in ultra-high-speed applications. However, the low cost, low power consumption and high packing density of MOS ICs have made them particularly attractive in recent years, especially in applications utilizing LSI (e.g., certain types of computer memories and electronic calculators).

17.6.   Hybrid ICs

Most hybrid integrated circuits consist of thin-film or thick-film conductor-resistor networks deposited on an insulating substrate, with semiconductor devices and MOS capacitors added in chip form. Another form of hybrid IC consists of monolithic bipolar ICs with thin-film resistors fabricated on the silicon oxide; this combination provides resistors which are more accurately controlled than monolithic diffused resistors.

Hybrid ICs have several advantages over pure monolithic ICs. Hybrid ICs often combine specifications that are unequaled by monolithic ICs. Hybrid ICs offer:

(a) much greater flexibility in circuit design.
(b) somewhat lower cost than monolithics in small quantities.
(c) improved circuit performance since passive component values (resistance and capacitance) can be trimmed to precision values.
(d) higher possible values of resistors and capacitors than are available in monolithic ICs.

Hybrid ICs can be divided into at least two categories: thin film and thick film.

Thin-film hybrid ICs

Thin-film technology is so named because conductors, resistors and capacitors (and, theoretically, transistors) are prepared as very thin films on the order of a few thousand angstroms thick. [An angstrom (Å) is 10^{-10} meters.] These thin films are generally deposited on substrates such as glass or alumina through a series of high-vacuum vapor-deposition processes. Resistors are formed by depositing nichrome, tantalum or tin oxide in strip form on the surface of the substrate. The resistance value, which is controlled by varying the length, width and thickness, ranges between 10 Ω and 1 MΩ. Gold-nichrome is often used for the interconnecting conductor films. Transistors, diodes and capacitors are added in chip form and secured to the substrate, usually with a conductive epoxy. Figure 17.6 illustrates a typical thin-film hybrid IC.

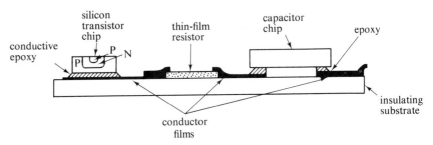

Figure 17.6.   TYPICAL THIN-FILM IC STRUCTURE

Thick-film hybrid ICs

Thick-film hybrid ICs are fabricated without the use of vacuum equipment and result in films which are on the order of a mil (0.001 inches) in thickness. A thick-film circuit is formed by printing or silk-screening a pattern onto an insulating substrate using as the "ink" a mixture of pulverized glass and aluminum (or another conductor-insulator mixture).

A wide range of resistor values, or a conductor pattern, may be produced by varying the proportions of the mixture. The thick-film network is then put in a high-temperature oven to be fired to the form of a stable circuit. Diodes, transistors and capacitor chips are then added to the substrate in the same manner as for thin-film ICs.

GLOSSARY

Integrated circuit: electronic circuit constructed as an inseparable assembly of components in a single structure.

Monolithic integrated circuit: electronic circuit in which all circuit elements are formed and interconnected on or within a single piece of silicon.

Diode isolation: method of electrically isolating components in a monolithic IC by creating reverse-biased P-N diodes between all components.

Silicon-dioxide isolation: method of isolation in which silicon dioxide acts as the insulator.

Hybrid integrated circuit: integrated circuit which combines thin- or thick-film passive components with silicon monolithic chips.

(See also Section 17.4.)

REFERENCES

Deboo, G. J. and C. N. Burrows, *Integrated Circuits and Semiconductor Devices: Theory and Application.* New York: McGraw-Hill Book Company, 1971.

Hnatek, E. R., *User's Handbook of Integrated Circuits.* New York: John Wiley and Sons, Inc., 1973.

Stern, L. S., *Fundamentals of Integrated Circuits.* New York: Hayden Book Company, Inc., 1968.

APPENDICES

Appendix I

Periodic Table of the Elements

Much valuable information about individual elements can be obtained from the periodic table of elements. For example, if we look at the element *carbon* in the table, we find that it is listed under the column numbered IV. This indicates that the carbon element has 4 valence electrons. Similarly, *nitrogen* in column V has 5 valence electrons. Referring to carbon again, we see that it is in the row numbered 2, indicating that the carbon atom has 2 shells of electrons. Similarly, *silicon* in the row numbered 3 has 3 shells of electrons.

The individual element squares contain information on atomic number, weight and electron-shell configurations. The carbon square is repeated in Figure I.1. The number 6 represents the atomic number, C is the symbol for carbon and 12 is the atomic weight (rounded off). The numbers 2–4 indicate the arrangement of electrons in the $K, L, M, N \ldots$ shells around the carbon nucleus. For carbon the K-shell contains 2 electrons and the L-shell contains 4 electrons. All the remaining shells are empty. Using this information, we can now construct the model of the carbon atom. The carbon nucleus has 12 particles since its atomic weight is 12. With an atomic number of 6, there are 6 protons in the nucleus along with 6 neutrons. The complete carbon model is shown in Figure I.2.

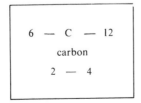

Figure I.1.

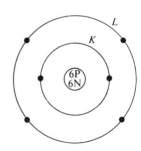

Figure I.2.

PERIODIC TABLE

Max. No. Electrons Per Shell	Valence Shells	I	II	III	IV	V
2	1	1–H–1 Hydrogen				
8	2	3–Li–7 Lithium	4–Be–9 Beryllium	5–B–11 Boron 2–3	6–C–12 Carbon 2–4	7–N–14 Nitrogen
18	3	11–Na–23 Sodium	12–Mg–24 Magnesium	13–Al–27 Aluminum 2–8–3	14–Si–28 Silicon 2–8–4	15–P–31 Phosphorus
32	4	19–K–39 Potassium	20–Ca–40 Calcium	21–Sc–45 Scadium	22–Ti–48 Titanium	23–V–51 Vanadium
		29–Cu–63 Copper	30–Zn–64 Zinc 2–8–18–2	31–Ga–69 Gallium 2–8–18–3	32–Ge–74 Germanium 2–8–18–4	33–As–75 Arsenic 2–8–18–5
32	5	37–Rb–85 Rubidium	38–Sr–88 Strontium	39–Y–89 Yttrium	40–Zr–90 Zirconium	41–Nb–93 Niobium
		47–Ag–107 Silver	48–Cd–114 Cadmium	49–In–115 Indium 2–8–18–18–3	50–Sn–120 Tin	51–Sb–121 Antimony 2–8–18–18–5
	6	55–Cs–133 Cesium	56–Ba–138 Barium	57–La–139 Lanthanum	72–Hf–180 Hafnium	73–Ta–181 Tantalum
		79–Au–197 Gold	80–Hg–202 Mercury	81–Ti–205 Tallium	82–Pb–206 Lead	83–Bi–209 Bismuth
	7	87–Fr–223 Francium	88–Ra–226 Radium	89–Ac–227 Actinium		

OF THE ELEMENTS

VI	VII		VIII		0
					2–He–4 Helium
8–O–16 Oxygen	9–F–19 Fluorine				10–Ne–20 Neon
16–S–32 Sulfur	17–Cl–35 Chlorine				18–A–40 Argon
24–Cr–52 Chromium	25–Mn–55 Manganese	26–Fe–56 Iron	27–Co–59 Cobalt	28–Ni–58 Nickel	
34–Se–80 Selenium 2–8–18–6	35–Br–79 Bromine				36–Kr–84 Krypton
42–Mo–09 Molybdenum	43–Tc–99 Technetium	44–Ru–102 Ruthenium	45–Rh–103 Rhodium	46–Pd–106 Palladium	
52–Te–130 Tellurium	53–I–127 Iodine				54–Xe–132 Xenon
74–W–184 Wolfram (Tungsten)	75–Re–187 Rhenium	76–Os–192 Osmium	77–Ir–193 Iridium	78–Pt–195 Platinum	
84–Po–210 Polonium	85–At–210 Astatine				86–Rn–222 Radon

Appendix II

Manufacturers' Data Sheets

The material presented on the following pages is done so through the courtesy of the following:

(a) Motorola Semiconductor Products Inc.
(b) Texas Instruments Incorporated.
(c) General Electric Company.
(d) Radio Corporation of America Laboratories.

—— Silicon Zener Diodes ——

1N3821 thru 1N3830

1 W
3.3 — 7.5 V

1N3821A thru 1N3828A USN/JAN

1N3821A thru 1N3828A HI-REL

CASE 52

Low-voltage, alloy-junction zener diodes in hermetically sealed package with cathode connected-to-case. Available as standard industrial types as well as for military and high-reliability applications.

MAXIMUM RATINGS

Junction and Storage Temperature: -65°C to +175°C.
D-C Power Dissipation: 1 Watt.(Derate 6.67 mW/°C above 25°C)

The type numbers shown have a standard tolerance on the nominal zener voltage of ±10%. A standard tolerance of ±5% on individual units is also available and is indicated by suffixing "A" to the standard type number.

ELECTRICAL CHARACTERISTICS (25°C Ambient $V_F = 1.5$ V @ $I_F = 200$ mA for all units)

TYPE NO.	Nominal Zener Voltage @ I_{ZT} (V_Z) Volts	Test Current I_{ZT} mA	Max Zener Impedance		Max DC Zener Current I_{ZM} mA	$I_R = 10\mu A$ Max @ Reverse Voltage V_R	Typical Zener Voltage/temp. Coeff. %/°C
			Z_{ZT} @ I_{ZT} ohms	Z_{ZK} @ $I_{ZK}=1.0$mA ohms			
1N3821	3.3	76	10	400	276	1	-.075
1N3821A	3.3	76	10	400	276	1	-.075
1N3822	3.6	69	10	400	252	1	-.065
1N3822A	3.6	69	10	400	252	1	-.065
1N3823	3.9	64	9	400	238	1	-.055
1N3823A	3.9	64	9	400	238	1	-.055
1N3824	4.3	58	9	400	213	1	-.040
1N3824A	4.3	58	9	400	213	1	-.040
1N3825	4.7	53	8	500	194	1	-.020
1N3825A	4.7	53	8	500	194	1	-.020
1N3826	5.1	49	7	550	178	1	+.005
1N3826A	5.1	49	7	550	178	1	+.005
1N3827	5.6	45	5	600	162	2	+.020
1N3827A	5.6	45	5	600	162	2	+.020
1N3828	6.2	41	2	700	146	3	+.035
1N3828A	6.2	41	2	700	146	3	+.035
1N3829	6.8	37	1.5	500	133	3	+.040
1N3829A	6.8	37	1.5	500	133	3	+.040
1N3830	7.5	34	1.5	250	121	3	+.045
1N3830A	7.5	34	1.5	250	121	3	+.045

Courtesy of Motorola Semiconductor Products Inc.

TUNNEL DIODE SPECIFICATIONS

1N3150

Outline Drawing No. 1

The 1N3150 is a germanium tunnel diode which makes use of the quantum mechanical tunneling phenomenon thereby attaining a unique negative conductance characteristic and very high frequency performance. This device is designed for low level switching and small signal applications with frequency capabilities up to 1.3 Kmc. It features closely controlled peak point current, good temperature stability and extreme resistance to nuclear radiation.

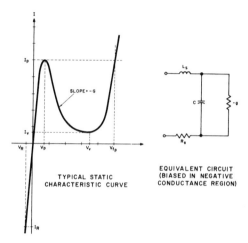

TYPICAL STATIC
CHARACTERISTIC CURVE

EQUIVALENT CIRCUIT
(BIASED IN NEGATIVE
CONDUCTANCE REGION)

SPECIFICATIONS

ABSOLUTE MAXIMUM RATINGS: (25°C)

Current

Forward (−55 to +100°C)		100	ma
Reverse (−55 to +100°C)		100	ma

Temperature

Storage	T_{STG}	−55 to +100	°C
Operating Junction	T_J	−55 to +100	°C
Lead Temperature 1/16″ ±1/32″ From Case for 10 seconds	T_L	260	°C

ELECTRICAL CHARACTERISTICS: (25°C) (1/8″ Leads)

		Min.	Typ.	Max.	
Peak Point Current	I_p	20	22	24	ma
Valley Point Current	I_v		2.9	4.80	ma
Peak Point Voltage	V_p		60		mv
Valley Point Voltage	V_v		350		mv
Reverse Voltage ($I_R = 22$ ma)	V_r			30	mv
Forward Peak Point Current Voltage	V_{fp}	450	500	600	mv
Peak Point Current to Valley Point Current Ratio	I_p/I_v		8		
Negative Conductance	$-G$		100×10^{-3}		mho
Total Capacity	C		60	150	pf
Series Inductance	L_s*		6		nh
Series Resistance	R_s		.15	1.0	ohm

*Inductance will vary 1-12 nh (10^{-9} henries) depending on lead length.

Courtesy of General Electric Company

PHOTOCELLS

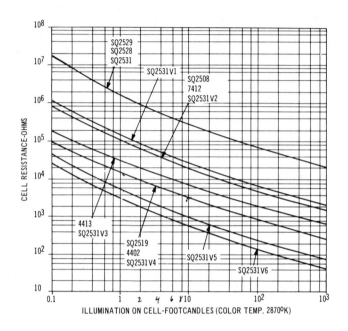

DC Cell Resistance of 1/4" – Diameter Broad-Area Photoconductive
Cells as a Function of Cell Illumination

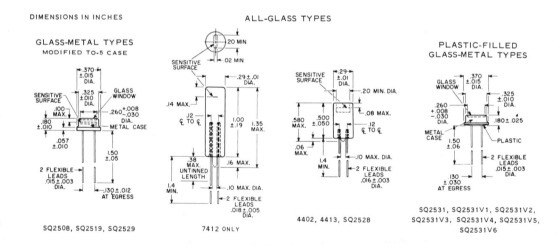

Courtesy of Radio Corporation of America Laboratories

PHOTOCELLS

1/4"-Diameter Broad-Area Cadmium-Sulfide Photoconductive Cells

RCA TYPES				MAXIMUM RATINGS			CHARACTERISTICS AT 25° C				
Glass-Metal Types[a]	All-Glass Types[a]	Plastic-Filled Glass-Metal Types[b]	Spectral Response	Voltage Between Terminals DC or Peak AC volts	Power Dissipation[c] watt	Photo-current ma	Voltage Between Terminals dc volts	Illumination[d] foot-candles	Photocurrent[e] ma		Max. Decay Current[f] μa
									Min.	Max.	
SQ2529	SQ2528	SQ2531	S-15	300	0.05	5	12	1	0.004	0.012	0.1
–	–	SQ2531V1	S-15	200	0.05	5	12	1	0.04	0.12	1
SQ2508	7412	SQ2531V2	S-15	200	0.05	5	12	1	0.065	0.275	1
–	4413	SQ2531V3	S-15	110	0.05	5	12	10	1.4	2.75	12
SQ2519	4402	SQ2531V4	S-15	300	0.05	5	12	10	1.6[g]	–	12
–	–	SQ2531V5	S-15	110	0.05	7	12	1	1	3	15
–	–	SQ2531V6	S-15	110	0.05	7	12	1	1.6	4.8	15

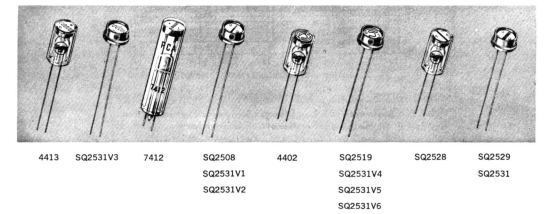

4413	SQ2531V3	7412	SQ2508	4402	SQ2519	SQ2528	SQ2529
			SQ2531V1		SQ2531V4		SQ2531
			SQ2531V2		SQ2531V5		
					SQ2531V6		

a The maximum ambient operating temperature range for these cells is −75° C to +60° C.

b The maximum ambient operating temperature range for these cells is −40°C to +60° C.

c In continuous service with sensitive surface of cell fully illuminated. The dissipation rating applies up to the maximum ambient operating temperature.

d For conditions where the light source is a tungsten-filament lamp operated at a color temperature of 2870° K.

e This characteristic is determined after the cell has been exposed for a period of 16 to 24 hours to 500 footcandle illumination (white fluorescent light).

f Measured 10 seconds after removal of incident-illumination level.

g This characteristic is determined after the cell has been exposed for a period of 16 to 24 hours to 50 to 100 footcandle illumination (white fluorescent light).

Courtesy of Radio Corporation of America Laboratories

TYPES 2N404, 2N404A
P-N-P ALLOY-JUNCTION GERMANIUM TRANSISTORS

High-Frequency Transistors for Computer and Switching Applications

Close parameter control and the JEDEC TO-5 welded package ensure device reliability and stable characteristics

environmental tests

To ensure maximum reliability, stability, and long life, all units are aged at 100°C for 100 hours minimum prior to electrical characterization. All transistors are thoroughly tested for complete adherence to specified design characteristics. In addition, continuous qualification tests are made comprising temperature-humidity cycling, shock, and vacuum leak testing under rigid in-process control procedures.

mechanical data

Metal case with glass-to-metal hermetic seal between case and leads. Unit weight is approximately 1 gram. These units meet JEDEC TO-5 registration.

All leads insulated from the case.

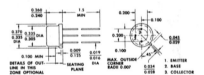

ALL DIMENSIONS IN INCHES

*absolute maximum ratings at 25°C free–air temperature (unless otherwise noted)

	2N404	2N404A
Collector-Base Voltage	25 v	40 v
Collector-Emitter Voltage (see note 1)	24 v	35 v
Emitter-Base Voltage	12 v	25 v
Collector Current	100 ma	150 ma
Emitter Current	100 ma	150 ma
Total Device Dissipation (see note 2)	150 mw	150 mw
Operating Collector Junction Temperature	85°C	100°C
Storage Temperature Range	−65°C to	−65°C to
	+100°C	+100°C

NOTES: 1. Punch-through voltage.
        2. For 2N404 derate linearly to 85°C free-air temperature at the rate of 2.5 mw/°C;
          For 2N404A derate linearly to 100°C free-air temperature at the rate of 2.0 mw/°C.
*Indicates JEDEC registered data.
        The maximum power dissipation at 25°C case temperature is 300 mw.

Courtesy of Texas Instruments Incorporated

TYPES 2N404, 2N404A
P-N-P ALLOY-JUNCTION GERMANIUM TRANSISTORS

electrical characteristics at 25°C free-air temperature (unless otherwise noted)

parameter		test conditions	2N404			2N404A			unit
			min	typ	max	min	typ	max	
I_{CBO}	Collector cutoff current	$V_{CB} = -12$ v, $I_E = 0$	—	−1	−5*	—	−1	−5*	μa
		$V_{CB} = -12$ v, $I_E = 0$ $T_A = 80°C$	—	−40	−90*	—	−40	−90*	μa
I_{EBO}	Emitter cutoff current	$V_{EB} = -2.5$ v, $I_C = 0$	—	−1	−2.5*	—	−1	−2.5*	μa
BV_{CBO}	Collector-base breakdown voltage	$I_C = -20\mu a$, $I_E = 0$	−25*	—	—	−40*	—	—	v
BV_{EBO}	Emitter-base breakdown voltage	$I_E = -20\mu a$, $I_C = 0$	−12*	—	—	−25*	—	—	v
h_{FE}	DC forward current transfer ratio	$V_{CE} = -0.15$ v, $I_C = -12$ ma	30	100	—	30	100	—	—
		$V_{CE} = -0.20$ v, $I_C = -24$ ma	24	110	—	24	110	—	—
V_{BE}	Base-emitter voltage	$I_B = -0.4$ ma, $I_C = -12$ ma	—	−0.26	−0.35*	—	−0.26	−0.35*	v
		$I_B = -1$ ma, $I_C = -24$ ma	—	−0.30	−0.40*	—	−0.30	−0.40*	v
$V_{CE(sat)}$	Collector-emitter saturation voltage	$I_B = -0.4$ ma, $I_C = -12$ ma	—	−0.08	−0.15*	—	−0.08	−0.15*	v
		$I_B = -1$ ma, $I_C = -24$ ma	—	−0.08	−0.20*	—	−0.08	−0.20*	v
V_{pt}	Punch-through voltage†	$V_{EBfl} = -1$ v	−24*	—	—	—	—	—	v
V_{EBfl}	Emitter-base floating potential	$V_{CB} = -35$ v	—	—	—	—	−0.2	−1*	v
h_{fe}	AC common-emitter forward current transfer ratio	$V_{CE} = -6$ v, $I_C = -1$ ma $f = 1$ kc	—	135	—	—	135	—	—
h_{ie}	AC common-emitter input impedance	$V_{CE} = -6$ v, $I_C = -1$ ma $f = 1$ kc	—	4	—	—	4	—	Kohm
h_{oe}	AC common-emitter output admittance	$V_{CE} = -6$ v, $I_C = -1$ ma $f = 1$ kc	—	50	—	—	50	—	μmho
h_{re}	AC common-emitter reverse voltage transfer ratio	$V_{CE} = -6$ v, $I_C = -1$ ma $f = 1$ kc	—	7×10^{-4}	—	—	7×10^{-4}	—	—
C_{ob}	Common-base output capacitance	$V_{CB} = -6$ v, $I_E = 0$ $f = 1$ mc	—	9	20*	—	—	—	pf
		$V_{CB} = -6$ v, $I_E = 1$ ma $f = 2$ mc	—	—	—	—	9	20*	pf
f_{hfb}	Common-base alpha cutoff frequency	$V_{CB} = -6$ v, $I_E = 1$ ma	4*	12	—	4*	12	—	mc

†V_{pt} is determined by measuring the emitter-base floating potential V_{EBfl} using a voltmeter with 11 megohms minimum input impedance. The collector-base voltage, V_{CB}, is increased until $V_{EBfl} = -1$ v; this value of $V_{CB} = (V_{pt} + 1)$. Care must be taken not to exceed maximum collector-base voltage specified under maximum ratings.

switching characteristics at 25°C free-air temperature

parameter		test conditions	2N404			2N404A			unit
			min	typ	max	min	typ	max	
t_d	Delay time	See Circuit 1	—	0.14	—	—	0.15	—	μsec
t_r	Rise time	See Circuit 1	—	0.20	—	—	0.27	—	μsec
t_s	Storage time	See Circuit 1	—	0.38	—	—	0.38	—	μsec
t_f	Fall time	See Circuit 1	—	0.19	—	—	0.24	—	μsec
Q_{sb}	Stored base charge	See Circuit 2	—	800	1400*	—	800	1400*	pcb

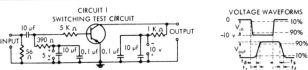

CIRCUIT 1
SWITCHING TEST CIRCUIT

VOLTAGE WAVEFORMS

NOTES: 1. Input pulse supplied by generator with following characteristics:
a. Output impedance: 50 ohms
b. Repetition rate: 1 KC
c. Rise and fall time: 20 nanoseconds maximum
2. Waveforms monitored on scope with following characteristics:
a. Input resistance — 10 megohms minimum
b. Input capacitance — 15 pf maximum
c. Risetime — 15 nanoseconds maximum
3. All resistors ±1% tolerance.

Courtesy of Texas Instruments Incorporated

TYPES 2N404, 2N404A
P-N-P ALLOY-JUNCTION GERMANIUM TRANSISTORS

COMMON-EMITTER COLLECTOR CHARACTERISTICS ... AS MEASURED ON TEKTRONIX 575 CURVE TRACER

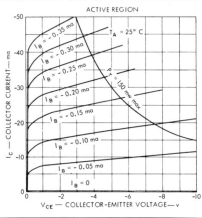

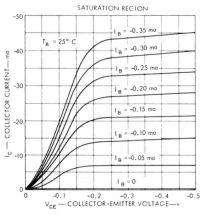

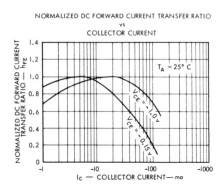

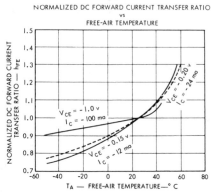

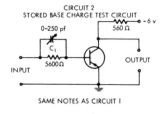

CIRCUIT 2
STORED BASE CHARGE TEST CIRCUIT

VOLTAGE WAVEFORMS

MEASUREMENT PROCEDURE

C_1 is increased until the t_{off} time of the output waveform is decreased to 0.2 μsec. Q_{sb} is then calculated by $Q_{sb} = C_1 V_{in}$.

SAME NOTES AS CIRCUIT I

Courtesy of Texas Instruments Incorporated

TYPES 2N404, 2N404A
P-N-P ALLOY-JUNCTION GERMANIUM TRANSISTORS

TYPICAL CHARACTERISTICS

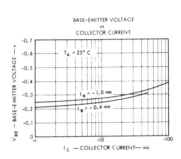

BASE-EMITTER VOLTAGE
vs
COLLECTOR CURRENT

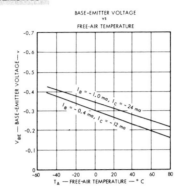

BASE-EMITTER VOLTAGE
vs
FREE-AIR TEMPERATURE

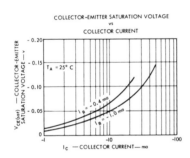

COLLECTOR-EMITTER SATURATION VOLTAGE
vs
COLLECTOR CURRENT

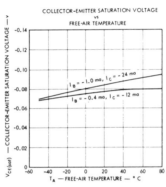

COLLECTOR-EMITTER SATURATION VOLTAGE
vs
FREE-AIR TEMPERATURE

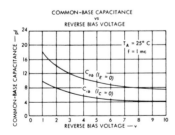

COMMON-BASE CAPACITANCE
vs
REVERSE BIAS VOLTAGE

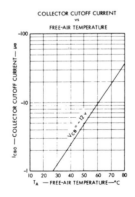

COLLECTOR CUTOFF CURRENT
vs
FREE-AIR TEMPERATURE

Courtesy of Texas Instruments Incorporated

TYPES 2N250A AND 2N251A
P-N-P ALLOY-JUNCTION GERMANIUM POWER TRANSISTORS

INTERMEDIATE-POWER TRANSISTORS
FOR
MILITARY AND COMMERCIAL APPLICATIONS

environmental tests

To ensure maximum integrity, stability and long life, all finished devices are heat aged at $+110°C$ for 100 hours and temperature cycled from $-55°C$ to $+100°C$ for four complete cycles over an 8-hour period prior to thorough testing for rigid adherence to specified characteristics.

mechanical data

The use of silver alloy to assemble the mounting base and the use of resistance welding to seal the can, provides a hermetically sealed enclosure. During the assembly process the absence of flux, combined with extreme cleanliness, prevents sealed-in contamination.
The mounting base provides an excellent heat path from the collector junction to a heat sink which must be in intimate contact to permit operation at maximum rated dissipation. The approximate weight of the unit is 18 grams.

*The transistors are in a JEDEC TO-3 case.

* THE COLLECTOR IS IN ELECTRICAL CONTACT WITH THE CASE

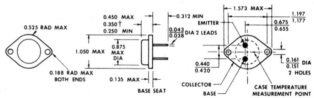

† THE TI SEATED HEIGHT IS 0.350 INCHES MAXIMUM.

DIMENSIONS ARE IN INCHES

absolute maximum ratings at 25°C case temperature (unless otherwise noted)

	2N250A	2N251A
Collector-Base Voltage	40 v	60 v
Collector-Emitter Voltage (see Note 1)	35 v	55 v
Emitter-Base Voltage	← 20 v →	
Collector Current	← 7 a →	
Base Current	← 2 a →	
Total Device Dissipation at (or below) 25°C Case Temperature (see Note 2) . .	← 90 w →	
Operating Collector Junction Temperature	← $+100°C$ →	
Storage Temperature Range	← -55 to $+100°C$ →	

*Indicates JEDEC registered data.

NOTES: 1. This value applies when base-emitter voltage, $V_{BE} = +0.2$ v.
    2. Derate linearly to $+100°C$ case temperature at the rate of 1.2 w/C°.

Courtesy of Texas Instruments Incorporated

TYPES 2N250A AND 2N251A
P-N-P ALLOY-JUNCTION GERMANIUM POWER TRANSISTORS

electrical characteristics at 25°C case temperature (unless otherwise noted)

PARAMETER		TEST CONDITIONS	TYPE	MIN.	MAX.	UNIT
BV_{CBO}	Collector-Base Breakdown Voltage	$I_C = -2$ ma, $\quad I_E = 0$	2N250A 2N251A	-40 -60		v
BV_{CEO}	Collector-Emitter Breakdown Voltage	$I_C = -500$ ma, $I_B = 0$ (see note 3)	2N250A 2N251A	-25^* -35^*		v
BV_{CEX}	Collector-Emitter Breakdown Voltage	$I_C = -2$ ma, $\quad V_{BE} = +0.2$ v	2N250A 2N251A	-35 -55		v
BV_{EBO}	Emitter-Base Breakdown Voltage	$I_E = -2$ ma, $\quad I_C = 0$	All	-20^*		v
I_{CBO}	Collector Cutoff Current	$V_{CB} = -10$ v, $\quad I_E = 0$	All		-500	μa
I_{CBO}	Collector Cutoff Current	$V_{CB} = -30$ v, $\quad I_E = 0$ $V_{CB} = -60$ v, $\quad I_E = 0$	2N250A 2N251A		-1.0^* -2.0^*	ma
I_{CBO}	Collector Cutoff Current	$V_{CB} = -20$ v, $\quad I_E = 0$ $T_C = 70°C$ $V_{CB} = -30$ v, $\quad I_E = 0$ $T_C = 70°C$	2N250A 2N251A		-5^* -5^*	ma
h_{FE}	Static Forward Current Transfer Ratio	$V_{CE} = -1.5$ v, $\quad I_C = -500$ ma	All	35^*		—
h_{FE}	Static Forward Current Transfer Ratio	$V_{CE} = -1.5$ v, $\quad I_C = -3$ a	All	25^*	100^*	—
V_{BE}	Base-Emitter Voltage	$I_B = -150$ ma, $\quad I_C = -3$ a	All		-1.0^*	v
$V_{CE(sat)}$	Collector-Emitter Saturation Voltage	$I_B = -150$ ma, $\quad I_C = -3$ a	All		-0.7^*	v
$\|h_{fe}\|$	Small-Signal Common-Emitter Forward Current Transfer Ratio	$V_{CE} = -2$ v, $\quad I_C = -1$ a, $\quad f = 80$ kc	All	2^*		—

NOTE: 3. If the transistor is tested without a heat sink, perform this test with a 100 msec current pulse and a duty cycle less than 2%.

*Indicates JEDEC registered data.

Courtesy of Texas Instruments Incorporated

TYPES 2N250A AND 2N251A
P-N-P ALLOY-JUNCTION GERMANIUM POWER TRANSISTORS

TYPICAL CHARACTERISTICS

COMMON-EMITTER COLLECTOR CHARACTERISTICS

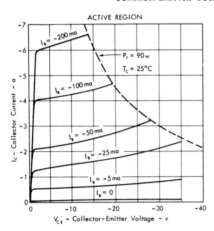

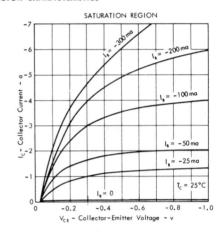

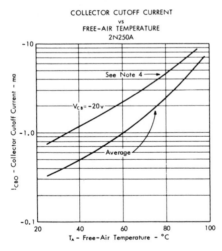

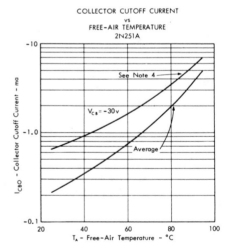

NOTE: 4. Typically, I_{CBO} for 90% of all units should be less than this value.

Courtesy of Texas Instruments Incorporated

TYPES 2N250A AND 2N251A
P-N-P ALLOY-JUNCTION GERMANIUM POWER TRANSISTORS

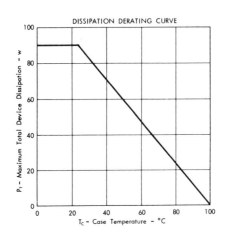

DISSIPATION DERATING CURVE

STATIC FORWARD CURRENT TRANSFER RATIO
vs
COLLECTOR CURRENT

$T_C = 25°C$

$V_{CE} = -1.5v$

See Note 5

P_T – Maximum Total Device Dissipation – w

T_C – Case Temperature – °C

h_{FE} – Static Forward Current Transfer Ratio

I_C – Collector Current – a

NOTE: 5. Typically, h_{FE} for 90% of all units should be greater than this value.

TYPICAL APPLICATION DATA

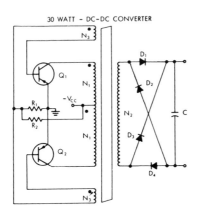

30 WATT – DC-DC CONVERTER

Performance Characteristics at 30 Watt Power
Output at T_A = 25°C

Input Current	3a
Output Voltage	290v
Overall Efficiency	83%
Input Voltage	12v
Frequency of Oscillation	350 cps

Q_1 & Q_2	=	TI 2N250A or TI 2N251A
C	=	4 μf
R_1	=	15 Ω, 2w
R_2	=	1500 Ω, 1/2w
D_1	=	1N2071
D_2	=	1N2071
D_3	=	1N2071
D_4	=	1N2071
Core	–	Magnetic Metals #500 172A or Equivalent
N_1	=	78i #16 AWG Bifilar Wound
N_2	=	2000T #29 AWG
N_3	=	30T #29 AWG
Note:		All resistance values ± 5% tolerance.

Courtesy of Texas Instruments Incorporated

Silicon
Unilateral Switch

| D13D1 |

(SUS)

The General Electric D13D1 is a diode thyristor with electrical characteristics that closely approximate those of an "ideal" four layer diode. The device uses a silicon monolithic integrated structure to achieve an 8 volt switching voltage, an on voltage of 1.75 volts at 200ma and temperature coefficient of switching voltage of less than 0.05%/°C. A gate lead is provided to give access to the circuit between the top of the zener and the base of the PNP.

Silicon Unilateral Switches are specifically designed and characterized for use in monostable and bistable applications where stability of the switching voltage is required over wide temperature variations. They are ideally suited for telephone switching, SCR triggering and for a variety of logic and memory applications.

EQUIVALENT CIRCUIT

DIMENSIONS WITHIN JEDEC OUTLINE TO-18

absolute maximum ratings:

25°C (unless otherwise specified)

Storage Temperature Range	-65 to $+200$	°C
Junction Temperature Range	-55 to $+150$	°C
Power Dissipation*	350	mw
Peak Reverse Voltage	-30	volts
DC Forward Current *	200	ma
Peak Recurrent Forward Current (1% duty cycle, 10 μsec pulse width, $T_A = 100$°C)	1.0	amp
Peak Non-Recurrent Forward Current (10 μsec pulse width, $T_A = 25$°C)	5.0	amps

*Derate linearly to zero at 150°C.

CIRCUIT SYMBOL

NOTE 1: Lead diameter is controlled in the zone between .050 and .250 from the seating plane. Between .250 and end of lead a max. of .021 is held

NOTE 2: Leads having maximum diameter (.019) measured in gaging plane .054 + .001 — .000 below the seating plane of the device shall be within .007 of true position relative to a maximum width tab.

NOTE 3: Measured from max diameter of the actual device

ALL DIMEN. IN INCHES AND ARE REFERENCE UNLESS TOLERANCED

electrical characteristics:

25°C (unless otherwise specified)

STATIC

	MIN.	TYP.	MAX.	UNITS
Forward Switching Voltage, V_S	6	8	10	V
Forward Switching Current, I_S			500	μa
Holding Current, I_H			1.5	ma
Reverse Current				
($V_R = -30V$ @ $T_A = 25$°C)			.1	μa
($V_R = -30V$ @ $T_A = 100$°C)			10.0	μa
Forward Current (off state)				
($V_F = 5V$ @ $T_A = 25$°C)			1.0	μa
($V_F = 5V$ @ $T_A = 100$°C)			20.0	μa
Forward Voltage Drop (on state)				
($I_F = 200$ ma)			1.75	V
Temperature Coefficient of Switching Voltage ($T_A = -55$°C to $+100$°C)			$\pm.05$	%/°C

DYNAMIC

	MIN.	TYP.	MAX.	UNITS
Turn-on Time, t_{on} (See Circuit 1)			1.0	μsec
Turn-off Time, t_{off} (See Circuit 2)			25.0	μsec
Peak Pulse Voltage (See Circuit 3)	3.5			V
Capacitance (0V., f = 1 MHz)		2.5		pF

Courtesy of General Electric Company

C30, C32

**Light Industrial and Consumer
Medium Current SCR
25 Amperes RMS Max.
Outline Drawing No. 4**

The C32 Silicon Controlled Rectifier is a three-junction semiconductor device for use in power switching and control applications requiring a blocking voltage of 400 volts or less and average load currents (full-wave rectified) up to 16 amperes. The C32 is provided in the popular "Press Fit" cup housing to insure ease of installation in large volume applications. Low cost makes this device suitable for high volume consumer and Light Industrial applications.

The C30 is the same as the C32 except that it is mounted on a $\frac{7}{16}$" hex and $\frac{1}{4}$"-28 stud as an added convenience to those manufacturers with production facilities more adaptable to "nut and bolt" type assembly.

- Flexibility of Mounting (Available in cup construction to solder, glue, or press fit, and as a stud device)
- One-piece Terminals
- High Surge Current Capabilities
- Low Power Required for Triggering

Type	Minimum Forward Breakover Voltage, $V_{(BR)FO}$*	Repetitive Peak Reverse Voltage V_{ROM} (rep)*	Nonrepetitive Peak Reverse Voltage (<5.0 Millisec) V_{ROM} (non-rep)*
C30U, C32U	25 V	25 V	35 V
C30F, C32F	50 V	50 V	75 V
C30A, C32A	100 V	100 V	150 V
C30B, C32B	200 V	200 V	300 V
C30C, C32C	300 V	300 V	400 V
C30D, C32D	400 V	400 V	500 V

MAXIMUM ALLOWABLE RATINGS

Repetitive Peak Forward Blocking Volatge, PFV...500 volts
RMS Forward Current, On-State...25.0 amperes (all conduction angles)
Average Forward Current, On-State...........13 amperes at 75°C case (half wave rectified sine wave)
19 amperes at 75°C case (full wave rectified sine wave)
Peak One Cycle Surge Forward Current I_{FM} (surge)...225 amperes
Peak Gate Power Dissipation, P_{GM}...5.0 watts
Average Gate Power Dissipation, $P_{G(AV)}$...0.5 watts
Peak Reverse Gate Voltage, V_{GRM}...5.0 volts
Max. Storage Temperature, T_{stg}...100°C
Max. Operating Junction Temperature...100°C
Stud Torque (C30 Only)...25 inch-pounds

CHARACTERISTICS

Test	Symbol	Min.	Typ.	Max.	Units	Test Conditions
Peak Reverse and Forward Blocking Current*	I_{ROM} and I_{FOM}					$T_J = 100°C$
C30U, C32U		—	1.0	10.0	ma	$V_{ROM} = V_{FOM} = $ 25 volts peak
C30F, C32F		—	1.0	10.0	ma	= 50 volts peak
C30A, C32A		—	1.0	7.0	ma	=100 volts peak
C30B, C32B		—	1.0	3.5	ma	=200 volts peak
C30C, C32C		—	1.0	2.3	ma	=300 volts peak
C30D, C32D		—	1.0	1.7	ma	=400 volts peak
Gate Trigger Current	I_{GT}	—	4.0	25.0	mAdc	$T_J = 25°C$, $V_{FX} = 6Vdc$, $R_L = 60$ ohms
Gate Trigger Voltage	V_{GT}	— 0.2	0.8 0.5	1.5 —	Vdc Vdc	$T_J = 25°C$, $V_{FX} = 6Vdc$, $R_L = 60$ ohms $T_J = 100$ C, $V_{FXM} =$ rated, $R_L = 1000$ ohms
Peak on Voltage	V_{FM}	—	1.30	1.5	V	$T_J = 25°C$, $I_{FM} = 50$ A peak, single half sine wave pulse, 2.0 millisec. wide
Holding Current	I_{HO}	—	10.0	50.0	mAdc	$T_J = 25$ C, anode supply = 24 Vdc

*Values apply for zero or negative gate voltage only. Maximum case to ambient thermal resistance for which maximum V_{ROM} (rep) ratings apply equals 18°C per watt.

Courtesy of General Electric Company

2N2646 & 2N2647

Unijunction Transistors—Silicon Types
Outline Drawing No. 10

The General Electric 2N2646 and 2N2647 Silicon Unijunction Transistors have an entirely new structure resulting in lower saturation voltage, peak-point current and valley current as well as a much higher base-one peak pulse voltage. In addition, these devices are much faster switches. The 2N2646 is intended for general purpose industrial applications where circuit economy is of primary importance, and is ideal for use in triggering circuits for Silicon Controlled Rectifiers and other applications where a guaranteed minimum pulse amplitude is required. The 2N2647 is intended for applications where a low emitter leakage current and a low peak point emitter current (trigger current) are required (i.e. long timing applications), and also for triggering high power SCR's.

ABSOLUTE MAXIMUM RATINGS: (25°C)

Power Dissipation (Note 1)............300 mw
RMS Emitter Current...................50 ma
Peak Emitter Current (Note 2)........2 amperes
Emitter Reverse Voltage................30 volts
Interbase Voltage......................35 volts
Operating Temperature Range −65°C to +125°C
Storage Temperature Range.. −65°C to +150°C

ELECTRICAL CHARACTERISTICS: (25°C)

PARAMETER		2N2646			2N2647			
		Min.	Typ.	Max.	Min.	Typ.	Max.	
Intrinsic Standoff Ratio ($V_{BB}=10V$)	η	0.56	0.65	0.75	0.68	0.75	0.82	
Interbase Resistance ($V_{BB}=3V$, $I_E=0$)	R_{BBO}	4.7	7	9.1	4.7	7	9.1	KΩ
Emitter Saturation Voltage ($V_{BB}=10V$, $I_E=50$ ma)	$V_{E(SAT)}$		2			2		volts
Modulated Interbase Current ($V_{BB}=10V$, $I_E=50$ ma)	$I_{B2(MOD)}$		12			12		ma
Emitter Reverse Current ($V_{B2E}=30V$, $I_{B1}=0$)	I_{EO}		0.05	12		0.01	0.2	μa
Peak Point Emitter Current ($V_{BB}=25V$)	I_P		0.4	0.5		0.4	2	μa
Valley Point Current ($V_{BB}=20V$, $R_{B2}=100\Omega$)	I_V	4	6		8	11	18	ma
Base-One Peak Pulse Voltage (Note 3)	V_{OB1}	3.0	6.5		6.0	7.5		volts
SCR Firing Conditions (See Figure 26, back page)								

NOTES:

1. Derate 3.0 MW/°C increase in ambient temperature. The total power dissipation (available power to Emitter and Base-Two) must be limited by the external circuitry.
2. Capacitor discharge—10 μfd or less, 30 volts or less.
3. The Base-One Peak Pulse Voltage is measured in the circuit below. This specification on the 2N2646 and 2N2647 is used to ensure a

minimum pulse amplitude for applications in SCR triggering circuits and other types of pulse circuits.
4. The intrinsic standoff ratio, η, is essentially constant with temperature and interbase voltage. η is defined by the equation:

$$V_P = \eta \, V_{BB} + V_D$$

Where V_P = Peak Point Emitter Voltage
V_{BB} = Interbase Voltage
V_D = Junction Diode Drop (Approx. .5V)

FIGURE 1

FIGURE 2
Unijunction Transistor Symbol with Nomenclature used for voltage and currents.

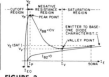

FIGURE 3
Static Emitter Characteristics curves showing important parameters and measurement points (exaggerated to show details).

G-E SCR's, RECTIFIERS, AND UNIJUNCTION TRANSISTORS WORK WELL TOGETHER!

G-E Controlled Rectifier (SCR) Type	G-E Rectifier Type	Compatible Rectifier				Compatible Unijunction Triggering Transistor
		Max On-Current $I_{F(AV)}$ at Stud Temperature	Peak 1 Cycle Sugre	Max. Rectifier Transient $V_{RM (non-rep)}$	Package	
C5, C7, 2N1595–1599	1N536–1N547	.75 A at 50°C Amb.	15 A	800 V	Double Ended Lead	2N2646
	1N2610–1N2615	.75 A at 50°C Amb.	30 A	775 V		
C8, C9	1N1341A–1N1348B	7 A at 140°C	150 A	800 V	7/16″ Hex	2N2646
C10, C11, C12	CAR A27	8.5 A at 150°C	240 A	∞	7/16″ Hex	
C20, C22, C30	A40 or A41 A44 or A45	15 A at 150°C	300 A	600 V	9/16″ Hex	2N2646
C35, C38, C40	1N1199A 1N1206A	12 A at 150°C	240 A	800 V	7/16″ Hex	2N2646
	1N2154–1N2160	25 A at 145°C	400 A	800 V	11/16″ Hex	
	CAR A38	22 A at 140°C	500 A	∞	11/16″ Hex	
C36,C37	1N1341A–1N1348A	7 A at 140°C	150 A	800 V	7/16″ Hex	2N2646
	1N1199A–1N1206A	12 A at 150°C	240 A	800 V	7/16″ Hex	2N2646
C45, C46, C50, C52, C55 C56, C60, C61, C150	1N3289–1N3293 (A70)	70 A at 150°C	1600 A	1300 V	1-1/16″ Hex	2N2647
	CAR A76	70 A at 150°C	1600 A	∞		
C80, C85	1N3736–3742 (A90)	200 A at 150°C	4500 A	1300 V	1-1/4″ Hex	2N2647
	CAR A92	200 A at 150°C	4500 A	∞		
6RW71	1N3736–3742 (A90)	200 A at 150°C	4500 A	1300 V	1-1/4″ Hex	2N2647
	CAR A92	200 A at 150°C	4500 A	∞		
	6RW62	390 A at 150°C	7000 A	1000 V	1-5/8″ Hex	2N2647

Stack assemblies available for all rectifier types listed.

Courtesy of General Electric Company

2N4220
thru
2N4222

2N4220A
thru
2N4222A

SILICON N-CHANNEL
JUNCTION FIELD-EFFECT TRANSISTORS

Depletion Mode (Type A) devices designed for general-purpose amplifier and switching applications.

- Low Transfer Capacitance — C_{rss} = 2.0 pF (Max)

- Low Input Capacitance — C_{iss} = 6.0 pF (Max)

- Low Gate Leakage Current — I_{GSS} = 100 pA (Max)

- Low Noise Figure — NF = 2.5 dB (Max) @ 100 Hz ("A" Versions)

**SILICON N-CHANNEL
JUNCTION FIELD-EFFECT
TRANSISTORS**

TYPE A

FEBRUARY 1968 — DS 5187 R1
(Replaces DS 5187)

TO-72

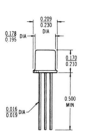

TO-72
Case 20 (3)

Drain and Source
may be interchanged

MAXIMUM RATINGS (T_A = 25°C)

Characteristic	Symbol	Rating	Unit
Drain-Source Voltage	V_{DS}	30	Vdc
Drain-Gate Voltage	V_{DG}	30	Vdc
Gate-Source Voltage	V_{GS}	30	Vdc
Drain Current	I_D	15	mAdc
Total Device Dissipation	P_D	300	mW
Derate Above 25°C		2.0	mW/°C
Operating Junction Temperature	T_J	175	°C
Storage Temperature Range	T_{stg}	-65 to +200	°C

Courtesy of Motorola Semiconductor Products Inc.

ELECTRICAL CHARACTERISTICS (T_A = 25°C unless otherwise noted)

Characteristic		Symbol	Min	Typ	Max	Unit
OFF CHARACTERISTICS						
Gate-Source Breakdown Voltage (I_G = -10 μAdc, V_{DS} = 0)		$V_{(BR)GSS}$	-30	-	-	Vdc
Gate Reverse Current (V_{GS} = -15 Vdc, V_{DS} = 0)		I_{GSS}	-	-	-0.1	nAdc
(V_{GS} = -15 Vdc, V_{DS} = 0, T_A = 150°C)			·	-	-100	
Gate-Source Voltage (I_D = 50 μAdc, V_{DS} = 15 Vdc)	2N4220, 2N4220A	V_{GS}	-0.5	-	-2.5	Vdc
(I_D = 200 μAdc, V_{DS} = 15 Vdc)	2N4221, 2N4221A		-1.0	-	-5.0	
(I_D = 500 μAdc, V_{DS} = 15 Vdc)	2N4222, 2N4222A		-2.0	-	-6.0	
Gate-Source Cutoff Voltage (I_D = 0.1 nAdc, V_{DS} = 15 Vdc)	2N4220, 2N4220A	$V_{GS(off)}$	-	-	-4.0	Vdc
	2N4221, 2N4221A		-	-	-6.0	
	2N4222, 2N4222A		-	-	-8.0	
ON CHARACTERISTICS						
Zero-Gate-Voltage Drain Current* (V_{DS} = 15 Vdc, V_{GS} = 0)	2N4220, 2N4220A	I_{DSS}*	0.5	-	3.0	mAdc
	2N4221, 2N4221A		2.0	-	6.0	
	2N4222, 2N4222A		5.0	-	15	
DYNAMIC CHARACTERISTICS						
Forward Transfer Admittance* (V_{DS} = 15 Vdc, V_{GS} = 0, f = 1.0 kHz)	2N4220, 2N4220A	$|y_{fs}|$*	1000	2500	4000	μmhos
	2N4221, 2N4221A		2000	3500	5000	
	2N4222, 2N4222A		2500	4500	6000	
Output Admittance* (V_{DS} = 15 Vdc, V_{GS} = 0, f = 1.0 kHz)	2N4220, 2N4220A	$|y_{os}|$*	-	-	10	μmhos
	2N4221, 2N4221A		-	-	20	
	2N4222, 2N4222A		-	-	40	
Drain-Source Resistance (V_{DS} = 0, V_{GS} = 0)	2N4220, 2N4220A	$r_{ds(on)}$	-	500	-	Ohms
	2N4221, 2N4221A		-	400	-	
	2N4222, 2N4222A		-	300	-	
Input Capacitance (V_{DS} = 15 Vdc, V_{GS} = 0, f = 1.0 MHz)		C_{iss}	-	4.5	6.0	pF
Reverse Transfer Capacitance (V_{DS} = 15 Vdc, V_{GS} = 0, f = 1.0 MHz)		C_{rss}	-	1.2	2.0	pF
Common-Source Output Capacitance (V_{DS} = 15 Vdc, V_{GS} = 0, f = 30 MHz)		C_{osp}	-	1.5	-	pF
Noise Figure (V_{DS} = 15 Vdc, V_{GS} = 0, R_S = 1.0 Megohm, f = 100 Hz)	2N4220A	NF	-	-	2.5	dB
	2N4221A		-	-	2.5	
	2N4222A		-	-	2.5	

*Pulse Test: Pulse Width = 630 ms, Duty Cycle = 10%

FIGURE 1 – EQUIVALENT LOW FREQUENCY CIRCUIT

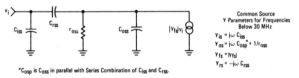

Common Source
Y Parameters for Frequencies
Below 30 MHz

$y_{is} = j\omega\, C_{iss}$
$y_{os} = j\omega\, C_{osp}^{*} + 1/r_{oss}$
$y_{fs} = |y_{fs}|$
$y_{rs} = -j\omega\, C_{rss}$

*C_{osp} is C_{oss} in parallel with Series Combination of C_{iss} and C_{rss}.

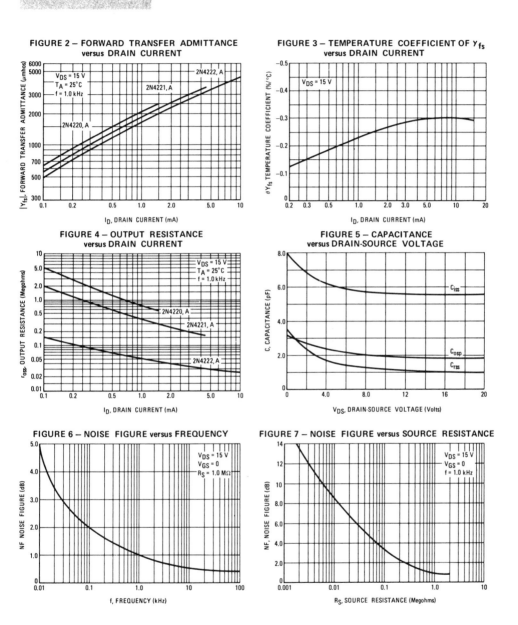

FIGURE 2 – FORWARD TRANSFER ADMITTANCE
versus DRAIN CURRENT

FIGURE 3 – TEMPERATURE COEFFICIENT OF Y_{fs}
versus DRAIN CURRENT

FIGURE 4 – OUTPUT RESISTANCE
versus DRAIN CURRENT

FIGURE 5 – CAPACITANCE
versus DRAIN-SOURCE VOLTAGE

FIGURE 6 – NOISE FIGURE versus FREQUENCY

FIGURE 7 – NOISE FIGURE versus SOURCE RESISTANCE

Courtesy of Motorola Semiconductor Products Inc.

Appendix III

Partial Table of Natural Logs ($\log_e n$)

n	$\log_e n$	n	$\log_e n$	n	$\log_e n$
	*	4.5	1.5041	9.0	2.1972
.1	7.6974	4.6	1.5261	9.1	2.2083
.2	8.3906	4.7	1.5476	9.2	2.2192
.3	8.7960	4.8	1.5686	9.3	2.2300
.4	9.0837	4.9	1.5892	9.4	2.2407
.5	9.3069	5.0	1.6094	9.5	2.2513
.6	9.4892	5.1	1.6292	9.6	2.2618
.7	9.6433	5.2	1.6487	9.7	2.2721
.8	9.7769	5.3	1.6677	9.8	2.2824
.9	9.8946	5.4	1.6864	9.9	2.2925
1.0	.0000	5.5	1.7047	10	2.3026
1.1	.0953	5.6	1.7228	i1	2.3979
1.2	.1823	5.7	1.7405	12	2.4849
1.3	.2624	5.8	1.7579	13	2.5649
1.4	.3365	5.9	1.7750	14	2.6391
1.5	.4055	6.0	1.7918	15	2.7081
1.6	.4700	6.1	1.8083	16	2.7726
1.7	.5306	6.2	1.8245	17	2.8332
1.8	.5878	6.3	1.8405	18	2.8904
1.9	.6419	6.4	1.8563	19	2.9444
2.0	.6931	6.5	1.8718	20	2.9957
2.1	.7419	6.6	1.8871	25	3.2189
2.2	.7885	6.7	1.9021	30	3.4012
2.3	.8329	6.8	1.9169	35	3.5553
2.4	.8755	6.9	1.9315	40	3.6889
2.5	.9163	7.0	1.9459	45	3.8067
2.6	.9555	7.1	1.9601	50	3.9120
2.7	.9933	7.2	1.9741	55	4.0073
2.8	1.0296	7.3	1.9879	60	4.0943
2.9	1.0647	7.4	2.0015	65	4.1744
3.0	1.0986	7.5	2.0149	70	4.2485
3.1	1.1314	7.6	2.0281	75	4.3175
3.2	1.1632	7.7	2.0412	80	4.3820
3.3	1.1939	7.8	2.0541	85	4.4427
3.4	1.2238	7.9	2.0669	90	4.4998
3.5	1.2528	8.0	2.0794	100	4.6052
3.6	1.2809	8.1	2.0919	110	4.7005
3.7	1.3083	8.2	2.1041	120	4.7875
3.8	1.3350	8.3	2.1163	130	4.8676
3.9	1.3610	8.4	2.1282	140	4.9416
4.0	1.3863	8.5	2.1401	150	5.0106
4.1	1.4110	8.6	2.1518	160	5.0752
4.2	1.4351	8.7	2.1633	170	5.1358
4.3	1.4586	8.8	2.1748	180	5.1930
4.4	1.4816	8.9	2.1861	190	5.2470

* Subtract 10 for $n < 1$. Thus $\log_e .1 = 7.6974 - 10 = -2.3026$.

Appendix IV

Vacuum Tube Devices

Since the advent of the transistor and semiconductor technology, the role of vacuum tubes in the fields of communications, radar and automatic control has been rapidly diminishing. Although they no longer enjoy the widespread use which characterized the first half of the century, vacuum tubes have maintained superiority in various areas including microwave and very high-power applications. For this reason, and because of the many similarities between vacuum tubes and certain semiconductor devices, a brief discussion of vacuum tube principles is presented below. For a more comprehensive treatment, the references cited at the end of this appendix should be consulted.

1

The first practical vacuum tube device was developed as a result of work done by Thomas Edison and J. A. Fleming at the turn of the century. The *vacuum tube diode* is comprised of a heated *cathode* surrounded by a metal *anode* (or *plate*) enclosed in an evacuated tube, usually made of glass or metal. Figure IV.1 shows a typical vacuum diode construction. In normal operation the cathode is heated to a temperature high enough to cause electrons to be emitted

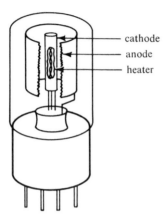

Figure IV.1. VACUUM DIODE CONSTRUCTION

from its surface, and a positive potential is applied to the plate to attract these electrons. These electrons moving from the cathode to the plate constitute a current. This current will only flow when the plate is positive with respect to the cathode; when the plate is negative relative to the cathode, the emitted electrons are repelled by the plate and no current flows.

Because it must be an efficient emitter of electrons, the cathode of a vacuum tube must be constructed of special material. Although free electrons in any conductor will escape from the solid if given enough heat energy, some materials are more suitable because of a greater emission efficiency or because of a greater resistance to high temperatures. The most modern cathode material is oxide-coated metal. The cathode consists of a metal sleeve coated with a mixture of barium and strontium oxides and is the most efficient emitter of electrons used to date. As shown in Figure IV.1, the oxide-coated cathode is usually indirectly heated by a separate heater or filament. The heater is usually heated by an AC current. The conventional circuit symbol for the vacuum tube diode is shown in Figure IV.2. Normally, the heater is omitted in the circuit symbol since it is not electrically connected with the diode.

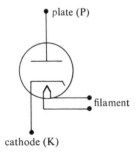

Figure IV.2. CIRCUIT SYMBOL FOR VACUUM DIODE

2

The electrical characteristics of a typical vacuum diode are shown in Figure IV.3 for two different cathode temperatures $T1$ and $T2$ (with $T1 > T2$). The cathode temperature is controlled by varying the filament power. The variation of plate current I_P with plate-to-cathode voltage E_P is similar for both temperature cases, with the curve for the higher temperature existing at higher plate currents. This seems reasonable since a higher cathode temperature will cause the emission of a larger number of free electrons and thus a greater current flow from cathode to plate.

Several significant features are evident in the curves of Figure IV.3. First, the plate current initially increases rapidly as the plate voltage is increased positively from zero. As E_P is increased further, the rate of increase in I_P becomes smaller and smaller until a region of saturation exists in which I_P remains virtually constant. This saturation region exists when all the electrons which the cathode can emit at a given temperature are collected by the plate and any further increase in plate potential cannot draw any additional electron current. Secondly, regardless of the cathode temperature, no plate current flows when E_P is negative. Thus the vacuum diode has the property of *rectification*; that is, it can conduct current in one direction only. This property is illustrated in Figure IV.4. Here, a voltage consisting of alternately positive and negative pulses is applied to a circuit consisting of a vacuum diode and a load resistor. The diode is said to *rectify* the input voltage since it allows current to flow only when the input makes the plate positive relative to the cathode, in this case when E_{in} is positive. When current flows, a voltage is developed across R_L. When E_{in} is negative the current is zero and E_L is zero.

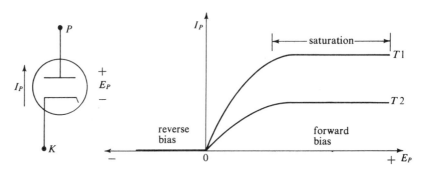

Figure IV.3.   TYPICAL VACUUM DIODE PLATE CHARACTERISTICS

It should be pointed out that in practical vacuum diodes the reverse current (plate negative relative to the cathode) is not exactly zero because of leakage currents. Typically, the diode's resistance is around 10 megohms in the reverse direction. This compares to a typical value of 100 ohms in the forward direction, making it a very good rectifier.

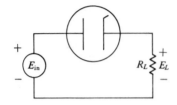

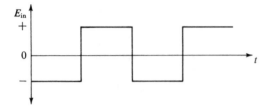

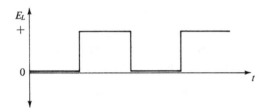

Figure IV.4.   VACUUM DIODE RECTIFICATION

3

One of the most important steps in modern electronics was the invention of the *triode*. In 1907 deForest added a third electrode to the vacuum tube diode to form a triode. This third electrode, called a *control grid*, opened up a whole new era in the field of radio communications. In the triode, the cathode and plate retain their functions as an emitter of electrons and a collector of electrons, respectively. The control grid is usually a fine wire helix located between them, nearer to the cathode, as shown in Figure IV.5.

The purpose of the control grid is to regulate the movement of electrons between the cathode and plate. The plate is still held at a positive potential as in the diode. However, unlike the diode, the flow of electrons from cathode to plate (plate current) depends not only on the plate-to-cathode voltage but also on the potential of the grid relative to the cathode. By varying the grid-to-cathode potential, the plate current may be increased or decreased even if the plate voltage is held constant. This control grid action can be better explained by referring to Figure IV.6 where the triode circuit symbol is shown with the electrode voltages properly labelled.

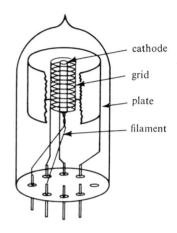

Figure IV.5.   CONSTRUCTION OF A TRIODE

Let us assume that E_P is positive since a negative E_P will produce zero plate current as in the diode. If the control grid were not present, the triode would behave essentially like the vacuum diode and a particular value of plate current would flow; call this value I_{PO}. With the grid present, as in the triode, we are at liberty to bias it at any potential relative to the cathode. With this potential, E_G, set to zero volts (that is, with $E_G = 0$), the grid will have little effect on the plate current and its value will remain I_{PO}. If the grid-to-cathode voltage is made negative ($E_G < 0$) a significant *decrease* in plate current will occur. This decrease is a result of the repelling effect which the negative grid has on the cathode-emitted electrons. As the grid is made more negative, the number of electrons which will travel from cathode to plate is decreased. In fact, a sufficiently negative value of E_G will cut off all of the plate current. In other words, even with a large plate-to-cathode positive voltage the plate current can be decreased to zero by making the grid voltage sufficiently negative. It is important to note that when E_G is negative, none of the traversing electrons collide with the grid. Instead, they are repelled by the negative potential and make their way to the plate by passing through the spacings in the grid mesh. Thus $I_G = 0$ when $E_G \leq 0$.

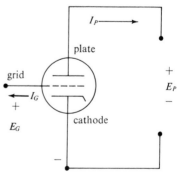

Figure IV.6.   TRIODE CIRCUIT SYMBOL

When the grid voltage is made positive, the opposite effect takes place; an increase in plate current occurs and grid current begins to flow as a result of the positive grid's attraction for the emitted electrons. That is, when $E_G > 0$ we have $I_P > I_{PO}$ and $I_G > 0$. The vacuum triode is usually not operated with a positive grid because the grid current can become excessive. There are applications, however, where the grid is allowed to go positive periodically without damaging the tube.

The above results can be summarized as in Figure IV.7, where some typical current and voltage values are also given. Notice that with $E_G = -8$ V, the plate current is decreased to zero. This condition is called *cutoff*. Notice also that a decrease in grid potential from 0 to -2 V causes a 60 percent decrease in plate current (10 mA to 4 mA). This is a good indication of the grid's control over the plate current.

Grid voltage E_G	Plate voltage E_P	Plate current I_P	Grid current I_G
No grid present	$+200$ V	10 mA	—
0 V	$+200$ V	10 mA	0 mA
-2 V	$+200$ V	4 mA	0 mA
-8 V	$+200$ V	0 mA	0 mA
$+2$ V	$+200$ V	16 mA	2 mA

Figure IV.7.   SUMMARY OF GRID CONTROL

4

Although the data in Figure IV.7 illustrates the behavior of a triode tube, it is usually necessary to use a more complete method of characterizing a given triode's operating characteristics. In particular, a set of curves, called the *plate characteristics*, are often utilized in analysis and design. Figure IV.8 contains the plate characteristics of the 6J5 triode. The plate characteristics consist of a family of curves each of which plots the variation of plate current with plate voltage for a given value of grid voltage. Let us now examine the 6J5 characteristics.

The curve labelled $E_G = 0$ is essentially the same curve that would be obtained if the grid were not present. In this case, plate current will flow as soon as the plate voltage is increased from zero. For example, with $E_P = 25$ V, a plate current of 1.7 mA flows (point A). Increasing the plate voltage to 50 V increases the current to 4.3 mA (point B). With $E_G = -2$ V the curve shifts to the right, indicating that for the same value of E_P a smaller I_P will flow than with $E_G = 0$. For example, at $E_P = 25$ V no plate current flows and, in fact, none

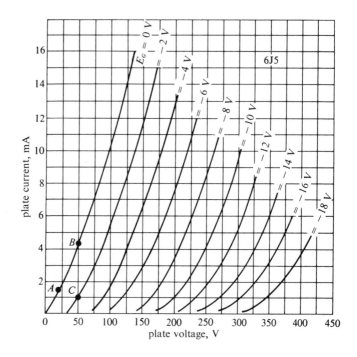

Figure IV.8. TRIODE PLATE CHARACTERISTICS

will flow until E_P is increased to about 35 volts. At $E_P = 50$ V the plate current increases to only 1 mA (point C; compare this with point B). Similar behavior takes place as E_G is made even more negative. Notice also that the separation of successive curves remains fairly uniform. This is an important feature of the triode that allows it to be used as a linear amplifier.

5

The majority of applications of the triode vacuum tube utilize its ability to *amplify* an input signal voltage. A typical 6J5 triode amplifier is shown in Figure IV.9. It consists of two bias supplies, E_{CC} and E_{BB}, a load resistor, R_L, the 6J5 triode and, of course, the input signal, e_S. The grid supply, E_{CC}, establishes the DC grid-to-cathode potential, E_G; that is, $E_G = -E_{CC}$. The plate supply, E_{BB}, provides the DC plate-to-cathode potential which allows plate current to flow. The load, R_L, is used to develop the output signal voltage across. The amplification process can easily be understood if we consider it in two separate steps. The first step is the establishment of the DC operating point (frequently called the Q point) which is determined with no signal applied ($e_S = 0$) by the values of E_{CC}, E_{BB} and R_L. Under the no-signal condition a certain DC plate current I_{PO}, plate voltage E_{PO} and grid voltage E_{GO} are thus established. This condition will persist indefinitely until a signal is introduced

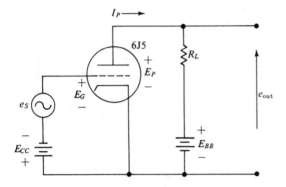

Figure IV.9.   TRIODE AMPLIFIER CIRCUIT

at the grid. The input signal, e_S, which we will consider to be pure AC, will serve to alter the grid-cathode voltage by alternately adding to and subtracting from the DC value $E_{GO} = -E_{CC}$. The resultant total grid-cathode voltage con-

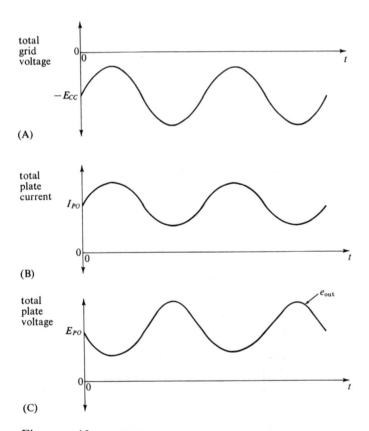

Figure IV.10.   TRIODE AMPLIFER WAVEFORMS

sists of an AC signal riding on a DC level (see Figure IV.10A). Since we have found that a change in grid voltage will produce a change in plate current, we can expect that the signal portion of the grid voltage will induce a signal portion of plate current. The resultant plate current waveform consists of the AC plate current riding on a DC level (I_{PO}) and is shown in Figure IV.10B. Notice that the plate current signal is *in phase* with the grid signal. This occurs since the portion of the grid signal which is going positive reduces the total grid-cathode negative potential, thus allowing an increase in plate current and vice versa.

Because of this AC component of plate current we can expect that the total plate voltage will also have a signal component produced across R_L. The resultant plate voltage waveform is shown in Figure IV.10C. Notice that the AC portion of plate voltage is 180° *out of phase* with the input signal. This phase shift occurs because, as the grid signal voltage goes positive, the plate current will increase, thus increasing the voltage across R_L. Since the sum of the voltage drops across the tube (E_P) and R_L must be equal to the voltage supply E_{BB}, this results in a decrease in plate voltage as the grid signal increases and vice versa, producing the 180° phase shift.

The output of the amplifier in Figure IV.9 is usually taken between plate and ground. In a well-designed amplifier the output signal e_{out} will have a magnitude several times greater than the input signal e_S. The ratio of these two signals is the voltage gain A_v. That is

$$A_v \equiv \frac{e_{out}}{e_S}$$

Typically, A_v is between 10 and 100 for this amplifier circuit. The value of A_v for a given amplifier depends on a very important property of a triode—the grid's ability to control the plate current. As we saw previously, a small variation in grid voltage can cause a relatively large change in plate current since the grid is very close to the cathode. A measure of this property is the triode parameter called *transconductance*, which is given by the symbol g_m. Mathematically, g_m is the ratio of the plate current change to the grid voltage change which produced it while holding plate voltage constant. That is,

$$g_m \equiv \frac{\Delta I_P}{\Delta E_G}\bigg|R_P = \text{constant}$$

The units of transconductance are mhos, the reciprocal of ohms. Typical values of g_m for a triode are in the range of 1 to 10 millimhos (10^{-3} mhos). The higher the value of g_m, of course, the higher the triode's amplifying capability.

Another important triode parameter is *plate resistance*, r_p. The triode plate resistance is a measure of the plate voltage's control over the plate current. Mathematically it is given as the ratio of a change in plate voltage to the resultant change in plate current for a given value of grid voltage. That is,

$$r_p \equiv \frac{\Delta E_P}{\Delta I_P}\bigg|E_G = \text{constant}$$

Typical values of r_p range from 2 kΩ to 20 kΩ. Because the control grid is much closer to the cathode than the plate, the plate's voltage control is much less than the grid's. For this reason the value of r_p is usually much greater than the *reciprocal* of g_m. We can express this as

$$r_p \gg \frac{1}{g_m}$$

or, equivalently, as

$$r_p \times g_m \gg 1$$

The product of r_p and g_m is given the special symbol μ (mu) and is called the *amplification factor*. That is,

$$\mu \equiv r_p \times g_m$$

The value of μ typically ranges from 10 to 200. It is essentially a measure of the relative control ability of the grid as compared to the plate; it is also a good indication of the voltage gain available in a given amplifier. For the amplifier in Figure IV.9, for example, the approximate value of voltage gain can be obtained from the expression

$$A_v \approx \frac{\mu R_L}{r_p + R_L}$$

This expression shows that a higher voltage gain can be obtained by using a higher μ triode or by increasing R_L.

6

A very important characteristic of a triode amplifier is its very high input impedance. As mentioned previously, with a negative grid voltage only a minute grid current flows. This means that the current drain from the signal source will be very small. A typical input impedance is 1 megohm and is determined mainly by an external grid-to-ground resistance needed for stability.

Despite its very high input impedance, the triode does not approach the ideal voltage amplifier mainly due to its high *output impedance*, which is essentially equal to r_p. However, for values of $R_L \gg r_p$ the triode is close to being a perfect voltage amplifier with $A_v \approx \mu$, independent of R_L.

7

The inherent capacitances that exist between the three electrodes form one of the principal drawbacks of the triode amplifier. Typically, in the 5 to 20 picofarad range, these *interelectrode capacitances* do not affect the amplifier operation at low and medium frequencies, but at high frequencies (above 50 kHz) their effect on input impedance and gain is noticeable.

In an effort to reduce the interelectrode capacitances, and thereby improve the high-frequency capabilities of the triode, the *tetrode* and then the *pentode* were developed. In the tetrode a *screen grid* was introduced to the basic triode construction between the control grid and plate. It did accomplish the reduction in capacitances but other undesirable effects brought about the development of the pentode. A typical example of pentode construction is shown in Figure IV.11 along with the pentode circuit symbol. A third grid, the *suppressor*

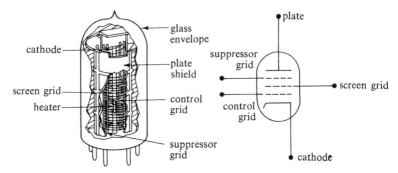

Figure IV.11.   PENTODE CONSTRUCTION AND ELECTRONIC SYMBOL

grid, is present between the screen grid and plate. The effects of the pentode construction on the electrical behavior of the tube are the lower interelectrode capacitances (fractions of pF), higher values of μ (100 to 1000) and higher values of r_p (100 kΩ to 1 megohm) than the triode. A simple pentode amplifier circuit is shown in Figure IV.12. Note that the screen grid is biased positively in relation to the cathode in order to help accelerate electrons toward the plate. The suppressor grid is normally connected to the cathode. The approximate voltage gain of this pentode amplifier is

$$A_v \approx \frac{u}{r_p} \times R_L = g_m \times R_L$$

and is typically higher than for a triode.

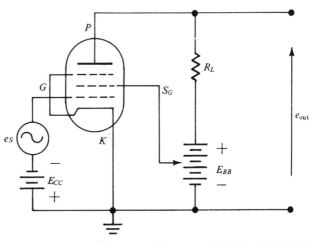

Figure IV.12.   PENTODE AMPLIFIER

A typical pentode plate family of characteristic curves is shown in Figure IV.13. In comparison to the triode curves, the pentode curves are flatter, indicating that the plate voltage has a smaller effect on plate current, which results in a higher r_p and μ. This is because of the effect of the screen and suppressor grids shielding the cathode from the plate.

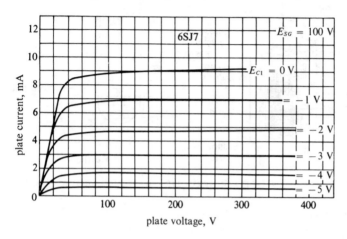

Figure IV.13.   PENTODE PLATE FAMILY

REFERENCES

Brazee, J. G., *Semiconductor and Tube Electronics.* New York. Holt, Rinehart and Winston, Inc., 1968.

Romanowitz, H. A. and R. E. Puckett, *Introduction to Electronics.* New York: John Wiley & Sons, Inc., 1968.

Appendix V

Device Symbols

Device	Circuit symbol	Alternate symbol
P-N diode; rectifier		
Zener diode		
Tunnel diode		
Photoconductive cell		
Photodiode		
Photovoltaic cell; solar cell		
LED		

Phototransistor		
NPN transistor		
PNP transistor		
Shockley (four-layer diode (SUS)		
SBS		
DIAC		
Silicon-controlled rectifier (SCR)		
Gate-controlled switch (GCS)		
Silicon-controlled switch (SCS)		
Light-activated SCR (LASCR)		
TRIAC		

Unijunction transistor (UJT)	$B2$ / E / $B1$	
Complementary UJT	$B1$ / E / $B2$	
Programmable UJT (PUT)	A G / K	
Junction field-effect transistors	D / G / N channel* S / *For P channel, reverse the arrow	
Depletion- MOSFET	D / G / S / N-channel	
Enhancement- MOSFET	D / G / S / N-channel	

Appendix VI

Checking Semiconductors
with an Ohmmeter

One of the most common ways for checking any semiconductor device is to use an ohmmeter to measure relative resistance between the device's different terminals. Ohmmeter checks are usually used to determine if the device has any gross defects (short, open, very high leakage current, etc.). In addition, the ohmmeter can often be used to determine the lead configuration of the device if it is not known; for example, to establish which lead is the emitter, which is the base and which is the collector of a transistor.

A typical ohmmeter applies a voltage through a series resistor to any device connected to its terminals. The amount of current which flows as a result is an indication of the resistance of the device. The ohmmeter scale is calibrated directly in ohms to give a reading of the device resistance. Since most semiconductor devices are polarity-sensitive, it is important to note how the ohmmeter is connected to the device. The ohmmeter terminal labeled *common*, or *ground*, is the *negative* terminal; the other one is the *positive* terminal.

The following table will describe the common ohmmeter checks for the various devices. These should be studied and understood thoroughly since they are an important part of a technician's laboratory skills.

TABLE VI.1

Device	Positive ohmmeter lead tied to	Negative lead tied to	Expected results
P-N diodes (also zener diodes, photodiodes and P-N junctions of any device)	anode (forward bias)	cathode	low resistance; usually 10–1000 Ω depending on type of diode and ohmmeter range used. Reading should be smaller on lower ohmmeter ranges.
	cathode (reverse bias)	anode	very high resistance; typically 1 MΩ or greater for germanium and 10 MΩ or greater for silicon.
Tunnel diodes	anode (forward bias)	cathode	very low resistance in either direction. Resistance is usually slightly lower when anode is negative and cathode positive.
	cathode (reverse bias)	anode	
Photoconductive cells	either end	either end	ohmmeter reading should be the same in either direction and depends on cell sensitivity and amount of ambient light. Cell resistance should increase considerably when cell is darkened.
Photodiodes, photovoltaic cells, and LEDs	SAME AS FOR P-N DIODES		
NPN transistor	emitter	base	high resistance (reverse-biased junction) unless ohmmeter voltage exceeds *E-B* breakdown voltage, BV_{EBO}.
	base	emitter	low resistance (forward-biased junction).
	collector	base	high resistance.
	base	collector	low resistance; usually not as low as *E-B* junction since collector is lightly doped.
	emitter	collector	high resistance in both directions but about 10 to 50 times less than *E-B* reverse resistance. Reading is also usually higher when emitter is negative and collector positive.
	collector	emitter	

TABLE VI.1 cont'd

Device	Positive ohmmeter lead tied to	Negative lead tied to	Expected results
PNP transistor	SAME AS FOR NPN, EXCEPT THAT ALL POLARITIES ARE REVERSED		
Four-layer diode and silicon unilateral switch (SUS)	anode (forward bias)	cathode	high resistance; > 1 MΩ.
	cathode (reverse bias)	anode	high resistance; usually greater than in other direction but may be impossible to detect with some ohmmeters.
SBS and DIAC	either end	either end	> 1 MΩ in either direction.
Silicon controlled rectifier (SCR), light-activated SCR, gate-controlled switch (GCS)	anode (forward bias)	cathode	> 1 MΩ (might be less for very high current SCRs).
	cathode (reverse bias)	anode	> 1 MΩ but usually greater than in forward direction.
	gate	cathode	Similar to P-N diode with low resistance when gate is positive and high resistance when gate is negative.
	cathode	gate	
	gate	anode	> 1 MΩ in either direction.
	anode	gate	
TRIAC	either anode 1 or 2	either anode 1 or 2	very high resistance; > 1 MΩ but may be less in very high current TRIACs.
	gate	anode 1	low resistance in both directions.
	anode 1	gate	
	gate	anode 2	high resistance in both directions.
	anode 2	gate	
Unijunction transistor	base 1	base 2	same resistance in either direction; typically 4 kΩ–10 kΩ.
	base 2	base 1	
	emitter (forward bias)	base 1	moderate resistance; usually in the 3 kΩ–15 kΩ range.
	base 1	emitter	very high resistance; > 1 MΩ.
	emitter (forward bias)	base 2	moderate resistance; usually in the 2 kΩ–10 kΩ range. *Usually less than the emitter-base 1 forward resistance.*
	base 2	emitter	very high resistance; > 1 MΩ.

TABLE VI.1 cont'd

Device	Positive ohmmeter lead tied to	Negative lead tied to	Expected results
Complementary UJT	SAME AS FOR UJT, EXCEPT THAT ALL POLARITIES ARE REVERSED		
Programmable UJT (PUT)	anode	cathode	high resistance; $> 1 \ M\Omega$.
	cathode	anode	
	anode	gate	low resistance (forward bias).
	gate	anode	high resistance.
	gate	cathode	high resistance in either direction.
	cathode	gate	
N-channel JFET	drain	source	same resistance in either direction; typically 500 Ω–5 kΩ.
	source	drain	
	gate	drain or source	low resistance (forward-biased P-N junction).
	drain or source	gate	high resistance; $> 10 \ M\Omega$ unless ohmmeter battery exceeds JFET breakdown voltage BV_{GDO}.
P-channel JFET	SAME AS FOR N CHANNEL, EXCEPT THAT ALL POLARITIES ARE REVERSED		
Enhancement MOSFET	drain	source	very high resistance; $> 10 \ M\Omega$.
	source	drain	
	gate	drain or source	very high resistance; $> 100 \ M\Omega$ for either direction.
Depletion MOSFET	drain	source	moderate resistance; in 500 Ω–5 kΩ range.
	source	drain	
	gate	drain or source	very high resistance; $> 100 \ M\Omega$ for either direction.
	drain or source	gate	

Answers To Selected Problems

CHAPTER 2: **2.1** Each has 0.5 A.

CHAPTER 3: **3.1** $V_{ac} = -6$ V; $V_{bc} = +12$ V; $V_{ab} = -18$ V. **3.5** $I_a = 15$ mA; 5 mA; 0 mA. **3.7** $I_a \simeq 5$ mA; $I_B \simeq 7$ mA.

CHAPTER 4: **4.19** No.

CHAPTER 5: **5.21** (a) 150 V; (b) 500 mA. **5.24** 2μA @ 36°C. **5.25** 8 mA and 0.7 V. **5.26** 5.3 mA. **5.27** $\simeq 0$ V; 50 V. **5.29** 9 mA; 11 mA. **5.32** PRV > 10 V. **5.33** (a) -2 V; (b) 2 V; (c) 4 V; (d) 6.7 V; (e) 6.7 V; (g) $I_{F(max)}$; V_{RDC}. **5.35** 65 mW. **5.37** 1 W. **5.41** (a) 0.2°C/mW; (b) 500 mA; (c) 48 Ω; (d) $P_{D(max)} = 425$ mW. **5.42** $I_{R(max)} = 4$ mA; $R_{min} = 5$ kΩ.

CHAPTER 6: **6.5** (a) 3 mA; (b) 20 V. **6.7** Yes. **6.9** (a) 5.6 V; (b) 5.684 V. **6.10** (a) 4.3 V; (b) 4.171 V. **6.11** (a) 833 mW. **6.13** 9.9 V ; 0 V. **6.15** $\sim$ 50 Ω. **6.18** 24 Ω < R_S < 6700 Ω. **6.19** (b) 650 Ω.

CHAPTER 7: **7.8** (a) $V_F \simeq 0$; $I_F = 6$ mA; (b) $V_F \simeq 0.5$ V; $I_F = 23.5$ mA **7.10** (a) $R_1 = R_2$ chosen between 273 Ω and 546 Ω; (b) $R_1 = R_2$ chosen ≤ 273 Ω but > 120 Ω; (c) $R_1 = R_2 = R_3$ chosen as in step (a).

CHAPTER 8: **8.1** Blue. **8.2** 4840 Angstroms. **8.3** 2.63×10^{12} eV. **8.7** 2 mA. **8.10** Approx. 1000 fc **8.11** Approx. 0.5 fc. **8.12** 1 MΩ. **8.15** 5.7 kΩ. **8.18** 276 Ω. **8.20** 12.2 percent. **8.22** $V_{OUT} = 1.4$ V at 1 fc **8.25** 25 mW.

CHAPTER 9: **9.10** (a) Active; (b) Cut-off; (c) Saturation; (d) Saturation. **9.11** (a) $I_E = 20$ mA. **9.13** For $I_E = 10$ mA, $I_C = 9.91$ mA and $I_B = 0.09$ mA. **9.22** $I_C \simeq I_E = 0.5$ mA. **9.23** (a) 5.3 mA. **9.24** (a) $I_C = I_E = 10$ mA, $V_{CB} = 20$ V; (b) 3.3kΩ. **9.25** (a) $I_E = 0.23$ mA, $I_C = 0.215$ mA, $V_{CB} = 8.91$ V; (b) —; (c) $I_E = 2.3$ mA, $I_C = 2$ mA, $V_{CB} \simeq 0$. **9.26** 1.8 V. **9.27** $I_C = 10$ mA, $I_B = 5$ mA. **9.33** 0.106 mA. **9.37** $R_C = 70$ Ω. **9.38** $I_B = 34.8$ μa; $I_E = 5.25$ mA; $V_E = 5.25$ V. **9.40** $I_E = 1.65$ mA; $V_E = -3.3$ V. **9.41** (a) 40 V; (b) 20 Ω; (c) 37.4 Ω. **9.43** (a) 25°C − $I_C = 2.3$ mA, $V_{CE} = 9.7$ V; (b) 75°C − $I_C = 5.34$ mA, $V_{CE} = 6.66$ V. **9.44** (a) 0.57 mA; 71 mA. **9.47** 14.4×10^6. **9.54** 22.7. **9.55** 36 mW; 1.6 mW. **9.56** $\simeq$ 1.4 fc.

CHAPTER 10: **10.4** $A_p = A_i = 10$. **10.5** $Z_{IN} = 25$ Ω, $Z_{OUT} = 200$ Ω. **10.6** For $R_L = 10$ kΩ, $G_v = 980$, $G_i = 9800$, $G_p = 96 \times 10^5$.

CHAPTER 11: **11.1** 153.6 mV. **11.5** 0.975. **11.10** (a) 2.5 mA, 2.5 mA, 15 V; (b) 1 V p-p; (c) $A_i \simeq 1$, $A_v \simeq 1000$, $A_p \simeq 1000$. **11.12** 10 Ω, 10 kΩ, 240. **11.14** (a) 1.5 mA, 5 V; (b) $v_{CB} = 7.5$ V p-p; (c) $G_v = 450$. **11.16** 0.43 mA, 15.7 V.

CHAPTER 12: **12.2** 50. **12.4** 2.5 kΩ. **12.5** (a) 2.6 mA, 24.6 V; (b) 0.8 V p-p; (c) $A_i = 75$, $A_v = 400$, $A_p = 30,000$. **12.7** $Z_{IN} = 750$ Ω, $Z_{OUT} = 3.6$ kΩ, $G_v = 176$, $G_i = 22$, $G_p = 3872$. **12.9** (b) $e_{out} = 2.3$ V p-p; $G_v = 115$. **12.11** Approx. 60 mV p-p. **12.12** (a) 9.65 mA, 10.35 V; (b) 14.5 mA, 5.5 V.

CHAPTER 13: **13.1** (a) $V_E \simeq 20$ V, $I_E \simeq 2$ mA $\simeq I_C$; (b) $h_{ib} = 12.5$ Ω, $h_{ie} = 1875$ Ω; (c) $Z_{ic} \simeq 33$ kΩ; (d) $e_{out} = 27.9$ mV; (e) $G_i = 0.65$, $G_p = 0.364$. **13.2** 63.5 Ω. **13.9** 200 kHz, 16.6 MHz. **13.15** (a) 60 W; (b) 0.7 mA; (c) 25. **13.16** (a) 0.91 μs; (b) 100; (c) 20. **13.17** 333.5. **13.18** 90 kHz. **13.19** 1215 pF.

CHAPTER 14: **14.6** $E_{ON} = 8.5$ V, $E_{OFF} = 2.5$ V. **14.10** $R_{max} = 8$ kΩ, $R_{min} = 7.3$ kΩ; $f_{min} = 1136$ Hz, $f_{max} = 1250$ Hz. **14.12** $R < 12.7$ kΩ; $C = 20 \times 10^{-3}/R$. **14.13** Approx. 7 V. **14.21** (a) 100 V; (b) 100 V; (c) 4 mA; (d) 0.8 V; (e) 10 mA; (f) 100°C; (g) 5 V; (h) 25 A. **14.24** 15 μs. **14.25** $R < 6.7$ kΩ. **14.28** 2.17 V. **14.30** 10 kΩ. **14.31** $\alpha = 9.2°$. **14.32** $\alpha = 20°$. **14.33** $R_T = 1$ kΩ. **14.34** 7.4 ms.

CHAPTER 15: **15.1** 9.5 V. **15.5** (a) 13.9 V; (b) 5 V. **15.6** (a) 16.7 V; (b) 45 Hz. **15.7** 0.62 Hz = 66.6 Hz. **15.9** 1.9 s. **15.11** 144 Ω. **15.17** $V_p = 19.7$ V; $f = 5.2$ kHz. **15.19** 54 s.

CHAPTER 16: **16.6** 3 V; 5 V; −5 V. **16.7** 10 mA; 6.4 mA; 3.6 mA; 1.6 mA; 0.4 mA; 0 mA. **16.9** 50. **16.10** 2.76. **16.14** (a) 4.1 mA, 22 V; (b) $A_v \approx 4$; (c) 10 MΩ, 2 kΩ. **16.15** 488 Ω. **16.16** 144 Ω; 0.4.

INDEX

INDEX